PROTEIN METABOLISM
OF THE BRAIN

STUDIES IN SOVIET SCIENCE

LIFE SCIENCES

1973

MOTILE MUSCLE AND CELL MODELS
  N. I. Arronet
PATHOLOGICAL EFFECTS OF RADIO WAVES
  M. S. Tolgskaya and Z. V. Gordon
CENTRAL REGULATION OF THE PITUITARY-ADRENAL COMPLEX
  E. V. Naumenko

1974

SULFHYDRYL AND DISULFIDE GROUPS OF PROTEINS
  Yu. M. Torchinskii
MECHANISMS OF GENETIC RECOMBINATION
  V. V. Kushev

1975

THYROID HORMONES: Biosynthesis, Physiological Effects, and
  Mechanisms of Action
  *Ya. Kh. Turakulov, A. I. Gagel'gans, N. S. Salakhova, A. K. Mirakhmedov,
  L. M. Gol'ber, V. I. Kandror, and G. A. Gaidina*

1977

THE EVOLUTIONARY ECOLOGY OF ANIMALS
  S. S. Shvarts
HEMATOPOIETIC AND LYMPHOID TISSUE IN CULTURE
  E. A. Luriya
STRUCTURE AND BIOSYNTHESIS OF ANTIBODIES
  R. S. Nezlin
PROTEIN METABOLISM OF THE BRAIN
  A. V. Palladin, Ya. V. Belik, and N. M. Polyakova

A Continuation Order Plan is available for this series. A continuation order will bring
delivery of each new volume immediately upon publication. Volumes are billed only upon
actual shipment. For further information please contact the publisher.

STUDIES IN SOVIET SCIENCE

PROTEIN METABOLISM OF THE BRAIN

A. V. Palladin, Ya. V. Belik, and N. M. Polyakova

Academy of Sciences of the Ukrainian SSR
Kiev, USSR

Translated from Russian by
Basil Haigh

Translation edited by
Abel Lajtha
New York State Research Institute
for Neurochemistry and Drug Addiction
Ward's Island, New York

Springer Science+Business Media, LLC

Library of Congress Cataloging in Publication Data

Palladin, Aleksandr Vladimirovich, 1885-1973
    Protein metabolism of the brain.

    (Studies in Soviet science)
    Translation of Belki golovnogo mozga i ikh obmen.
    Includes bibliographical references and index.
    1. Brain chemistry. 2. Protein metabolism. I. Belik, Iakov Vasil'evich, joint
author. II. Poliakova, Nina Mikhailovna, joint author. III. Title. IV. Series.
QP376.P3313   599'.01'88   77-2307
ISBN 978-1-4684-1619-0   ISBN 978-1-4684-1617-6 (eBook)
DOI 10.1007/ 978-1-4684-1617-6

The original Russian text, published by Naukova Dumka in Kiev in 1972, has been
corrected by the authors for the present edition. This translation is published
under an agreement with the Copyright Agency of the USSR (VAAP).

БЕЛКИ ГОЛОВНОГО МОЗГА И ИХ ОБМЕН
А. В. ПАЛЛАДИН, Я. В. БЕЛИК, Н. М. ПОЛЯКОВА

BELKI GOLOVNOGO MOZGA I IKH OBMEN
A. V. Palladin, Ya. V. Belik, and N. M. Polyakova

© 1977 Springer Science+Business Media New York
Softcover reprint of the hardcover 1st edition 1977
Originally published by Consultants Bureau, New York in 1977

Foreword

I have had the privilege of conversing with Professor Palladin on
numerous occasions during the last 15 years of his life. In these
years, after the age at which most people retire, he was still a
most admirable man, with an active interest in science, and a wide
and wise approach to its many facets. I was always impressed by
his enthusiasm for, and knowledge about, studies of the nervous
system, both his own and those of others. He was interested in
myriads of details, which he was able to synthesize into concepts
and systems. The present book reflects, in addition to his warmth
and wit and wisdom, Dr. Palladin's approach and style: how he
and his collaborators came to their conclusions after careful consideration of all the facts. The references in this book, for example,
show the great number of works that were found worthy of quoting
and including. The reader can see the stepwise emergence of a
concept as he reads the pages of this book. Facts are not given for
facts' sake, but to try to present a picture based on a large number
of observations, by most of the scientists who studied this aspect
of the nervous system. The work emerges as from a community;
there is no attempt to minimize results of others or to maximize
those of Palladin's laboratories.

Palladin's working lifespan covered over six decades, and
many stages in biochemistry. He was present at the beginning of
neurochemistry, and of the use of isotopes in biochemistry; thus
this book, like Palladin, bridges classical and modern approaches
to our understanding of the nervous system. The historical summary forms a valuable part of the book.

We tend to be concerned mainly with the latest results and the
current views on problems, and rarely have the time or the incli-
nation to consider the development of concepts in historical per-
spective. It is clear that methodology constantly improves, and
that all the experiments of yesteryear can now be performed more
precisely or in a more meaningful way, but we are apt to ignore
or discard the many valuable results and well-established findings
of the past. This book can, by its balanced presentation of the
facts, rekindle our interest in viewing problems in a larger per-
spective, and teach us to appreciate the use of methods such as
electrophoresis or extraction of proteins that are not used as often
at the present time. The book is by no means concerned exclusive-
ly with the past; the heuristic description of the facts usually brings
into focus fruitful approaches. The need and promise for further
great contributions by neurochemistry is expressed in many pages.

Protein metabolism is treated on a wide basis, as it should be.
Amino acid pools and changes in amino acid composition are discussed; and in addition to protein synthesis and breakdown, such
factors as enzymes and other structures participating in protein
metabolism are discussed. The literature dealing with alterations
of protein metabolism (development, drug effects, and effects of
hypothermia) is also treated in detail.

This summary of neural protein metabolism comes from a
laboratory that has contributed in a most significant way to our
understanding in this area. It illustrates and clarifies the dynamic
aspects of cerebral metabolism, and also the plasticity of brain
metabolism reacting to influences and changing requirements; it
shows how very complex problems can be studied and understood
by the scientific community.

                                                    Abel Lajtha

Preface

All functions of the cell and all physiological phenomena involve conversions of proteins. The elucidation of the biochemical basis of the various physiological functions of any cell, including the most highly organized and specialized nerve cell, is therefore impossible without knowledge of the composition of its proteins, their physicochemical and biological properties, and the principles governing their conversions during cell function.

The experimental material now available on the proteins of nerve tissue and, in particular, of the brain and on their metabolism in various functional states has not been adequately reflected in the modern literature, especially in the USSR. Experimental treatises and surveys by Western authors have dealt chiefly with recent research into individual problems concerned with the study of brain proteins, and have drawn insufficiently on results obtained by Soviet researchers.

Nevertheless protein metabolism in nerve tissue has been studied for the last 20 years in the Department of Biochemistry of the Nervous System, Institute of Biochemistry, Academy of Sciences of the Ukrainian SSR, and in other Institutes in the Soviet Union. For example, workers in the authors' Department have studied the protein content in various functional states of the body, the fractions of soluble brain and peripheral nerve proteins, the metabolic heterogeneity of proteins in the subcellular structures of nerve tissue at different stages of postnatal development of the brain, and the activity of intracellular proteolytic enzymes.

The authors have set themselves the task, when writing this monograph, of describing the principal results of research by both

Soviet and Western scientists into the proteins of the nervous system and thus of presenting a complete picture of our present knowledge of brain proteins and their metabolism, of examining historically the stages of development of this research in the USSR and elsewhere, and of including the fullest possible bibliography of the world literature.

Particular attention was paid to protein metabolism in the subcellular structures of the brain during the various stages of postnatal development, when the composition of the protein fractions of nerve tissue is established. Data are given on the composition of the brain protein fractions, their metabolic heterogeneity, the correlation between the metabolic activities of the proteins of the brain and its functional state, and the regional and subcellular localization and age changes in the activity of the intracellular proteinases (acid and neutral) — enzymes responsible for the catabolic phase of protein metabolism in nerve tissue. Primacy is given to a full account of the investigations into the functional biochemistry of cell structures conducted by the authors. In this field, which has been so fruitful in the last two decades, the authors' investigations occupy a significant place.

Throughout this monograph the biochemistry of the brain proteins is presented under the three principal subdivisions of analytical, dynamic, and functional biochemistry.

It can rightly be said that this monograph represents a chapter in neurochemistry, namely, the neurochemistry of the brain proteins.

Academician A. V. Palladin

Contents

Introduction

In the higher animals the nervous system plays a leading role in the regulation of physiological functions and of the biochemical processes on which they are based. The regulatory mechanisms of the nervous system are the most perfect and the most complex among the whole diversity of biological control systems.

The specific functions of nerve tissue such as the generation, conduction, and synaptic transmission of nervous impulses, the electrical activity of nerve cells, the excitation and inhibition of their activity, and the utilization and storage of information are all linked with proteins. The brain proteins also play an active part in such unique mechanisms of regulation of biochemical processes as ammonia formation and fixation, the homeostatic functions of the blood—brain barrier, and the spatial coupling of some and uncoupling of other metabolic reactions with the aid of complex systems of protein—lipid membranous structures. Inborn or acquired abnormalities of function of the nervous system and disorders of mental activity probably depend on disturbances of protein metabolism in nerve tissue.

The primary role of proteins in the functions of the nervous system is not in dispute. It is therefore a remarkable fact that until recently the proteins of nerve tissue had received far less study than, for example, other biologically important substances of the brain or the proteins of the blood, muscles, and other organs.

All physiological functions and the biochemical processes on which they are based take place with the active participation of proteins [80, 135, 399, 488, 509]. Some proteins perform enzymic functions in the body, others have hormonal, protective, transport, structural, and other functions [135, 196, 305, 657, 706, 1327].

1

The predominant role of enzymes and hormones of protein nature in the performance and regulation of the basic metabolic reactions of tissue is well known [135, 196, 255, 644]. Many enzymes and hormones have been isolated and purified, their physicochemical and biological properties studied, and in certain cases the amino acid sequence determined (ribonuclease, trypsin, chymotrypsin, insulin, vasopressin, oxytocin, etc.) [1108, 1109, 1717-1720, 1743].

The structure and biochemical functions of proteins concerned with oxygen transport (particularly hemoglobin and myoglobin) have now been studied in detail [93, 859, 994]. Much progress has also been made in the study of the physicochemical and biochemical properties of the contractile proteins of muscle tissue, proteins with mechanochemical functions [241, 390, 675].

Much is now known about the role of proteins in immunochemical reactions [42, 135, 404] and in the processes of blood clotting – an important protective reaction of the animal against injury [39-41, 305, 354, 808, 1264].

Considerable attention is being paid at present to the study of the physicochemical properties of proteins of cell membranes [142, 279, 630, 707, 1078, 1214, 1267, 1268, 1439, 1742]. For a long time a mainly plastic function has been ascribed to these proteins. However, it has been established more recently that these proteins are heterogeneous with regard to their physicochemical properties and that they take part enzymatically in many vitally important reactions including active membrane transport of metabolites at the tissue, cellular, and subcellular levels [48, 181, 244, 1563], and also in the performance of the mechanochemical functions of membrane structures [317, 512].

Considerable progress has been made in the study of protein metabolism in animal tissues [488, 1327]. Development has been particularly rapid in the study of protein biosynthesis [178, 282, 366] and of the role of enzymes and nonenzymic proteins in the regulation of intracellular reactions [90, 229, 255, 407, 666]. The most valuable results in this field have been obtained by investigating the metabolism and physiological functions of the "soluble" proteins of blood plasma and other tissues. The study of the "insoluble" proteins, the structural elements of the numerous tissue membranes, and the "nonstructural" proteins firmly bound to the membranes is in fact only just beginning [142, 918, 919, 1074, 1078, 1267, 1323].

Against the background of these achievements in the study of the physicochemical and biological properties of proteins and of their structure and metabolism in various tissues, the growth of our knowlege about proteins of the nervous system until very recently has been negligible. Hardly anything was known of the composition of the protein fractions of nerve tissue, the content of the various proteins in functionally different macro- and microstructures of the nervous system, or their role in the functions of the cell and its organelles. The molecular mechanisms of intracellular synthesis and breakdown of the tissue proteins, the features of these processes specific to nerve tissue, and the dependence of their intensity on the functional activity of the cell remained uninvestigated. Especially little was known about the specific proteins of nerve tissue concerned with functions such as memory and the generation of the nervous impulse, its conduction, and its synaptic transmission.

Since protein structures play an active part in the regulatory functions of the nervous system, including those under conscious control, research into the metabolic conversions of nerve tissue proteins and the intensity of their synthesis and breakdown must be not only of the greatest scientific interest for neurochemistry, but also of exceptional importance on the general biological plane.

Chapter 1

The Principal Stages and Basic Trends of Research into Proteins and Protein Metabolism in the Nervous System

In the century and a half or more of the history of research into the proteins of nerve tissue it is possible to distinguish two principal periods, the boundary between which is formed by the fifth decade of this century. In that decade radioactive methods, especially the use of amino acids labeled with radioactive isotopes, began to be used extensively for the study of the metabolic conversions of proteins in the nervous system.

The first period, which began in 1811 and which can be called the "preisotope" period, lasted about 140 years and ended with the discovery of the chemical topography of proteins in the tissues of the nervous system. For several decades in the 19th century the attention of scientists was concentrated chiefly on the study of the lipid and carbohydrate composition of nerve tissue; proteins received far less study. The chief reason for this was that the histochemical methods existing at that time and the biochemical methods that followed soon after were chiefly for studying lipid and carbohydrate tissue components. In addition, it was a long time before the important role of proteins in the activity of the nervous system was recognized at all. A mainly plastic function was ascribed to them.

Even after the work of Danilevskii and his pupils [179, 183, 184, 708], in which for the first time the important functional role of proteins in nerve tissue was emphasized, interest in the study of these compounds remained as low as ever. The isolated investigations of Petrovskii, Lents, Slovtsov, Halliburton, Koch, McGregor,

5

and others [443, 529], undertaken during the 19th and the first
quarter of the 20th centuries, were devoted chiefly to the study of
the content of total proteins and their fractions in tissues of phylogenetically and functionally different parts of the brain.

Later, Palladin and co-workers [443] described the results
of their investigations to determine the content of total and protein
nitrogen and of total proteins and their individual fractions in
various parts of the central and peripheral nervous system. These
investigations led to the formulation of the following general principle for the distribution of proteins in nerve tissue: The protein
content is higher in phylogenetically younger and functionally more
complex parts of the nervous system and lower in parts appearing
in the earlier stages of phylogeny; the composition of the proteins
also varies in different parts of the nervous system. There is
a much higher content of water-soluble proteins in the gray matter
of the cerebral hemispheres than in the white matter.

The next stage in the study of the proteins of nerve tissue is
characterized by the determination of the composition of protein
fractions chiefly by methods based on the differential solubility of
proteins in water, in aqueous solutions of neutral salts, acids, and
alkalies, and in organic solvents. This path of research, which
began with the early work of Danilevskii, Lents, Halliburton, and
others, underwent further development in the laboratories of Palladin [441, 442, 451, 477, 480], Vladimirov [63, 68, 103, 108, 649],
Éngel'gardt [327, 330, 729, 730], and others [280, 1018, 1019, 1021,
1058, 1273, 1707]. The results of these investigations were supported by observations revealing the complexity of the protein
composition of nerve tissue and its differences in the structures
of the central and peripheral nervous systems. They also revealed
a number of hitherto unknown protein fractions and protein complexes with specific physicochemical properties, differing in their
metabolic activity, and performing different functions in the tissues.
The existence of protein complexes with nucleic acids, lipids,
carbohydrates, and other substances in nerve tissue was demonstrated. These compounds include ribonucleoproteins, deoxyri-
bonucleoproteins, phosphoproteins, lipoproteins, proteolipids,
phosphatidopeptides, and cholesterol – protein and glycogen–
protein complexes.

During the second, or "isotope," period, although shorter in
duration than the first, considerable progress has been made studying the role of proteins in the functions of nerve tissue, and in dis-

covering the laws governing their metabolic conversions in nerve cells and subcellular structures. Progress has also been made in this period in separating nerve tissue proteins into many fractions, isolating individual proteins in a highly purified form, and studying their composition and physicochemical and biological properties The results of these investigations, in conjunction with data obtained by modern methods of fractionation of intracellular structures, has shown the key role of the individual cell organelles in metabolic reactions and has brought us significantly closer to an understanding of the biochemical basis of nervous activity.

Because of the high sensitivity and specificity of the isotopic indicator method it is now possible to detect and to determine quantitatively conversions of matter of such small magnitude in the living organism as were previously beyond the limit of sensitivity of earlier methods. Another distinguishing feature of the isotope method is that the transformations of biologically important substances, especially proteins, on an extremely small scale can be detected in the intact organism without any significant disturbance of the functions of the organs of the experimental animal. This fact has made the use of labeled compounds particularly promising in the study of metabolic processes in nerve tissue which, because of the anatomical characteristics of the nervous system and the extremely rapid alternation of excitation and inhibition that occurs in it [44, 428], proved to be extremely difficult to study by earlier methods. It must also be emphasized that because of the extremely high sensitivity of radioactive labeling, it can be used to study metabolic conversions of substances in extremely small quantities of tissue. By using autoradiographic studies the exact localization and intensity of processes taking place in the microstructures of the cell can be determined.

The isotope method has proved to be virtually irreplaceable as a means of determining the rate of renewal of biologically important macromolecular compounds, including proteins, in the adult animal in vivo. Since the content of any substance in a tissue depends ultimately on the relationship between the rates of its synthesis and breakdown, an increase or decrease in the content of a substance in a tissue reflects nothing more than the degree of predominance of synthetic over catabolic reactions or vice versa. In the case of dynamic equilibrium between these two processes, as is observed in the tissues of adult animals, the total protein content in a tissue does not vary appreciably and, for that reason, the methods available previously could not provide information about the rates of protein synthesis or breakdown. Under these conditions only the isotope method can give comparatively satisfactory characteristics of the velocities of these processes. In developing animals, unlike the adult, there is a steady increase in the total content of proteins, nucleic acids, carbohydrates, lipids, and other tissue components, which can easily be determined by the ordinary chemical methods. Even in this case, however, the use of labeled precursors has certain advantages, for in conjunction with other modern methods it can be used to determine not only a change in the content of proteins, carbohydrates, and other substances in the tissues, but also the intensity of the metabolic conversion of these substances, their individual fractions, and individual components at all stages of ontogenetic development.

The most important advances in the investigation of the composition and functional role of proteins in nerve tissue have been made chiefly during the last 15 to 20 years. As a result of the extensive use of electron microscopy and polarization-optical and x-ray structural methods, the role of proteins in the structural organization and function of the various membrane structures of neurons and other cells has been elucidated [666, 707, 1313, 1545].

The use of modern analytical methods of protein chemistry has revealed [467, 469, 526-529, 847] some characteristics of the quantitative and qualitative composition of proteins in different parts of the CNS and PNS undetectable by the less sensitive methods previously available.

Investigations [374, 469, 473, 854, 979, 1057, 1750] have shown that the composition of the protein fractions of nerve tissue is established in ontogeny. Some phylogenetic variations in the quantitative and qualitative composition of the brain proteins have been discovered [374, 375]. Nerve tissue, unlike blood serum, contains chiefly globulin-like proteins [467, 469, 555, 844, 847, 945]. Water-soluble brain proteins can be divided into fractions that differ in certain physicochemical properties, such as solubility, electrophoretic mobility, enzymic activity, and other features [110, 204, 535, 843, 946, 955, 1183, 1199, 1397, 1546, 1547, 1722, 1795]. Certain proteins have been isolated from nerve tissue in a purified form and some of their physicochemical and biological properties have been studied. These proteins include, for example, intracellular acid proteinase [332, 333, 536], adenosine deaminase [472], a protein with coronary dilator action [132, 133], the structural

protein of microsomal membranes [1074], basic proteins with encephalitogenic properties [892, 975, 988, 992, 1191-1193, 1307,
1357, 1421], and other basic proteins [299, 300, 303, 304, 1684],
and also acid proteins [816, 1187, 1396, 1485]. One fraction of
acid proteins, known as S-100 protein, is considered to be specific
for nerve tissue [1153, 1317, 1396-1398, 1400, 1705]. This protein
fraction has been found to be heterogeneous on electrophoresis on
polyacrylamide or on mixed polyacrylamide−agarose gel [1064,
1065, 1667]. An organ-specific protein was recently isolated from
membranes of the microsomal fraction of the brain [1691]. This
protein is basic (it migrates toward the cathode during electrophoresis) and is species specific. An immunospecific glycoprotein
[1750, 1752] and several organ-specific antigens of protein nature
[151] have also been discovered in brain tissue.

In experimental investigations of the biochemistry of the nervous system, including protein metabolism, three principal move-
ments can be distinguished: Depending on the problems to be
solved, these may be designated the static, dynamic, and functional
biochemistry of the nervous system.

The earliest biochemical investigations of nerve tissue were
chiefly analytical, i.e., determinations of the chemical composition of the whole brain and of various anatomical structures of the
nervous system. With the improvement of older methods and the
introduction of methods of great sensitivity and specificity, there
was a gradual swing to research in dynamic biochemistry, followed
by a further movement toward functional biochemistry of the nervous system. Research in these two new categories sought to reveal the mechanisms of conversion of biologically important substances in the tissues of the nervous system and to establish links
between these processes and functional states of the organism as
a whole and of the nerve tissue and its anatomical and microscopic
structures in particular, i.e., to study the biochemical basis of
nervous activity.

Although dynamic and functional biochemistry of the nervous
system have occupied leading positions in research in the last
decades, the role of analytical biochemistry as a means of establishing the biochemical basis of nervous functions must not be under-
estimated. With the appearance of techniques that have greatly
increased the scope of investigations of this type, a change has
taken place in the level of study of the composition of nerve tissue.
Whereas previously these investigations were carried out with

whole tissue and its principal anatomical structures, nowadays it is possible to study the chemical composition and structure of individual cell groups, types of cells, intracellular organelles, and the structures composing them. Whereas formerly the study of the composition of nerve tissue was limited to determination of the content of the principal chemical classes (proteins, carbohydrates, lipids, nucleic acids, etc.), now we can investigate the content of separate fractions of these as well as of some individual components.

One of the most important and productive fields in the functional biochemistry of the nervous system, and one which has developed successfully during the last two decades, is the functional biochemistry of cellular structures. Progress in this branch of research followed the introduction of methods of differential centrifugation of tissue homogenates conjoined with electron-microscopic study of the isolated cell components and with modern microquantitative techniques for the analysis of their chemical composition and biochemical organization. The wide use of these methods has allowed the ultrastructure of the cell organelles to be studied in detail, and their role in basic metabolic processes regulating cell functions to be examined. Our ideas regarding functional interactions between these structures during intracellular metabolism, including protein biosynthesis, have thereby been broadened.

Fractionation of the Tissue Proteins of the Nervous System. Physicochemical and Biological Properties of Some Protein Fractions

The First Stages in the Study of the Proteins of the Nervous System

The study of the chemical composition of the central nervous system began in the second half of the 18th century, when the work of Henzig was published in 1779. In 1811, Vauquelin listed the chief chemical components of brain tissue, including protein.

In the descriptions of investigations conducted in the first half of the 19th century, most of which are now of purely historical importance, the data given relate chiefly to the content of water, salts, and lecithin in the brain. Protein content is rarely mentioned [529].

The first paper devoted to the study of the proteins of the gray and white matter of the brain was published in 1873 by Petrowsky [1487]; his experiments were carried out in Hoppe-Seyler's laboratory. Petrowsky determined the total content of proteins in the brain tissue and studied their properties. To judge from these properties, the proteins which he called "albuminous substances" were really globulins. According to Petrowsky's findings, more than half of the dry weight of the gray matter of the brain consists of proteins whereas cholesterol and lipids account for only one-quarter. The gray matter contains twice as much protein as the white.

The same distribution of proteins and lipids in the brain is given in a monograph by Thudichum [1679], who carried out the fullest investigation of the chemical composition of the human brain. According to his observations the protein content in the gray matter is 2.5 times greater than the lipid content; in white matter, on the other hand, the lipid content is 2.5 times greater than the protein content. However, Thudichum attached great importance to the brain lipids whereas he regarded the proteins "as a stroma or skeleton of the bioplasm, in which the specific elements are housed." Like Thudichum, other workers who studied the components of the brain for a long time devoted their attention chiefly to the lipids and very little to the investigation of proteins.

In the early stages of biochemistry, progress in the study of brain proteins was due almost entirely to the work of Russian scientists and, in particular, to Danilevskii [183, 184], who demonstrated conclusively the important physiological role of proteins in the nervous system. He wrote that "when the relationship between the chemical composition of a tissue and its vital functions is studied, attention must be paid chiefly to the study of the nature of its protein forms .... [T]he high irritability, the reflex behavior, the specialization of the biological role of the nerve cell — all these depend to a high degree on particular protein substances contained in the cell."

Danilevskii [183] found that brain tissue contains a globulin which differs from the globulin of muscles by its high phosphorus content, and which he called neuroglobulin; he also found another brain protein — neurostromin.

The study of the brain proteins begun by Danilevskii was continued by his pupils. Gutnikov [179] determined for the first time the content of water, phosphorus, nitrogen, and sulfur in the white and gray matter of the human brain. He studied the brain not only of the adult, but also of embryos at various ages, and of infants at birth and at the age of 2 months. Shkarin [708] determined the content of various proteins in the cerebral cortex of man and animals at different ages. The proteins of the gray matter of the dog's brain were studied by Fleisher [659].

Continuing the study of the brain proteins commenced by Danilevskii, Lents [319] set out to extract all the neuroglobulins from the brain, to extract the neurostromin from the residual brain mass, and to establish the identity of the proteins he extracted with the neuroglobulin and neurostromin described by other workers.

Before publication of the work of Danilevskii and his pupils a paper was published in 1885 by Baumstark [800] on the chemical investigation of horse brain tissue. Having isolated proteins from the brain, Baumstark postulated that they were completely identical with the proteins extracted from muscle tissue.

Extensive investigations of the proteins of brain tissue were carried out by Halliburton [865, 1102, 1103]. He investigated the composition of nerve tissue quantitatively, paying particular attention to the protein content in different parts of the nervous system and to the isolation and study of the properties of the proteins contained in nerve tissue. He found that proteins account for more than 50% of the dry weight of the gray matter, rather less in the white matter, and less still in nerves.

Halliburton isolated three proteins from the gray matter that differed in their coagulation temperature and in the ease with which they were precipitated by neutral salts and by acetic acid; two of them were globulins whereas the third contained phosphorus, and Halliburton called it nucleoalbumin.

Ultimately the idea took hold that brain tissue contains chiefly four groups of proteins. The first group consists of proteins soluble in water; the content of this group of proteins (presumably albumins) in the brain is very small and most investigators disregarded it in their analyses and did not describe it in the results of their work. The second group of proteins consists of globulins, or neuroglobulins. The third group is the neurostromins and, finally, the fourth group consists of proteins insoluble in weak acids or alkalies, alcohol, and ether. This group was called the "residue." Its most important components are neurokeratin and nuclein or, more correctly, a nucleoprotein that is a component of the cell nuclei.

Unlike Halliburton, Koch [1207] distinguished three groups of proteins in the brain: simple proteins, nucleoproteins, and neurokeratin.

By using various methods to fractionate the proteins of brain tissue, McGregor [1319] found three proteins in animal brain tissue, two of which contained phosphorus and iron.

In 1919 Danilevskii [185] published a paper in which he postulated that the neurostroma is the excitable substance of nerve cells; he cited the observations of Shkarin who found neuroglobulins and neurostromins in the brains of animals and men at different ages. In the same year, continuing his investigations into the chemical composition of the gray matter of the brain, Lents published a paper [320] in which he proposed a new method of extracting neuroglobulin from brain tissue, by means of which he considered that the neuroglobulin content in different parts of the nervous system could be determined.

Slovtsov and Georgievskaya [598] investigated the chemical composition of the gray and white matter of the human brain and also found a higher protein content in the gray than in the white matter.

The foundations of the study of brain proteins in the Department of Biochemistry of the Nervous System, Institute of Biochemistry, Academy of Sciences of the Ukrainian SSR, were laid by the work of Palladin and Rashba [474] on the determination of the total nitrogen content in parts of the brain of various animals. They found that in cows, dogs, rabbits, rats, guinea pigs, pigeons, and lizards, the total nitrogen, most of which consists of protein nitrogen, is highest in the gray matter of the cerebral hemispheres, lower in the cerebellum, and lowest of all in the white matter of the hemispheres. These differences are most clearly marked in mammals; in birds, i.e., in animals with a less highly differentiated nervous system, the differences are less.

To confirm the conclusion that the complexity of functions performed by the structures of nerve tissue bears a relationship to their protein content, the chemical composition of the cortical gray matter, basal ganglia, cerebellum, and spinal cord was studied [475]. The phylogenetically youngest but functionally most complex part of the gray matter of the brain, the cerebral cortex, was found to have the highest protein content; the gray matter of the cerebellar cortex and of the basal ganglia contains less protein, and the gray matter of the spinal cord — the functionally least complex and phylogenetically oldest part of the gray matter — has less protein still.

The study of the chemical composition of the spinal ganglia, certain parts of the autonomic nervous system, and the conducting pathways of the PNS showed [475, 476] that here also the phylogenetically youngest parts have the highest content of nitrogenous substances. In the course of these investigations the total protein content was determined in different parts of the nervous system. However, it could be claimed that structurally and functionally different parts of the nervous system contain proteins specific for each particular part. Accordingly the proteins extracted from nerve tissue were divided into four fractions (extractable with water, with 4.5% potassium chloride solution, and with 0.1 N sodi-

um hydroxide, and also the insoluble residue) and the content of
these fractions was determined in the gray and white matter of the
cerebral hemispheres.

The proteins of these various fractions, which differed in their
isoelectric points, were present in different quantities in the gray
and white matter of the brain. Proteins of the water-soluble fraction account for about 30% of the nitrogenous substances in the
gray matter and about 19% in the white matter. The gray matter
is thus richer in water-soluble proteins than the white matter.
Proteins extractable with potassium chloride solution are found in
almost equal amounts in the gray and white matter. However, the
gray and white matter differ sharply in their content of protein
insoluble in the solvents mentioned above; this fraction accounts
for about 5% of the total nitrogen in the gray matter and 20-22% in
the white matter.

These investigations thus showed that the gray and white
matter of the brain differ not only in their total protein content but
also in their content of the individual protein fractions: The gray
matter contains more of the water-soluble proteins but less of the
insoluble protein residue than the white matter. Chemical fractionation as a means of studying the brain proteins has also been
used by other workers [950, 1282].

We also investigated the structural proteins of the brain [480].
Having isolated from the gray matter of the brain a protein of the
nucleoprotein category, we showed that it is evidently a ribonucleoprotein and determined its content in the cortical gray matter of
the rat, rabbit, and cow. We isolated from the nuclei of the gray
and white matter of the cerebral hemispheres a nucleoprotein consisting mainly of deoxyribonucleoprotein with very small quantities
of RNA [477]. These results suggested that Danilevskii's neuroglobulin is nuclear deoxyribonucleoprotein, whereas neurostromin
is ribonucleoprotein.

Other workers also have isolated individual fractions of the
brain proteins [63, 68, 1607].

Brain tissue contains both ribonucleoproteins and deoxyribo-
nucleoproteins. The nucleoproteins may exist as complexes with
lipids, such as the liponucleoprotein isolated by Folch and Uzman [1020].

Folch and Lees [1019] found substances of lipoprotein nature,
consisting of lipid and protein moieties and called proteolipids, in
the white and gray matter of the brain and in other tissues although
not in blood plasma. They differ from ordinary lipoproteins in

that, whereas the latter are soluble in water or in salt solutions, the proteolipids are insoluble in water but soluble in chloroform — methanol mixtures. Folch and Lees isolated three proteolipids from the white matter of the brain, one of which, containing 50% each of protein and lipid, was isolated in crystalline form. These workers considered the lipid moiety in the lipoprotein molecule to be surrounded by protein, whereas in the proteolipid molecule the protein is surrounded by lipid. Proteolipids participate in the formation of the myelin sheath and they are resistant to proteolytic enzymes.

Another proteolipid was isolated by Wolfgram [1785] from the white matter of the brain; it is present in myelin, and it has less tryptophan but more histidine and tyrosine than the proteolipid of Folch and Lees. A similar proteolipid can be obtained from peripheral nerves.

Fractionation of the proteolipids in the subcortical white matter was studied by Lewin and Hess [1290] and others [1265].

Mokrasch [1393] determined the distribution of proteolipid among the subcellular particles of the brain and found that its concentration is highest in the myelin-rich unpurified mitochondrial fraction.

Neurokeratin, a component of the sheaths of the nerve fibers, is one of the group of nerve tissue proteins that are insoluble in water and resistant to hydrolysis by proteolytic enzymes. In its amino acid composition it differs from the ordinary keratins. Neurokeratin was first isolated in 1877 by Ewald and Kühne [990] from brain and myelinated nerves. It is insoluble in alcohols, ether, chloroform — methanol mixtures, and dilute potassium hydroxide solution and it is not digested by gastric or pancreatic juice. In its amino acid composition, especially in its total content of glycine, alanine, leucine, and isoleucine, it closely resembles the proteolipids. Expressed per wet weight of tissue the content of neurokeratin in the gray matter of the hemispheres is 0.327% and in the white matter 2.243% [1236]. Neurokeratin also was studied by Le Baron and Folch [1274] and by Wolfgram and Rose [1788].

Nerve tissue contains collagen and elastin, similar in their physicochemical properties to the collagen and elastin of other organs [1550], and also proteins bound with glycogen [667]. The glycogen in glycogen—protein complexes may be bound to either water-soluble or water-insoluble proteins [668].

The protein cerebrocuprein, which contains two copper atoms per molecule, is found in very small quantities in the brain. There is more of this protein in the cerebral cortex than in the white matter [1499].

The phosphoproteins, containing phosphorus in the form of phosphorylserine, constitute a special group of brain proteins. A method of determining phosphoprotein in animal tissues and, in particular, in brain tissue, was developed by Schmidt and Thannhauser [1590] and later modified by Vladimirov [108] and Heald [1114]. The phosphoprotein phosphorus is easily removed as inorganic phosphate in an alkaline medium. The phosphoproteins are among the most active of the protein fractions of brain tissue; they have a very high rate of renewal, much higher than the rate of renewal of phospholipids and nucleic acids [330].

Study of the Composition of Soluble Protein Fractions of Nerve Tissue by Electrophoresis

The fractional composition of brain proteins, especially the soluble proteins, has been determined by the method of zonal electrophoresis, a technique that has been used with great success for the fractionation and analysis of proteins from blood serum, liver, and other organs.

When the proteins of blood serum (like those of other tissue fluids) are fractionated by electrophoresis on paper no preliminary treatment of the serum is needed. These protein mixtures can be applied directly to the paper strips for electrophoresis, though of course tissue proteins must first be extracted from the tissues. To isolate soluble proteins from a given tissue the conditions of extraction must be such that the proteins will be extracted as fully as possible but not denatured. In addition, extracts with a high protein concentration, about 2% or more, must be obtained.

One difficulty in investigating tissue proteins by electrophoresis on paper is the fact that other substances, adversely affecting the electrophoretic fractionation of the proteins on paper, pass from the tissues into the extract along with the proteins. For example, the extract contains many lipids that are adsorbed on the paper, thereby preventing migration of the proteins in the electric field.

Difficulties of this nature have been encountered by research workers using electrophoresis on paper for the fractionation of brain tissue proteins in which the lipid content is high. When tissue proteins are obtained from the brain or other parts of the nervous system for subsequent fractionation by electrophoresis special attention must therefore be paid to the choice of methods and conditions of extraction of the proteins from the nerve tissue.

The method of zonal electrophoresis on paper was first used to fractionate the soluble brain tissue proteins by Demling and co-workers [945]. They studied proteins soluble in salt solutions from various rat tissues in order to determine the site of formation of the blood serum proteins. The proteins were extracted from the tissue by autolysis followed by high-speed centrifugation. Among other organs these workers also studied the brain. In their paper they give one illustration of the electrophoresis of rat brain proteins, an extract of which was prepared by homogenizing the tissue and centrifuging the homogenate at 22,000 rpm. After comparing the results of electrophoresis of the brain and serum proteins these workers concluded that most of the brain proteins are globulins and the albumin content is small; extracts of brain proteins were separated by electrophoresis into six fractions.

Electrophoresis on paper also was used by Kaps [1182] when studying the pathogenesis of brain swelling. It is clear from the illustrations in his paper on electrophoresis of the soluble proteins of the gray and white matter of the human brain that he fractionated the proteins into five or six fractions, although the line of demarcation between them was not sharp. Comparison of the results of electrophoresis of the brain and serum proteins shows the almost complete absence of the fraction with the mobility of serum albumins from the brain proteins.

The methods of extraction of soluble proteins from brain tissue used by Demling and Kaps do not extract all the proteins, and they probably cause some degree of denaturation.

To find the most suitable methods and conditions for protein extraction comparative tests were carried out in which proteins were extracted from brain tissue by means of different solvents (buffered solutions) with different hydrogen ion concentrations, followed by electrophoresis of the resulting protein extracts.

These investigations showed that the higher the pH of the solvent the more proteins passed into solution. Veronal — medinal (i.e., barbital — barbital sodium) and borate buffers extract more proteins than physiological saline, solutions of potassium chloride, and distilled water. Acetate buffer with pH 5.6 and 5.0 extracts considerably less protein. Extraction from brain tissue with succinate buffer with pH 4.2 and 3.6 solubilizes only a negligible amount of protein.

The same experiments [533] showed that solutions with a high protein content obtained by extraction of brain tissue with alkaline solvents do not give clear separation of the proteins into individual fractions; this is because alkaline solvents extract large quantities of lipids which are adsorbed on the paper and interfere with the separation of the proteins during electrophoresis.

The most complete electrophoretic separation of protein extracts from nerve tissue is obtained by first freezing the homogenate with liquid air. Under these circumstances almost transparent solutions are obtained, and the separation of the proteins into individual fractions is correspondingly easier [533].

These observations in our laboratory were confirmed by Le Baron and Folch [1275] who found that the pH and ionic strength of different solvents affect the quantity of proteins extracted from brain tissue.

This same method of protein extraction [533] was also used by Palladin and Polyakova [467] in the systematic study of soluble brain proteins by electrophoresis on paper. They studied the soluble proteins of the gray and white matter of the cerebral hemispheres, cerebellum, and spinal cord of cats and found that they contained six or seven principal protein fractions with the electrophoretic mobility of blood serum globulins. The content of proteins with the electrophoretic mobility of blood serum albumins was comparatively low. Electrophoresis revealed definite differences both in the quantity of the individual protein fractions and in their relative proportions in different parts of the nervous system.

Generally similar results were obtained [527] in a study of soluble proteins from the gray and white matter of the cerebral hemispheres, cerebellum, spinal cord, and medulla of the cow by electrophoresis on paper. The proteins of these parts of the nervous system separated into five or six fractions; in this case also the protein content in the fraction with electrophoretic mobility corresponding to that of serum albumins was very small. Most of the proteins found migrated in the electric field with the characteristic speed of blood serum globulins.

These investigations showed that among the soluble proteins of the spinal cord and peripheral nerves there are electrophoretic zones of proteins that migrate toward the cathode during electro-

phoresis [526, 527]. Similar results were obtained later by other
workers [883, 1131, 1156, 1190, 1420, 1528, 1546, 1553, 1685] who
fractionated the soluble proteins of the brain.

It was interesting to compare the soluble brain proteins with
the soluble proteins of peripheral nerves. Investigations of the
soluble proteins of the sciatic nerve of the cat and cow undertaken
by Polyakova [526, 527] showed a high content of proteins with the
electrophoretic mobility of serum albumins in the peripheral nerves;
this finding was confirmed by other workers [1391] with the
soluble proteins of the sciatic, vagus, and optic nerves.

Since connective-tissue proteins contain "albumin," it may be
postulated that the albumin discovered during electrophoresis of
nerve proteins belongs to the connective tissues of the sheaths and
not to the nerve fibers. To study this problem Polyakova and
Kabak [534] tested whole nerve, isolated bundles of nerve fibers,
and connective tissue dissected from nerve. They found that the
albumin detected by electrophoresis of proteins of the whole nerve
also was present both in the isolated nerve fibers and in the connective tissue of the epineurium and perineurium. These experiments thus showed that albumin is present in the nerve fibers
themselves and not only in the connective tissue of the nerve trunk.
In electrophoretic mobility and other properties, nerve albumin
resembles serum albumin [528]. If a mixture of serum albumin
and nerve albumin is subjected to electrophoresis, a single peak
is found at the same place as the peak during electrophoresis of
serum albumin alone. Nerve and serum albumins dissolve in water
and in weak salt solutions. Both are precipitated by ammonium
sulfate between 50 and 80% saturation and both have an isoelectric
point at pH 4.5. The identity of nerve albumin and serum albumin
was finally confirmed by immunoelectrophoresis [530].

Since the nerve was carefully separated from most of the blood
vessels and well washed with physiological saline in the experiments described above, the albumin found among the nerve proteins
cannot be regarded as serum albumin. Li and Sheng [1292] also
concluded that nerve albumin and blood serum albumin are one and
the same substance.

Albumin is found not only in the nerves of adult animals, but
also in the nerves of embryos (calf embryos at the age of 2, 4,
and 7 months). The albumin content in nerve rises during embryonic development while the composition of the globulin fractions
becomes more complex [473].

Large quantities of albumin are found among the proteins of both unmyelinated and myelinated nerves; in this way the soluble proteins of nerve fibers differ from the soluble proteins of the CNS [528].

The method of electrophoresis of proteins on paper has the disadvanatage that filter paper adsorbs considerable quantities of proteins, particularly lipoproteins, in which nerve tissue is rich; this prevents migration of the proteins in the electric field and interferes with their separation on the paper strips.

Because of this fact, agar-agar gel, 98.5% of which consists of buffer solution, can be used as the supporting medium for electrophoresis instead of paper for it does not prevent migration of the proteins and does not adsorb them. The value of this method was emphasized originally by the workers who introduced the method of electrophoresis on agar [1070, 1071]. Accordingly the method of electrophoresis on agar has been used on an increasingly wide scale for the fractionation of blood serum proteins, especially after development of the immunoelectrophoresis technique by Grabar and his collaborators [168, 1075].

Palladin and Polyakova [468] used the method of electrophoresis on agar-agar to fractionate the soluble proteins of nerve tissue on the assumption that more complete separation of the brain proteins could be achieved in this way. Though initially the conditions of electrophoresis used were the same as those described by Grabar [1075], Ilkov and Nikolov [243], and others [1060, 1061, 1791] for separating blood serum proteins, these were later somewhat modified [468].

By fractionating bovine brain proteins on agar gel, 16 protein fractions are found instead of the six to eight fractions obtained by electrophoresis on paper, and the fractions, moreover, are more clearly separated. Among the soluble brain proteins identified by this method, a prealbumin zone was observed [468]. This fact has also been reported by other workers [1183].

In 1960, Vladimirov et al. [107] undertook the comparative electrophoretic fractionation of soluble brain proteins on paper and on agar gel and confirmed the superiority of the latter over the former.

Allegranza and Marobbio [758] fractionated soluble proteins from the rat brain and from various parts of the human brain by electrophoresis and found 9 to 12 protein zones, three of which were lipoproteins and eight glucoproteins. Other workers [614, 1516] found 16 or 17 electrophoretic zones (four of which were "prealbumin").

A comparison of the results of electrophoresis of soluble proteins from the normal brain and from brain tumors [394] revealed substantial quantitative differences in the distribution of the protein by zones but no qualitative differences between the fractions.

Yordanov [1795] investigated water-soluble brain proteins by a combined autoradiographic and electrophoretic method and found that in most of the electrophoretic zones the intensity of labeling was proportional to the protein concentration; in addition, darkening of the x-ray film was found even in the absence of corresponding protein zones. In his opinion the explanation of this fact is that some zones with a small protein content cannot be detected because of the low sensitivity of the method used for staining the protein.

During investigations of the cytoplasmic proteins of the brain by a slightly modified version of horizontal electrophoresis in starch gel, which was suggested by Smithies, up to 15 protein zones were found, five or six of which migrate toward the cathode. Differences also were found in the soluble brain proteins from animals of different species and in the protein spectra from different parts of the central nervous system [783].

By vertical electrophoresis in starch gel the water-soluble proteins of the rat brain were separated into 12 zones, and proteins of white matter from human brain into 14-15 zones [833]. The white matter of the human brain contains a small quantity of albumin-like proteins. Later [796], in a study of the properties of esterases in human brain the water-soluble proteins were separated into 20-25 zones by the same method.

Fractionation of the soluble brain proteins with a particularly high degree of definition has been achieved by electrophoresis on the substance PAAG [938, 939], combining the advantage of the agar gel mentioned above with the high resolving power of starch gel. To improve the extraction of the proteins from the brain the nonionic detergent Triton X-100 was used [1722].

Later, as the study of soluble brain proteins by fractionation was expanded, Polyakova and Lishko [535] combined the fractional salting out of the proteins with ammonium sulfate with electrophoresis on agar. Their experiments showed that the proteins salted out with ammonium sulfate at different degrees of saturation could be separated by subsequent electrophoresis on agar into varying numbers of protein zones with sharply different electrophoretic patterns.

When proteins salted out with ammonium sulfate at complete saturation were dialyzed against veronal−medinal buffer and then

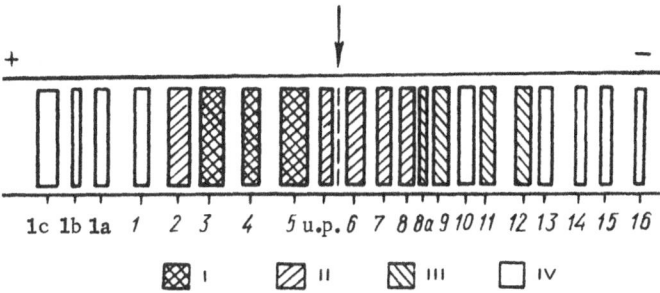

Fig. 1. Diagram of electrophoretic zones obtained by separation of nerve tissue proteins in agar gel. The arrow marks the point of application, "u.p." equals unfractionated proteins. The protein content of the zones is indicated by cross hatching: I > II > III > IV.

subjected to electrophoresis on agar, they separated into 12 fractions, whereas if distilled water was used for dialysis, 16 fractions were obtained [535]. Although much of the protein was precipitated and removed by dialysis against water, the number of protein zones obtained was increased and better separations were obtained than with the original undialyzed solutions.

Since the largest number of electrophoretic fractions (16) was obtained from protein solutions dialyzed against distilled water, these fractions were given identification numbers (Fig. 1) which were later used when describing the composition of fractions obtained with other extracts. The smallest number was given to the fraction migrating most rapidly toward the anode. As a result of these investigations a spectrum of soluble brain proteins was established.

The results showed that by the method described above, based on a combination of fractional salting out with ammonium sulfate and electrophoresis of the soluble brain proteins on agar, the largest quantity of a particular protein fraction can be obtained with a particular degree of saturation with ammonium sulfate. This is an important matter for the future investigation of these fractions and their individual protein components.

This method of fractionation of soluble brain proteins was used by Palladin and Kudinov [466] in a study of the proteins from the gray and white matter of the cerebral hemispheres to establish whether these soluble proteins differ in composition, since different workers had obtained conflicting results.

Their investigations showed hardly any qualitative difference between the proteins of the gray and white matter of the cerebral

hemispheres. Only by electrophoresis of proteins precipitated at
0.6-0.7 saturation with ammonium sulfate were differences found in
the protein spectra; the seventh electrophoretic fraction was not found
in white matter but was present among the proteins of gray matter.

Marked quantitative differences are found with the proteins in
individual electrophoretic fractions. For example, the first fraction of gray matter contains 50% more protein than the corresponding fraction from white matter.

The method of fractional salting out with ammonium sulfate
combined with electrophoresis on agar was later used for a comparative study of the proteins of the gray and white matter of the
spinal cord and of sciatic nerve [302]. No differences were found
between the proteins of gray and white matter of the spinal cord.
Comparison of the protein spectra of spinal cord with that of brain
fractions revealed some differences: Zone 16, which is always
observed among proteins of the cerebral hemispheres, is absent
from the proteins of the spinal cord.

The fractionation of sciatic nerve reveals very few proteins
with high electrophoretic mobility, i.e., those of electrophoretic
zones 1 and 2. This distinguishes the soluble nerve proteins from
those of the brain and spinal cord. The amount of protein in electrophoretic zone 3 (albumin) in the soluble nerve protein, however, is four times greater than in the soluble protein of the cerebral
hemisphere and spinal cord. Far fewer of the components not
readily separated by electrophoresis are found in nerve protein
than in spinal cord protein. Electrophoresis of the soluble protein
of the sciatic nerve failed to reveal any protein analogous to zone
16 of brain proteins.

This technique thus reveals electrophoretic zones in the soluble protein of the spinal cord and sciatic nerve that are not found
in the soluble protein of the cerebral hemisphere.

Fractionation of Proteins Extracted
from Subcellular Structures of Brain
Tissue by Detergents

Until recently the modern methods of separation of complex
protein mixtures were used [263, 466, 468, 529, 614, 1183] mainly
to study the proteins readily extracted from tissue with water or
salt solutions. These extracts contain chiefly proteins of the
cytoplasm of the cells and intercellular substance. The possibility
cannot, of course, be ruled out that this soluble fraction may also

contain soluble proteins from disintegrated cell organelles as well
as proteins extracted from intracellular structures during the
homogenization and fractionation of the tissue. The soluble fraction of the brain accounts for about 20% of the tissue protein.
Progress in the fractionation of soluble tissue proteins and the
study of their properties can be attributed mainly to the fact that
these proteins exist in the free state in the mixtures extracted:
either as simple proteins or as comparatively uncomplicated
supramolecular lipoprotein and nucleoprotein complexes.

Only a few years ago nothing was known about the composition
or properties of the proteins forming the complex membranous
structures of the cell or of the proteins of the cell organelles
weakly bound with the membranes, neither of which passes into the
soluble fraction during homogenization in isotonic solutions.
Nevertheless the role of these proteins in intracellular metabolism
and in the transport of metabolites through the cell membranes is
no less important for the functions of the cell than the role of the
water-soluble proteins of the hyaloplasm. Because nerve tissue
contains such a high proportion of lipids, lipoprotein, and proteolipid complexes, additional difficulties arise during the isolation
and study of the physicochemical and biological properties of its
water-insoluble proteins. It is not surprising, therefore, that the
first attempts to separate these proteins by electrophoresis in
agar and polyacrylamide gel [32, 478, 918, 1722] and on paper [393]
and also by chromatography on columns with DEAE-cellulose
[844] and hydroxyapatite [870] have only recently been undertaken. In
these investigations the proteins were extracted from whole tissue
homogenates [393, 1722] or from fractions of subcellular structures
[32, 478, 844, 870, 918] with the aid of detergents or urea or by
a combination of detergent treatment and irradiation with ultrasound.

We extracted proteins that could not be solubilized with the
usual solvents from brain mitochondria and from specific microstructures of nerve tissue such as myelin and nerve endings and
investigated the composition of these extracts. Of the methods of
solubilization tested (treatment with ultrasound, with Triton X-100,
and with sodium deoxycholate) treatment with the nonpolar detergent Triton X-100 was found to be the best. About 40% of the
protein from fractions of mitochondria and nerve endings, but only
about 17% of the protein of the myelin fraction, obtained by a modified [33] version of Whittaker's method [1769], can be solubilized
with a 0.1% solution of this detergent (primary Triton extracts).
By repeated treatment of the insoluble residues with 0.1% Triton

X-100 (secondary Triton extract) a further 15-20% of the protein from the fractions of mitochondria and nerve endings and 10-12% of the protein of the myelin fraction can be solubilized. Further treatment of the insoluble residues, but with a 1% solution of the detergent, solubilizes an additional 20-30% of the original protein of all the fractions. Some of the proteins insoluble in a 1% solution of the detergent can be extracted with 5% Triton X-100.

Sodium deoxycholate, in concentrations of 0.25 and 0.5%, solubilizes 60-70% and 80-85% of the protein of the subcellular fractions respectively. Not more than 15-25% of protein of the subcellular fractions tested could be solubilized with the aid of ultrasound.

Since sodium deoxycholate introduces substantial difficulties into the electrophoretic fractionation of the extracted proteins in agar gel and since ultrasound solubilizes too little protein for investigation, all subsequent solubilization of subcellular structures was carried out with solutions of Triton X-100. This substance extracts sufficient protein for analysis without interfering with the electrophoretic separation.

After appropriate treatment [32], the soluble proteins and the Triton extracts of the subcellular fractions of cat brain were fractionated by electrophoresis in agar gel in the usual way [243, 468]. The protein zones were identified by the numbers (Fig. 1) given previously [535] to the analogous fractions of water-soluble brain proteins.

The primary Triton extracts of mitochondrial and nerve ending fractions, like water-soluble brain proteins, consistently separate into 15 electrophoretic zones: five migrating toward the anode (1-5) and 10 toward the cathode (6-15). Additional zones, 8a and 1b, with a very low protein content, are sometimes found migrating toward the cathode during electrophoresis of water-soluble brain proteins.

Two additional zones of acidic proteins were found [32] during electrophoresis of the primary Triton extracts of the nerve ending fraction. These migrate toward the anode in front of zone 1 but are clearly separate from it and each other. These zones were called 1a and 1b (Fig. 1) and could not be found in Triton extracts of the myelin and mitochondrial fractions. Though it appeared that the strongly acid proteins of zones 1a and 1b might be specific components of nerve endings, this hypothesis was not confirmed experimentally. Quite clear zones 1a and 1b were found not only in aqueous extracts of gray matter from cerebral hemispheres, regions very rich in nerve endings, but also in the water-soluble proteins extracted from white matter, regions consisting chiefly of fibers, and also in extracts of corpus callosum, a structure composed entirely of a well-developed bundle of axons joining neurons of the right and left hemispheres.

More detailed information on the regional distribution, heterogeneity, and metabolic activity of the strongly acid proteins of 1a and 1b is given later in this chapter.

It must be emphasized that the proteins of the primary Triton extract of the myelin fraction are not as well separated in agar gel as the corresponding protein extracts from mitochondria and nerve endings [32]. The myelin fraction obtained by Whittaker's method [1769] is highly contaminated with soluble proteins and membranes of the endoplasmic reticulum. It is therefore possible some of the electrophoretic zones found in Triton extracts of the myelin fraction are of nonmyelin origin. This hypothesis was confirmed [478] by an investigation of the composition of the purified myelin obtained by Autilio's method [779].

The protein of the secondary Triton extracts of the subcellular structures differs significantly in composition from that of the corresponding primary Triton extracts. During electrophoresis in agar gel a single wide band migrates toward the anode in the region of zones 3-5. The protein extracted by 1% and 5% solutions of Triton X-100 from the residue of the subcellular structures cannot be separated electrophoretically.

The mechanisms of interaction between detergents and the protein–lipid membrane components have not yet received adequate study [143], and it is therefore not possible to give a detailed account of the protein composition of the fractions that we obtained by extracting these subcellular structures with Triton X-100 in different concentrations. It is certain that the primary Triton extracts contain not only the soluble proteins but also some enzymes fixed in situ on the membranes. This conclusion stems not only from the comparatively high content of protein in the extracts but is confirmed by other workers [870] who found more than 10 enzymes in mitochondrial Triton extracts. These were obtained after preliminary extractions of some of the protein from the mitochondria first with 0.02 M Tris buffer, pH 8.0, and later with distilled water. These initial extracts removed about 20% of the protein of the mitochrondrial fraction, whereas the Triton extract studied contained 33%.

During further treatment of the subcellular fractions with 0.1% solution of Triton X-100 part of the structural protein of the

membranes is evidently extracted. These proteins are comparatively homogeneous in their physicochemical properties for only
one extensive zone is formed during electrophoresis in agar.

The protein extractable from subcellular fractions with 1%
and 5% solutions of Triton X-100 also undoubtedly contain membrane structures. These proteins are not separated by electrophoresis in agar gel. After disintegration of the lipoprotein complexes of membrane structures with high concentrations of Triton
X-100, conformational changes possibly arise spontaneously in
the detached proteins, since the nonpolar detergents do not bind
with the protein molecules [1155, 1505] and probably do not affect
their structure. Another possibility is that these proteins cannot
be separated electrophoretically because they are in the form of
complexes, consisting of large fragments of membrane structures.

The quantitative analysis of protein extracts of the myelin
fraction obtained by Whittaker's method [1769], as was mentioned
above, was unreliable because of the large quantity of nonmyelin
structures present as impurities. Accordingly the protein composition of purified myelin isolated from the white matter of bovine
brain by Autilio's method [779] was next studied. In contrast to the original method, we did not separate the myelin preparations into "light"
and "heavy" myelin, but electron-microscopic investigation demonstrated the morphological homogeneity of the preparations (Fig. 2).

Proteins soluble in 0.5% Triton X-100 and in 0.1% sodium
dodecylsulfate were obtained successively from lyophilized preparation of purified myelin previously treated with acetone. The
preparation was treated with each detergent three times. The
Triton extract contained 22 ± 3% (mean of five experiments) of the
protein of the original myelin preparation, whereas the dodecylsulfate
extract contained 79 ± 3% (mean of six experiments) of the protein
insoluble in Triton X-100. Differences in the degree of extraction
of the protein with Triton X-100 and sodium dodecylsulfate are
evidently explained by differences in the degree of disintegration
of the myelin structures induced by these detergents, i.e., by the
different mechanisms of their interaction with the protein—lipid
components of the membranes. The nonpolar detergent Triton
X-100 probably causes changes in the membrane structure as a
result of hydrophobic interaction with the lipid—protein complexes.
Dodecylsulfate ions are evidently bound to the molecules of the
protein complex both by electrostatic interaction and by hydrophobic bonds, forming a hydrophobic structure around the molecule
[1155, 1505].

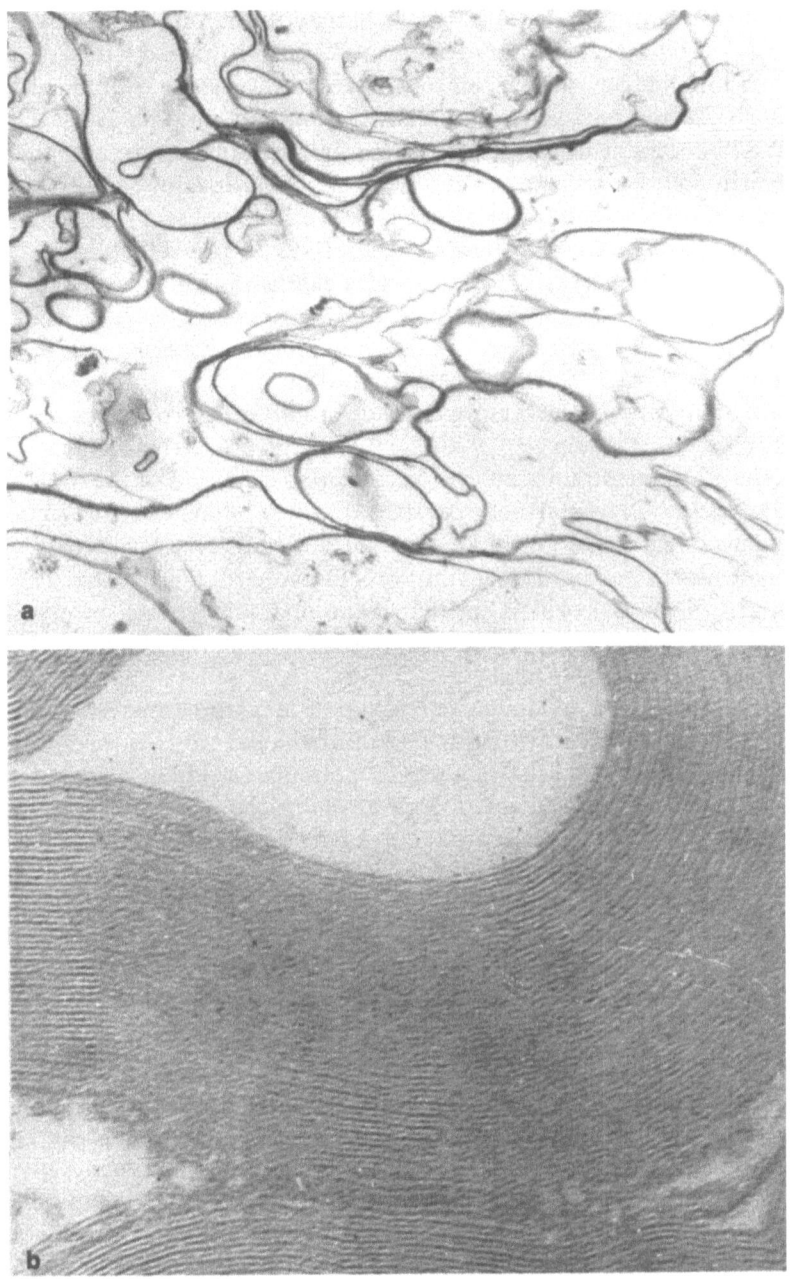

Fig. 2. Electron micrographs. a) Myelin fraction (14,000×) isolated from white matter of brain by Autilio's method [779]; b) fragment of myelin structure (100,000×).

The protein of a dodecylsulfate extract of myelin is separated [478] on a column of Sephadex G-200 equilibrated with 0.05 M
Tris glycine buffer, pH 8.3, into two peaks: The first peak contains
about 20% of the protein eluted from the column, and the second
about 80%. The proteins of the Triton and dodecylsulfate extracts
of myelin and the proteins of the two peaks obtained by gel-filtration of the dodecylsulfate extracts appeared microheterogeneous
when run through a column of Sephadex G-200. These proteins
were separated by vertical disc electrophoresis in PAAG as described by Davis [938, 939] and modified by Broome [867]. Electrophoresis was carried out in gels of various concentrations made
from Cyanogum-41 in Tris glycine buffer, pH 8.3.

The proteins of the Triton extract of myelin were fractionated
[478] in 4, 7.5, 11, and 15% PAAG. Better fractionation of the
proteins of the Triton extract of myelin was obtained in the 7.5%
and 11% gels. Five distinct protein zones were found. After prolonged washing of the gel in acetic acid, an additional five weakly
stained protein bands of uncertain origin could also be seen. The
possibility cannot be ruled out that these proteins are not of myelin
origin. Three distinct protein zones are found in the 4% gel and
four in the 15% gel.

Proteins of the dodecylsulfate extract of myelin were fractionated
in 7.5, 11, and 15% PAAG. These proteins separate into three distinct protein zones in the 7.5 and 11% gels but into four weakly
stained zones in the 15% gel. Much of the protein of the dodecylsulfate extract does not penetrate the gel but forms a deeply stained
protein band at the starting line.

The proteins of each of the fractions of peaks I and II obtained
by filtering the dodecylsulfate extracts of myelin through Sephadex
also were found to be heterogeneous on electrophoresis in PAAG.
The proteins of the first and second peaks separated in 7.5% gel
into two distinct bands, indicating that at least four protein fractions are present.

The protein spectra of Triton and dodecylsulfate extracts of
myelin thus differ significantly from each other. The nonpolar
detergent Triton X-100 and the polar detergent sodium dodecylsulfate evidently extract proteins with different physicochemical
properties and also, probably, with different functions, from myelin.
Differences in the protein composition of extracts obtained by
treating tissue membrane structures with various detergents and
salt solutions have also been described by other workers [844, 1729].

The proteolipid isolated from a chloroform—methanol brain extract is known [660, 1019] to be resistant to the action of trypsin, pepsin, papain, and erepsin, but to be split by pronase. We tested the resistance of the proteins of Triton and dodecylsulfate extracts of myelin to the action of trypsin and pronase. The results [478] showed that all the proteins of these extracts electrophoretically separated in PAAG were fully hydrolyzed by pronase and trypsin. The sensitivity of these proteins to hydrolysis by trypsin is propably explained by the dissociating action of the detergents both on the protein—lipid complex and on the protein liberated from it, as a result of which these fragments lose the resistance characteristic of the protein component of the original proteolipid.

Cotman and Mahler [918, 919] also detected a complex set of protein fractions during the fractionation of extracts of membrane structures from fractions of myelin, synaptic vesicles, synaptosomes, mitochondria, and various components of the microsomal fraction (membranes, microsomal vesicles, etc.), by the method of disc electrophoresis in PAAG. These workers obtained protein extracts of these membrane structures with a solution containing phenol, acetic acid, and water (2 : 1 : 1) and with 8 M urea. Altogether 17 electrophoretic bands were found, for example, in the protein spectrum of the synaptosome extracts. A similar composition was found also in extracts of the other subcellular fractions tested. Protein extracts of myelin, which separate in PAAG into four bands, two of them similar to proteolipids in their solubility, are an exception. Corresponding differences in the protein composition of myelin proteins soluble in detergents, on the one hand, and other subcellular structures (nuclei, mitochondria, synaptosomes), on the other hand, have been found by other workers [1729].

It may be noted here that by means of 0.9% NaCl solution about 75% of proteins separable electrophoretically in starch gel can be extracted from mitochondria of the rat brain. Eight bands of protein are seen, two of which migrate toward the cathode [648]. A complex spectrum of basic mitochondrial proteins has recently been discovered [1490] by a combination of salting out, gel filtration, and electrophoresis in polyacrylamide gel.

The proteins of the membrane components of the microsomal fraction extractable with 0.5% sodium deoxycholate solution can be separated [844] by column chromatography on DEAE-cellulose. Four fractions were eluted with 20 mM Tris HCl buffer, pH 7.6, followed by a concentration gradient of KCl (0.1-0.33-1.0 M) in the

same buffer. During starch gel electrophoresis of the deoxycholate extracts of the microsomal membranes a protein zone with
high cathodic mobility was found. This component characteristically is not found in the soluble fraction of the brain or in deoxycholate
extracts of mitochondria.

Protein extracts obtained by treating the gray matter of rabbit
brain with ultrasound in the presence of Triton X-100 (final concentration 0.5%) can be separated in PAAG of different concentrations
(from 4 to 15%) into 22-25 fractions [1722]. Since the gels of
different concentrations have bands that do not coincide, the authors cited suggested the extracts contained at least 30 protein
fractions. A very complex protein spectrum also was obtained
during separation in PAAG of the insoluble brain proteins isolated
with solvents containing K_2CO_3, urea, mercaptoethanol, and Triton
X-100 [1297], and also of Triton X-100-soluble proteins of subcellular brain fractions — mitochondrial, nerve endings, myelin,
and soluble [940]. However, it must be remembered that besides
membrane proteins, these extracts also contained all the soluble
proteins of the hyaloplasm and cell organelles, which are presumably responsible for these highly complex protein spectra.

The results described in this section are evidence that structural components of the cell and their membranes, like the soluble
tissue proteins, are highly complex although less heterogeneous
mixtures of proteins with varied physicochemical and biological
properties. The many approaches to separating these as yet
uninvestigated proteins of the membrane structures encourage
us to hope that more will soon be known of their structures and
functions. This type of approach is a prerequisite to elucidating
their role in the structural, metabolic, and functional organization
of the cell and its organelles.

Fractionation of the Basic Proteins

of Nerve Tissue

The intensified interest shown recently in the basic proteins,
i.e., those migrating toward the cathode during electrophoresis at
pH 8.6, results from investigations showing that these proteins are
not only structurally important components of the cell but they
also play an important role in its metabolism.

Of all the basic proteins, the histones have been studied the
most and have been chosen frequently by investigators as substances which might influence various biochemical processes. The
second group of proteins most commonly studied is the protamines.

The composition and the physical, chemical, and biological properties of the basic proteins show that they are highly heterogeneous and are represented in the cell not only by histones and ribosomal proteins, but also by many other basic proteins such as the cytoplasmic basic proteins and the soluble proteins of the nuclear sap.

One of the most interesting properties of the basic proteins is their ability to react with a very wide range of compounds (proteins, nucleic acids, lipids, carbohydrates). The complexes thus formed are highly labile. This high reactivity of the basic proteins also explains the fact that they affect many biological processes such as protein synthesis, the permeability of cell membranes, the growth of malignant tumors, processes of excitation, and so on.

As yet the basic proteins of the nervous system have received comparatively little study. In 1956, Polyakova [526], studying the soluble proteins of the nervous system by electrophoresis, first discovered basic proteins among the soluble proteins of spinal cord and later among the soluble nerve proteins [527].

Bailey and Heald [783] studied the cytoplasmic proteins of the brain by electrophoresis in starch gel and found five or six electrophoretic zones migrating toward the cathode, but they did not describe any additional details.

In a study of basic brain proteins, Honegger [1136] used the brain of a person aged 73 years removed 6 h after death. Extracts were fractionated on columns with Sephadex G-25 and G-100 and by electrophoresis in starch gel. He found that the gray and white matter of the brain contain up to 18 different basic proteins, five of which are characteristic of gray matter only, and six of white matter only, the rest being common to both parts of the brain. The basic proteins of white matter had higher mobility and lower molecular weight than those of the gray matter.

Six basic proteins were isolated from an acid extract of bovine spinal cord [1421] by fractionation on CM-cellulose and gel-filtration with Sephadex G-25. All six proteins were homogeneous on electrophoresis in acrylamide gel. The molecular weight as determined by gel filtration was 37,000. Three proteins had encephalitogenic activity.

Polyakova and Lishko [535] had found earlier that of the 16 zones obtained by electrophoresis of soluble brain protein on agar, fractions 14, 15, and 16 migrated toward the cathode (veronal-medinal buffer, pH 8.6), evidence of their basic character.

Accordingly Kudinov and Polyakova [300] set out to study
fraction 15 as a basic protein, to isolate it from the brain, purify
it, and study its composition and properties. For this purpose an
extract of water-soluble proteins was obtained from bovine cerebral hemispheres and fractionated with ammonium sulfate. The
fraction salted out at 0.7 saturation, was dialyzed, lyophilized, and
chromatographed twice on DEAE-cellulose and then on Sephadex
G-100. A protein of electrophoretic zone 15 was thus obtained
which proved to be homogeneous on electrophoresis in agar, re-
chromatography on DEAE-cellulose and Sephadex G-100, and ultracentrifugation.

Analysis of the amino acid composition of the protein and
microelectrophoretic studies confirm the basic character of this
protein.

The study of its amino acid composition by Moore's method
[1401] with an amino acid analyzer revealed 15 amino acids, of
which aspartic and glutamic acids were present in the largest
quantities; their combined content was greater than that of lysine,
histidine, and arginine together. The analysis also revealed a high
concentration of ammonia, indicating that some of the dicarboxylic
amino acids were formed during hydrolysis from glutamine and
asparagine. Determination of the content of amide ammonia
showed that about half of the total quantity of aspartic and glutamic
acids is contained in the protein in amide form. The ratio between
the combined content of basic amino acids and the total content of
dicarboxylic amino acids containing a free carboxyl group is 1.49,
further evidence of the basic nature of the protein. Almost the
same content of basic amino acids and almost the same ratio between basic and acidic amino acids are observed in the basic protein isolated from the soluble protein of the nuclei [1746].

The protein of zone 15 has a high content of leucine, valine,
and proline, i.e., of amino acids with lyophobic groups. In its
amino acid composition this protein differs from the proteins of
nuclear ribosomes, histones, and basic proteins of thymus nuclei.

Determination of the isoelectric point of the protein of
zone 15 showed that it lies between pH 9 and 9.4. Its molecular
weight was found to be 31,500, corresponding to the sedimentation
coefficient of 3.25 S found for this protein in a concentration of
0.74% [299].

Electrophoresis of this protein in PAAG demonstrated its
microheterogeneity: It separates into seven bands [299]. It contains seven N-terminal amino acids. All this indicates that the basic

proteins of electrophoretic zone 15 of the brain differ both in their amino acid composition and physicochemical properties from the basic proteins described previously (histones, ribosomal proteins, etc.).

Besides the results of investigations into the fractionation of basic proteins from different parts of the nervous system and isolation of the separate fractions, data suggesting their possible biological roles can also be found in the literature. For example, McIlwain et al. [925, 1321, 1322, 1352] studied the ability of brain cells in vitro to respond to electrical stimulation in a medium containing basic protein.

To examine the biological role of the basic brain protein that we had isolated, its effect on ATPase, on adenylic acid deaminase, and on guanosine triphosphatase was studied [303]. These enzymes were chosen for the following reasons. There is evidence in the literature that the ATPase participating in active ion transport is under the influence of basic proteins such as histones and protamines. The study of the effect of basic brain protein on adenylic acid deaminase is particularly interesting because this enzyme is sensitive to changes in phospholipid metabolism and, as evidence in the literature shows, is under the influence of some basic proteins. Guanosine triphosphatase is linked with the incorporation of amino acids into proteins. Our experiments showed that the protein of zone 15 activates Na^+, K^+-ATPase but inhibits Mg^{++}-ATPase as well as adenylic acid deaminase and guanosine triphosphatase.

The main object of the subsequent investigations of basic proteins of the nervous system was to discover protein fractions that induced allergic encephalomyelitis.

If the white matter of bovine brain is freed from lipids by treatment with a mixture of chloroform and methanol and subsequently treated with water and dilute HCl, about one-third of the protein passes into the acid extract [1359, 1360]. Fractionation of this acid extract by ion-exchange chromatography, gel-filtration, and disc electrophoresis yields a number of protein fractions. Most of the protein extractable by acid is insoluble in a neutral medium and consists of large molecules of nonbasic protein. However, the two protein fractions which are soluble at neutral pH are of low molecular weight and high basicity. These two fractions account for 1% of the wet weight of the white matter of the brain; they are rich in glycine and basic amino acids and they can induce experimental allergic encephalomyelitis.

Many basic protein and proteolipid fractions isolated from the brain and spinal cord possess encephalitogenic activity [1196,

1216, 1241, 1552, 1739]. Soluble collagen-like proteins also possess encephalitogenic activity [1550, 1551]. One of these proteins was isolated in a pure form and its homogeneity demonstrated by electrophoresis on paper and by ultracentrifugation [1551].

A highly purified encephalitogenic protein was obtained by Kibler from spinal cord [1192]. On electrophoresis in PAAG it separates into three zones with molecular weights below 10,000.

Later, Kies [1195] isolated two encephalitogenic basic proteins with a molecular weight of 35,000-40,000 from the white matter of the bovine spinal cord. She considers these two proteins identical and classes them among the "large" encephalitogenic proteins, in contrast to the protein described by Kibler.

The brains of embryonic and neonatal animals contain no encephalitogens [1480]. Other workers [1643] state that the proteins of the immature brain also have no encephalitogenic activity. During an investigation of human brains (taken at autopsy) at different stages of development (starting from 6 weeks before birth) by electrophoresis no encephalitogenic protein could be found at the 4th and 6th weeks before birth or at the 7th week after birth or, indeed, probably up to the 8th month [976-978].

Encephalitogenic peptides with a molecular weight of 1500-3500 were isolated from spinal cord. These peptides were fractionated on Sephadex G-75 into a number of fractions, two of which induced encephalomyelitis. Since a "mild" method was used to isolate the polypeptides, the encephalitogenic peptides were assumed to be present in the spinal cord and are not breakdown products produced during the fractionation.

Tests of the encephalitogenic activity of 173 different protein fractions from nerve tissue, including proteins, proteolipids, peptides, and compounds containing gangliosides, led to the conclusion [1311] that encephalitogenic activity is due to basic polypeptides with a molecular weight of 4000-6000, which can form complexes with lipids, phosphatides, and gangliosides.

A basic protein with molecular weight of 16,500 and polypeptides with molecular weights of about 3500 were isolated from bovine spinal cord [884]. Similar fractions were also isolated from guinea pig brain by a method including defatting, acid extraction, and repeated chromatography on Sephadex G-50. The basic protein and polypeptides thus obtained induce experimental allergic encephalomyelitis in guinea pigs. Encephalitogenic polypeptides are evidently formed from the basic protein of myelin as a result of the action of acid proteinase during extraction with acid.

An encephalitogenic protein was also isolated from bovine
spinal cord by Nakao [1422]. On electrophoresis in PAAG the acid
extract from almost completely defatted spinal cord separates
into seven components migrating toward the cathode. Five of them
were obtained in a relatively pure form and two induced allergic
encephalomyelitis in guinea pigs.

Three fractions can be obtained from spinal cord homogenates
by high-speed centrifugation: supernatant, gel, and residue. Although all these fractions have antigenic activity, the gel has the
strongest encephalitogenic properties [1278].

Chao and Einstein [1291] isolated an encephalitogenic fragment
obtained by enzymatic degradation of the encephalitogenic protein
of the spinal cord with acid proteinase from nerve tissue and
determined its chemical and physical characteristics.

An encephalitogenic protein isolated [991, 1110] in a homogeneous state from bovine spinal cord contains 38 moles basic amino
acids and only 7 moles of acidic amino acids per mole of protein.
It also contains one tryptophan and two methionine residues and differs
from histones in its higher content of histidine, glycine, and serine
and its lower content of alanine. The molecular weight is about
16,000. This protein is highly resistant to denaturation.

Closely linked with these investigations of the encephalitogenic
basic proteins of nerve tissue are the many recent studies of the
structure and properties of myelin. A basic protein contained in
myelin, as was stated above, has encephalitogenic activity. Surveys by Schmitt and Finean [998, 1591] containing information
about myelin were published in 1957. More recently Autilio and
co-workers have studied the isolation of purified myelin from the
CNS and its properties [777, 779].

A protein accounting for 20% of the myelin protein was isolated
from the white matter of the human and bovine brain [1307]. The
protein was extracted from the myelin preparations with 0.2 N
sulfuric acid; on electrophoresis in starch gel it migrated as a
single band toward the cathode. Amino acid analysis showed that
more than 20% of this basic protein is composed of dibasic amino
acids, with only a little methionine and no cysteine whatever. On
injection into guinea pigs this protein induced experimental allergic
encephalomyelitis.

Palladin, Terletskaya, and Belik [478] used the method of extracting proteins with various detergents to study the composition of
myelin. White matter from bovine brain was used, and myelin obtained from it by Autilio's method which gives a high yield of

myelin (5-6% of the wet weight of tissue) with a high degree of
purity. The myelin was freeze-dried and the purity verified under
the electron microscope. First some of the lipids were removed
from the myelin by means of acetone, after which the proteins
were extracted with the nonpolar detergent Triton X-100 and the
anionic detergent sodium dodecylsulfate.

Proteins of the dodecylsulfate extract, like proteins of the
Triton extract of myelin, are completely hydrolyzed by proteolytic
enzymes.

Our next paper [479] dealt with a basic protein of myelin that
induces experimental allergic encephalitis. This protein proved
to be homogeneous on gel-filtration, ion-exchange chromatography,
and ultracentrifugation; its molecular weight is 17,000. On electrophoresis in PAAG the protein showed high cathodic mobility
and gave one zone in an alkaline medium but three or five bands in an
acid medium depending on the gel concentration. The encephalitogenic material isolated is probably a mixture of molecules with
closely similar molecular structures.

That this protein has a higher content of basic than acidic
amino acids was determined by amino acid analysis. The total
content of basic amino acids is twice that of dicarboxylic amino
acids. The protein has a high content of serine, glycine, alanine,
and phenylalanine. Its amino acid composition is almost identical
to that of the encephalitogenic proteins isolated by other workers
from brain tissue. At the same time its amino acid composition
differs from the histone fraction obtained from brain nuclei, chiefly
in the lower content of histidine, serine, and glycine in the histones.

Comparison of the composition of myelin from the CNS and
PNS of anthropoid primates showed [1139] a similar molecular
structure but also differences in chemical composition. These
problems have been studied by other workers [1634, 1692, 1786].
Wolfgram and Kotorii [1786], for instance, isolated a myelin fraction by Autilio's method from spinal roots, after removal of the
membranes, and showed [1374] that the myelin obtained from a
peripheral nerve differed from the myelin of CNS white matter in
its higher content of valine, tryptophan, serine, lysine, and
aspartic acid and its lower content of cystine, cysteine, phenylalanine, alanine, histidine, proline, and glycine. The protein of nerve myelin was insoluble in a mixture of chloroform and
methanol and was hydrolyzed to a greater degree by trypsin and
pepsin. Myelin from peripheral nerve evidently does not contain

the Folch-Pi proteolipid protein, and its basic protein component is of a different nature.

After studying the protein composition of myelin from different parts of the nervous system Wolfgram and Kotorii [1786] concluded that the amino acid composition of the myelin is not identical with the amino acid composition of Folch-Pi proteolipid protein and that myelin contains an additional protein fraction rich in dicarboxylic acids.

The quantitative content of protein in myelin is another disputed problem. It can be concluded from investigations by several workers that the protein content in myelin varies between 19 and 30% [1447].

Having isolated myelin by Autilio's method [779] and having investigated the composition of its proteins, Gonzalez-Sastre [1069] obtained three different protein fractions. The first fraction was obtained as the insoluble residue after extraction of the myelin with neutral chloroform – methanol mixture. This fraction is hydrolyzed by pepsin and its amino acid composition is similar to that of the proteolipid protein studied by Wolfgram [1785]. The second fraction was precipitated by the addition of various electrolytes to the chloroform – methanol extract. This fraction was a basic protein identical in its electrophoretic mobility and amino acid composition with the basic protein obtained by Martenson and Le Baron [1360]. The third fraction, remaining in solution in the chloroform – methanol mixture, contained a protein resistant to trypsin with an amino acid composition similar to the proteolipid of Folch and Lees [1019].

There is evidence that myelin contains neurokeratin [1787] and also activities of various enzymes, notably cathepsin, aminopeptidase, and monoamine oxidase [743, 745, 1105].

The properties of myelin and its component proteins have been investigated by Martenson and co-workers [1356] who compared highly basic proteins isolated from bovine and rat brain [1357]. A study of two protein fractions obtained by electrophoresis of acid extracts from the rat brain suggested that the myelin contains at least two strongly basic proteins [1194].

The basic myelin proteins extracted from the white matter of the bovine brain have lower electrophoretic mobility than the two basic proteins from rat brain, which also are evidently components of myelin. Separation of the basic myelin proteins from the rat CNS into two fractions after repeated gel-filtration on Sephadex

G-100 was later reported [1355]. Each fraction was found to be homogeneous on gel-filtration and electrophoresis in 5% and 15% PAAG in an acid medium. These two fractions with molecular weights of about 17,000 and 14,000 [943] differed in their amino acid composition and encephalitogenic activity. Further investigations showed that the basic myelin protein isolated from CNS tissue can be separated into six components by electrophoresis in PAAG at an alkaline pH in the presence of urea [1353]. The relative quantities of the components obtained by electrophoresis [1357] are affected by the duration of treatment of the original rat brain homogenate with the chloroform—methanol mixture.

Fractions with similar molecular weights were found by electrophoresis in the basic myelin protein isolated from the CNS of four species of mammals (man, ox, rabbit, guinea pig) [1354, 1357].

This protein separated into two components, one of which, with a molecular weight of 17,000, is similar in molecular weight, amino acid composition, and encephalitogenic activity to the myelin proteins found in other species. The other component, with a molecular weight of 14,000, is not found in the myelin of any species so far studied other than those indicated. Two electrophoretically similar basic myelin proteins have also been found by other investigators [918, 988, 1373].

Chao and Einstein [893] used various physicochemical methods to study a highly basic encephalitogenic protein isolated from bovine spinal cord. Its molecular weight is about 18,000-20,000. The protein, in the presence or absence of a denaturing agent, is monodisperse, existing in a random configuration and behaving like a polyelectrolyte in neutral aqueous solutions.

Recently Kies [1194] published a survey of the latest research on myelin.

Fractionation of the Acid Proteins

of Nerve Tissue

Most proteins specific for nerve tissue so far discovered have been acid proteins. In the last few years they have been studied in many laboratories in various countries.

One such acid protein specific for nerve tissue was discovered in 1965 by Moore and McGregor [1397] by fractionation of water-soluble brain proteins on a DEAE-cellulose column and electrophoresis of the resulting fractions on starch gel. In one of the chromatographic fractions with high affinity for cellulose these workers found a protein zone with a high electrophoretic mobility toward the anode. This protein zone was not present in liver proteins. Later Moore [1396] showed that on electrophoresis in starch gel with the discontinuous buffer system described by Poulik [1501], this protein migrated toward the anode ahead of the buffer front as a separate band in front of the other acid proteins. Moore isolated it in a pure form, confirmed its specificity for the nervous system, and showed that this protein is present in the whole brain of 17 species of vertebrates (pig, rat, rabbit, mouse, dog, monkey, man, etc.) and also in different parts of the CNS and PNS (white and gray matter of the brain, cerebellum, brain stem, spinal cord, vagus and sciatic nerves, retina). This protein was not found in other organ tissues of the rat: liver, kidneys, heart, muscles, lungs, blood serum, erythrocytes.

The protein discovered by Moore was called S-100 protein because of its ability to remain in solution in the presence of 100% saturation with ammonium sulfate at neutral pH.

The method of isolating S-100 protein from the brain of cows, rabbits, and pigs as developed by Moore [1396] consists of extracting the soluble proteins from the brain tissue by 5 mM Tris phosphate buffer (pH 7.2), salting out with ammonium sulfate at complete saturation in acid medium (pH 4.2), and fractionation on a column of DEAE-cellulose by gradient elution. The final stage of purification of the protein was by chromatography on Sephadex G-200 and DEAE-Sephadex A-50.

Amino acid analysis of S-100 proteins from bovine and rabbit brain revealed the similarity of their amino acid composition and their high content of dicarboxylic amino acids (up to 30 mole % glutamic and aspartic).

The molecular weight of S-100 protein, determined by different methods, varies between 15,000 and 30,000 [1396, 1399, 1721]. Recently a molecular weight of 21,300 was established [932] by the method of high-speed equilibrium centrifugation as described by Yphantis [1799] in the presence of β-mercaptoethanol and it was observed that after electrophoresis in acrylamide gel the molecular weight was 7000. These results suggest that S-100 protein consists of three equal subunits and that the molecular weight of the native S-100 protein is 21,000.

After an antiserum against S-100 protein became available, a quantitative method of determination of this protein was developed on the basis of the C'-complement fixation test. It was confirmed by the use of this method that S-100 protein is present in the nerve tissue of all species of vertebrates studied and in some species of invertebrates [1289, 1399]. The distribution of S-100 protein varies from one part of the brain to another, with the cerebellum having the highest content. There is more in the white matter of the cerebral hemispheres than in the gray, but its content is almost the same in the white and gray matter of the cerebellum. Although considerable variations in the S-100 content are found in different parts of the adult human brain, the content in each part is consistent from one experiment to another [1396, 1400].

In 1966 Mandel and co-workers [1064] suggested another, more rapid method of obtaining S-100 protein. They filtered soluble brain proteins through millipore filters and adsorbed them on hydroxyapatite; the S-100 protein was removed from the hydroxyapatite by 0.02-0.2 M potassium phosphate buffer (pH 6.8). The concentrated filtrate was electrophoresed in starch gel by Smith's method [1624] and the pure protein eluted from the gel.

S-100 protein appears fairly early in ontogeny [1806]. By the 10th week the concentration of this protein can be reliably determined in the phylogenetically older parts of the human brain. By the 20th week the level of S-100 protein in all parts except the frontal cortex is half that found at birth. It may be extremely important that S-100 protein appears in the frontal cortex a few weeks before birth and that its level rises rapidly since its formation in the cortex of the frontal lobe coincides with the appearance of electrical activity of the brain.

Mainly two groups of workers have studied the cellular localization of S-100 protein. Hydén and McEwen [1153] studied a large neuron and the surrounding glia in Deiters' nucleus of the rabbit. By simple precipitation in a capillary tube with agar combined with the fluorescent antibody method they found that more S-100 protein is located in the neuroglia and less in the nucleus of the neuron.

In their study of the cellular localization of S-100 Moore and co-workers used Wallerian degeneration of the sciatic [1485] and optic [1484] nerves of the rabbit. They concluded from their experiments on the sciatic nerve that S-100 protein in the PNS is located in the axon rather than in the myelin and Schwann cells [1485]. However, after considering the results of their observations on the optic nerve [1484] these workers postulated that S-100 protein in the PNS is synthesized by the Schwann cell [1484]. During investigations of retrograde degeneration of posterior thalamic neurons, a parallel was observed [903] between the content of S-100 protein at the various stages of retrograde degeneration of the neurons and the degree of glial hypertrophy. As regards the type of glial cells in which this protein is localized, the results of these experiments indicate that most of it is in the astrocytes although the presence of S-100 protein in the oligodendroglia cannot be ruled out. The predominant localization of S-100 protein in the glia is also confirmed by the fact that an astrocytoma, which can induce the formation of glial tumors in young rats [811], contains up to 2% of S-100 while developing in tissue culture.

Although S-100 protein behaves as if homogeneous on ultracentrifugation and gel-filtration on Sephadex [1396], its heterogeneity has been revealed by other methods. For instance, Hydén and McEwen [1317] separated S-100 protein by electrophoresis in PAAG into three or four fractions. These results could be explained by the formation of complexes of S-100 protein with the components of the buffers or by aggregation of the protein. The hypothesis of protein aggregation was rejected on the basis of the sedimentation constants of the pure S-100 protein in four different buffers, which in all cases were very close [1065, 1067, 1668, 1721]. Determination of the N-terminal amino acids supplied additional evidence of the heterogeneity of S-100 protein. Terminal glutamic and aspartic acids, glycine, serine, alanine, threonine, lysine, isoleucine, and valine were found [1667].

An important contribution to the study of the heterogeneity of S-100 protein was made by the French workers Mandel et al. [997, 1705] using a technique of electrophoresis in mixed polyacrylamide—agarose gel developed by Uriel [1704]. They found that S-100 protein consists of two components — one migrating quickly, the other slowly. The ratio between these components in different parts of the bovine brain varies. In the gray matter of cerebral hemispheres and in the cerebellum the fast-migrating fraction is predominant, but in the white matter of the hemispheres, the cerebellar peduncles, and thalamus the slower fraction is predominant. Other evidence of the heterogeneity of S-100 protein was obtained by studying the effect of Ca^{++} ions on the protein's electrophoretic pattern in PAAG. Pure S-100 protein from bovine brain, although apparently homogeneous in starch gel and PAAG (without Ca^{++}), is found to be heterogeneous on chromatography, centrifugation in a sucrose density gradient, and electrophoresis in PAAG in the presence of Ca^{++} in low concentrations (0.01-0.1 mM) [881]. A study of the possible mechanism of action of Ca^{++} led to the conclusion that Ca^{++} ions bind with the S-100 protein and induce limited conformational changes, as a result of which the protein molecules become more open in structure. The action of Ca^{++} ions on S-100 protein is specific and evidently takes place at a site containing one tryptophan residue, several tyrosine and phenylalanine residues, and two of the three cysteine residues present.

Further investigations [1568] gave support to the view that the two SH groups of the cysteine residues in the protein molecule are oxidized during prolonged incubation in the presence of Ca^{++} ions with the formation of disulfide bridges, as a result of which the protein is stabilized in the modified conformation. During incubation one form of protein changes into the other, thus explaining the appearance of two forms of S-100 protein during electrophoresis in the presence of EDTA.

The localization of S-100 protein in the nerve tissue of only some species of animals, its unequal distribution in different parts of the nervous system, and its strong acidity, high mobility on electrophoresis, and solubility in concentrated salt solutions all suggest that S-100 protein may be linked with specific functions of the nervous system. Naturally, therefore, many different aspects of this protein are being closely studied in many laboratories of the world in attempts to discover its specific role in the functions of the brain.

There is increasing evidence to show that proteins and RNA participate in the mechanism of memory. This conclusion is supported by the intensified protein and RNA synthesis in nerve tissue when activity of the nervous system is increased in response to external stimulation, including irradiation experiments [623, 1157, 1561], and by disturbances of memory formation after administration of inhibitors of protein and RNA synthesis [794].

Among others, Hyden [674, 1151, 1152] has drawn attention to the role of protein synthesis in memory. As a result of his investigations he concluded that the protein synthesis taking place during learning is essential for the formation of long-term memory. Naturally, after the discovery of S-100 protein and of its specificity, he set out to study whether it was connected with the process of learning. To examine this problem Hydén studied the formation of behavioral responses in rats and chose as his test object the pyramidal neurons in the region of the cornu Ammonis of the hippocampus. He chose this region on the basis of observations on man which showed that bilateral destruction of the hippocampus leads to gross defects of memory and inability to learn [1392, 1454]. Hydén developed new behavior in rats by teaching them to use the left paw when feeding. The neurons were removed for study 15 min after the last period of training. Once developed, the new behavior persisted for a long time. Under halothane anesthesia an injection of 60 μCi leucine-^{3}H was given into the ventricles of both hemispheres 30 min before the beginning of the last training period and the specific activity of the proteins of the neurons determined. Hyden concluded [1152] that during learning the intensity of protein synthesis rises in hippocampal neurons and that this increase is particularly marked during the initial learning period and during a short period of retraining on the 14th day. Separating the hippocampal proteins of the control and of repeatedly trained animals and identifying the protein fractions suggested that during training the hippocampal neurons synthesized an acid S-100 protein specific to the brain much more intensively than did hippocampal neurons of the control animals. Apparently [1152, 674] the proteins participating in behavior formation are synthesized either during, or in the first few minutes after, training.

Richter [1523] observed the effect of stimulation on the loss of radioactivity of previously labeled rabbit brain protein. He found a relatively rapid fall in the specific activity of protein-bound glutamic acid in the first few days after administration of carbon-labeled glucose. These findings indicate that the brain contains a very rapidly metabolized protein with a high glutamic acid content. This protein is conjecturally acid S-100 protein.

The fact that the brain contains acid specific proteins is also confirmed by the results of immunochemical investigations. By immunoelectrophoresis on agar, 11 antigens were found among the soluble brain proteins [1217]; six of them were nonspecific, for they were found in other organs as well (bone marrow, liver, spleen, testicles), whereas five were specific for the brain; their precipitation arcs in the agar are still present even after the brain antiserum has first been exhausted with antigens from other tissues. Immunoelectrophoretic analysis showed that three of these five brain antigens have high mobility toward the anode, i.e., that they are acid proteins.

Other workers [582] also found five specific protein antigens among the soluble proteins of the adult rat brain. It is interesting to note that no specific proteins could be found among the soluble proteins in the brain of newborn rats. Specific protein appeared in the zone of the α- and β-globulins on the 5th day after birth. By the 7th to the 15th day, three specific antigens had appeared in the zone of the albumins and an additional protein antigen appeared in the prealbumin zone in the brain of adult rats. These specific proteins are presumably formed as brain functions become more complex. Other investigations [151, 1325] also indicate that at least five or six specific proteins can be detected by immunochemical methods among the soluble brain proteins.

Hydén and Mihailovic [674] studied antigenic differences between the glia and neurons of Deiters' nucleus in rabbits and found that glial antigens of Deiters' nucleus were nonspecific whereas the neuronal antigens, on the other hand, exhibited high specificity. Judging from the specific fluorescence data, the neuronal antigens are located at the periphery of the cell and in its thin processes, indicating a connection between the antigens and the plasma membranes

Besides S-100 protein, Moore and co-workers [1399] isolated a second protein specific for nerve tissue from the bovine and human brain and called it "acid protein 14-3-2" because of its characteristics during chromatographic purification. This protein is widely represented in the CNS and PNS of mammals and birds and, unlike S-100 protein, its content in the gray matter of the brain is higher than in the white. The study of the content of protein 14-3-2 in the optic nerve during Wallerian degeneration [1484] and in the thalamus during retrograde degeneration [903] led to the conclusion that this protein is present mainly in neurons and only in much smaller quantities in the glial cells. The protein readily forms aggregates consisting usually of two equal subunits. The molecular weight of the monomer is about 38,000 [1076].

Another acid specific brain protein, known as "antigen a," was isolated in 1968 by Bennet and Edelman [816]. Antigen a was obtained in the pure form from rat brain. Investigation of the isolated preparation by disc electrophoresis in PAAG revealed two bands. Studying this protein in various electrophoretic systems in the presence of urea and also by ultracentrifugation at different pH values in the presence or absence of β-mercaptoethanol led to the conclusion that antigen a consists of equal subunits whose molecular weight is about 39,000. Under certain conditions these subunits aggregate to form dimers with a molecular weight of the order of 83,000. Aggregation takes place with the participation of sulfhydryl groups which form disulfide bridges. Amino acid analysis of antigen a showed [816] that it contains more proline than S-100 protein but much less valine and phenylalanine; the content of dicarboxylic amino acids (glutamic and aspartic) is twice that of diaminomonocarboxylic acids.

Comparison of the physicochemical properties of protein 14-3-2 and antigen a shows that they are identical. The two pro-

tein preparations are possibly really the same protein obtained in
different ways by the authors who described them.

Nerve tissue also contains some specific acidic glycoproteins.
Acidic glycoprotein 10B, specific for brain tissue and containing
hexoses, hexosamine, and neuraminic acid, has been studied now
for several years [841]. This protein was first isolated by Bogoch
[842] from a glial tumor of the human brain, in which its content
was 69 times greater than in healthy human brain. The content of
this protein is increased sevenfold in Tay—Sachs disease. Protein
10B is evidently a glial protein.

Bogoch [842] developed a method of extracting and fractionating brain glycoproteins by which their total content in the brain
could be determined and the individual glycoproteins of the 10B
type (10BI, 10BII, 11A) could be isolated. These three proteins
differ in their structure and in the content of their carbohydrate
components. On the basis of his observations on quantitative
changes in certain proteins of the 10B type in pigeons during training and on the specific localization of glycoproteins in synaptic
junctions, Bogoch [842] postulated that glycoproteins are involved
in the biochemical reactions taking place during training and learning.

Yet another glycoprotein with a high content of neuraminic
acid and specific for the brain was found in 1967 in human white
and gray matter of cerebral hemispheres, spinal cord, and PNS
[1747, 1751]. This α_2-glycoprotein has a high carbohydrate content
but contains no gangliosides. It appears in the 26th-27th week of
development, at the same time as myelination of the glia [1748,
1752]. The isolation of this protein in a relatively pure form was
accomplished only in 1970 [1749]. The protein has a high affinity
for DEAE-cellulose and is eluted from the column by 0.5 M sodium
phosphate buffer, pH 6.5, in 1 M NaCl. Apparently this protein
does not induce encephalomyelitis but it may play an important
role in the formation and regeneration of myelin.

Of the other acid proteins containing carbohydrates, mention
must be made of three prealbumin glycoproteins extractable from
whole rat brain by buffer of low ionic strength. These proteins,
which appeared specific to brain [812], were purified by column
electrophoresis at three pH values and were analyzed for their content
of sialic acid, hexoses, amines, and protein. The ratio of sialic
acid to protein was found to decrease from the most mobile to the
least mobile fraction, with the relative content of fucose, hexose,
and hexosamines corresponding to the sialic acid distribution.

Singh and Talwar [1619] described a high rate of incorporation of radioactive amino acids, injected intracisternally, into the proteins of the occipital region of the cortex in monkeys and rabbits exposed to stimulation by flashes. In monkeys subjected to the same treatment, a higher incorporation of labeled lysine into the cortical proteins was observed than in monkeys kept in darkness. An increase in specific radioactivity was found both in proteins of the soluble fraction and in membrane-bound proteins. These results suggest the existence of a group of acid proteins of low molecular weight whose synthesis is intensified in animals exposed to flashes. These proteins give a crossed immunologic reaction with antiserum against S-100 protein.

There is another interesting group of organ-specific protein antigens found in animals of only one species. These antigens were first found in the rat brain by MacPherson and Liakopoulou [1326] in 1966. Later the same workers found two other antigenic proteins of the same type in bovine brain [1111]. Both antigens isolated from bovine brain have the electrophoretic mobility of β-globulins. One more species-specific antigen of this type, named SRANT (species restricted antigen of nervous tissue) [1293] has now been isolated from rat nerve tissue and some of its properties have been determined. It exists in two molecular forms and has the electrophoretic mobility of α-globulin and albumin. Only one, the more acid form of the protein with the mobility of albumins, has been isolated in the pure form. The purified protein with a molecular weight of about 70,000 forms a single band in 9% PAAG. The concentration of the protein in the rat brain increases during development, and in adult animals is found in all parts of the brain. The SRANT protein is also found in the spinal cord and sciatic nerve of the rat.

In 1970 the Canadian workers Hatcher and MacPherson [1112] described the isolation in a highly purified form of yet another acid protein described as α_1-BASNT (bovine antigen specific to nervous tissue) protein. On immunoelectrophoresis in agar this protein has the mobility of an α_1-globulin. This protein is also found in rat and human brain and in bovine spinal cord but not in bovine or rat sciatic nerve. It is formed in the first stages of development and is a glycoprotein with a molecular weight, determined by gel-filtration on Sephadex G-100, of about 84,000. This protein has only weakly acid properties: about 14% of its amino acids are basic.

As mentioned earlier, during electrophoretic fractionation of freshly prepared solutions of lyophilized water-soluble brain proteins which have not been kept for a long time, and also of primary Triton extracts of the nerve ending fraction [32], two additional zones of protein (1a and 1b) were found migrating toward the anode in front of zone I (Fig. 1). The proteins of these zones are found in the soluble fraction of whole brain tissue, the gray and white matter of the cerebral hemispheres, the corpus callosum, and the spinal cord as well as in soluble liver proteins but not in bloom serum. With proteins from the gray matter of the cerebral hemispheres more protein is always found in zone 1a than in zone 1b. In the white matter from the cerebral hemispheres, on the other hand, zones 1a and 1b contain about equal quantities of protein because zone 1b for white matter contains much more protein than the corresponding zone of the gray matter. The content of these highly acid proteins is higher in rabbit, cat, and bovine spinal cord than in the white matter, and much more so than in the gray matter of the brain. Zone 1b in spinal cord tissue always contains much more protein than zone 1a. Thus the content of strongly acid proteins (1a and 1b) is highest in the spinal cord with lesser amounts in the white matter and least of all in the gray matter of the cerebral hemispheres; i.e., higher concentrations are found in regions richer in conducting structures. In the cat brain, the protein content of 1b and 1a is characteristically lower than in the rabbit and bovine brain. A lower content of 1a and 1b in the cerebral hemispheres than in the spinal cord also was observed by Kudinov and Polyakova [302].

It has been shown [32] that proteins of zones 1a and 1b are present in the protein salted out by ammonium sulfate at saturations between 0.6 and 1.0, but not at saturations 0-0.2, 0.2-0.4, and 0.4-0.6. On repeated electrophoresis in agar gel, protein zones 1a and 1b occupy the same position in the gel and are indistinguishable from the original proteins either in the width of the bands or in the sharpness of their edges.

On fractionation of the proteins of zones 1a and 1b by vertical disc electrophoresis in PAAG by the method of Davis [938, 939] as modified by Broome [867], considerable microheterogeneity was found (Fig. 3). The proteins of zone 1a, for instance, separated in 15% gel into nine clearly defined bands, although in five of them the protein content is very small. The highest protein content is found in the second and third bands, migrating rapidly toward the anode.

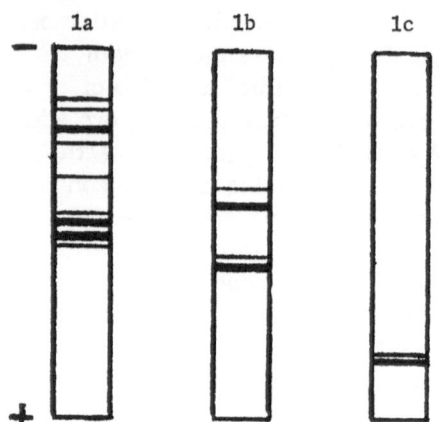

Fig. 3. Scheme of fractionation of proteins of electrophoretic zones 1a and 1b in 15% PAAG and of zone 1c in 7.5% PAAG.

The proteins of zone 1b separated into four bands, of which only the first and third contain comparatively large quantities of protein.

If a large quantity (4-5 mg) of the protein salted out in 0.6-1.0 saturation is subjected to electrophoresis in agar, one more previously unknown fraction of strongly acid proteins is found (Fig. 1, zone 1c). Its electrophoretic mobility is much higher than that of zone 1b. The substance contained in zone 1c has been obtained [603] in a purified form. The proteins salted out were first fractionated by ion-exchange chromatography on a column (32 × 2.8 cm) of DEAE-cellulose (capacity 0.47 meq/g) followed by filtration through Sephadex G-100. The proteins were eluted from the DEAE-cellulose column by borate buffer, pH 8.3, with ionic strength of 0.005 followed by 0.2, 0.34, and 1.2 M NaCl solutions in the same buffer, and finally 0.1 M NaOH solution. Electrophoresis of the eluates in agar gel showed that protein zone 1c occurs only in the eluate of the 1.2 M NaCl. On filtration of this eluate through columns (140 × 1.2 cm) with Sephadex G-100 (elution with 0.1 M NaCl solution made up in borate buffer, pH 8.3, with ionic strength 0.005) two peaks are obtained, containing 40 and 60%, respectively, of the protein applied to the column. Electrophoresis in agar showed that peak I contains proteins from electrophoretic zones 3, 4, and 5, whereas peak II contains only protein zone 1c.

The protein content in a purified, lyophilized preparation of zone 1c was determined by the double-wave spectrophotometric method of Groves [1083]. RNA was determined by Meibaum's orcine reaction [371], DNA by Burton's diphenylamine reaction [877], and phosphorus by Chen's method [895]. The preparation contained no DNA, and its content of RNA and protein was in the proportion of 2.68 : 1 by weight. Phosphorus amounted to 10.7% of the total RNA in the preparation. It can be concluded on the basis of these findings that RNA is the only phosphorus-containing compound in the isolated preparation.

During electrophoresis of a purified zone 1c in 7.5% PAAG (Tris glycine buffer, pH 8.3, ionic strength 0.005, current 5 mA to the tube, duration of electrophoresis 1 h) two components migrating toward the anode very close together were found. The band with the greater mobility does not stain with protein dyes and is evidently RNA. Alongside it was a very narrow but deeply stained protein band.

A protein of the second electrophoretic zone has been isolated from brain and partly purified. On electrophoresis in agar it migrates in front of the serum albumins. If electrophoresis is in 14.5% starch gel with Poulik's discontinuous system of buffers [1501] the protein migrates immediately after the buffer front. By disc electrophoresis in PAAG of different concentrations it separates into two to five protein bands. The molecular weight, determined in a thin layer of Sephadex G-200, is about 25,000. Large quantities of aspartic and glutamic acids were found by amino acid analysis after hydrolysis in 6 N HCl.

Recently a protein with a low molecular weight (about 10,000), extremely rich in glutamic acid [1511], has been isolated from monkey brain.

It is interesting to note certain special features distinguishing the acid protein fractions of nerve tissue. First, nerve tissue contains five prealbumin zones: 1c, 1b, 1a, 1, and 2 (Fig. 1), i.e., three more than blood serum. Other workers [193, 535, 555, 614, 1183] found only two or three prealbumin protein zones in tissues from various parts of the nervous system by electrophoresis in agar gel. Only Yordanov [1795] was sometimes able to detect four prealbumin protein zones by electrophoresis although he clearly distinguished a fifth prealbumin fraction by autoradiography. Bondy and Perry [844], who fractionated these proteins from rabbit by chromatography on DEAE-cellulose, also observed a high content of acid proteins in the brain tissues, a much lower content in liver and muscle, and none at all in blood serum.

The second distinguishing feature is that none of the strongly acid proteins of electrophoretic zones 1a, 1b, and 1c is evidently identical with the acid protein S-100, first isolated by Moore [1396] and at present under detailed investigation in many laboratories [1064, 1065, 1153, 1317, 1398, 1400, 1485, 1667, 1705]. Protein S-100 is soluble in saturated ammonium sulfate solution whereas the proteins of electrophoretic zones 1a, 1b, and 1c are precipitated by a saturated solution of the salt.

The third distinguishing feature of the acid proteins of nerve tissue is of great importance for the determination of the biological role of the strongly acid proteins of zones 1a and 1b, and that is that the protein content, particularly for 1b, is higher in those parts of the nervous system with a predominance of conducting structures.

The fourth property of these proteins is the microheterogeneity of the protein composition of electrophoretic zones 1b and, in particular, 1a.

The fifth distinguishing feature is that the strongly acid proteins of electrophoretic zones 1a, 1b, and 1c are metabolically less active (see Chapter 5) than proteins salted out first in the saturation range 0.6-1.0 and proteins of electrophoretic zones migrating more slowly toward the anode (zones 1, 2, and 3).

The study of specific brain proteins and their properties is extremely important not only for establishing their functional role in brain activity, but also for an appraisal of the view that certain mental and neurological diseases are manifestations of a state of autoimmunity to brain antigens. Antiserum against the caudate nucleus of the brain, if injected into the cerebral ventricles, reduces the electrical activity of this zone [1163, 1383], whereas antiserum against soluble brain proteins induces psychopathic states in man [1116] and leads to a disappearance of training effects in animals [1382]. It has also been shown that monospecific antiserum against protein S-100 causes destruction of isolated nerve endings in vitro, whereas if injected intracellularly into a mollusk neuron it causes a gradual lysis of the neuron [1530].

Chapter 3

The Blood–Brain Barrier and Amino Acid Transport Through Cell Membranes

The free amino acids of the blood are the chief source from which the amino acid supply of the brain tissue is formed. Knowledge of the mechanisms controlling amino acid transport between the blood stream and nerve cells is accordingly essential to understanding the many processes in which amino acids behave as irreplaceable substrates and particularly for understanding the special character of protein metabolism in the brain tissue. Unfortunately no general analysis has yet been made in the literature of the considerable experimental material that has accumulated as a result of the study of this problem. The existing monographs, surveys, and collections [144, 156, 221, 254, 545, 550, 714, 715, 952, 973] give an adequate picture chiefly of the morphological substrates of the blood–brain barrier (BBB) and its principal functions in various physiological and pathological states and also of the history of our knowledge of the BBB. Within the last few years general accounts have been given on some aspects of the mechanisms of membrane transport in nervous tissue and their role in determining the amino acid composition of the metabolic reserves of the brain [26, 195].

Accordingly, in this chapter we shall examine the fundamental structures and functions of the BBB and discuss the experimental data on exchange of amino acids between blood and brain in vivo and between the incubation medium and tissue slices. Data on membrane transport of amino acids in other organs will be examined in this chapter so far as they shed light on variations in tissue distribution and mechanisms of transport through cell membranes.

Basic Information on the BBB

History and Terminology. The study of the BBB began more than 80 years ago when in 1885 Ehrlich first found [254, 952] that certain aniline dyes, if injected into the blood stream, stain virtually all the tissues of the body except those of the CNS. Because of these findings and later evidence showing that other histologically detectable substances do not penetrate the brain or do so only very slightly, the view gradually took shape that a barrier, subsequently called the blood—brain barrier, existed between the blood stream and the brain tissue. With the discovery of new facts concerning the functions of the BBB, specialists in different fields became interested. It was studied by morphologists, physiologists, pathophysiologists, clinicians, and later by biochemists. From a problem of theoretical interest, the study of the BBB became a subject of great practical importance.

During the first six decades, research on the BBB was concentrated chiefly on the principles governing the penetration of foreign materials such as thiocyanate, silver, trypan blue, and other dyes into the brain tissues. Later the permeability to pharmacological and biologically active substances, such as bacteria, viruses, and toxins, was studied. Only in the last two decades have substances occurring naturally among the components of nerve calls been studied for BBB permeability. This stage coincides with the wide use of labeled compounds in biochemical research, as a result of which much progress has been made in elucidating the functions and role of the BBB. Normal activity of the CNS depends essentially on the proper functioning of three barriers: blood $\rightleftharpoons$ CSF, CSF $\rightleftharpoons$ brain, and blood $\rightleftharpoons$ brain. The relative constancy of the internal environment of the CNS tissue is maintained by functioning barriers in both directions. An important role is played by the cell membranes and their transport system, which also carry materials in two directions.

Two morphologically distinct but functionally interdependent barrier mechanisms thus participate in the transport of free amino acids from blood to brain: One is the BBB itself, controlling the transport from blood into the intercellular tissue medium (and vice versa), and the other is the cell membranes acting between the tissue medium and cells. Characteristically these two barrier transport systems are of great importance in regulating rates of metabolism, including protein metabolism, in all tissues but particularly in brain, which differs from other organs in the great

structural and functional variety of its cells, its very complex
system of intercellular connections, and several other features
(see Chapter 4).

It must be emphasized, however, that the separation of a single barrier into its components is purely conventional. While ad-
mitting this artificial division, investigators are well aware of
the fact that the BBB, like any other type of tissue barrier, acts
as a single, harmoniously functioning system selectively trans-
ferring metabolites from the blood into the cells and vice versa
while maintaining the concentration gradients characteristic of
each metabolite. Usually in animal experiments one determines
the net result of all the components of this barrier system, which
is affected by the functional state both of the organ itself and of
the body as a whole. The influence of morphological substances
and metabolic demands will be considered below.

As was discovered later [254, 713], a barrier between the
blood and the tissue medium exists in some form or other in
practically every animal organ, so that the wider concept of tissue—blood barriers (TBBs), in the sense of "the physiological
mechanism controlling and protecting the relative constancy of
the composition and properties of the direct internal medium of
cells, tissues, and organs" [716], was introduced. In their writings,
Shtern and his pupils have emphasized over and over again that
the barrier between the blood and brain is not a simple morphological concept but a functional unit including both an anatomical
substrate and a wide range of mechanisms for the transport of
different metabolites. In Kassil's monograph "The Blood—Brain
Barrier" [254] a definition of the BBB taken from Herlin is given:
"the concept of barrier must include all phenomena preventing,
diminishing, retarding, or even facilitating the penetration of substances into the central nervous system. This penetration can
take place through dialysis, ultrafiltration, osmosis, Donnan equilibrium, electrical changes, lipid solubility, special tissue affinity, or metabolic activity."

The definitions of tissue barriers, like many others, have im-
perfections, of course, although in general they reflect the physiological nature of barrier mechanisms correctly. It is evidently
a practical impossibility to give a definition of tissue barriers
that would completely reflect all their distinguishing features at
all levels of organization of living matter — from the capillary permeability of an organ, its specialized areas or functional units,
down to membrane barriers at the cellular and subcellular level.

Some Distinguishing Features of Tissue Bar-
riers. It will be clear even from this very short survey of the
study of TBBs in general and the BBB in particular that the protective and regulatory functions must be distinguished. The latter,
which is more important, is exhibited throughout the life of the
organism and ensures the relative constancy of the tissue medium in accordance with the functional demands of the organ at
every stage of its development. The protective function, guarding
the internal medium of each organ and of each of its cells and
subcellular structures against the harmful action of foreign substances, including bacteria, toxins, and antimetabolites, is essential to the normal activity of every organ. However, it does not
follow from this conventional distinction between regulatory and
protective functions that each results from independent mechanisms working in isolation.

One characteristic feature of tissue barriers is their selectivity, i.e., their ability to allow certain substances to pass into
the tissue (or vice versa) but to hold back others. This property,
established initially during investigation of BBB function, was
found to be shared by the other TBBs. Frequent attempts by research workers [155, 156, 254, 545, 714, 952] to establish general principles governing the degree of permeability of tissue barriers with respect to various substances have been unsuccessful.
Neither differences in the anatomical substrate of the barrier nor
the chemical nature of the substance transported could explain the
experimental facts or clinical observations demonstrating the se-
lective passage of ions, dyes, and other substances through the
tissue barriers. Similarly there is no satisfactory explanation of
the unequal ability of substances with closely similar chemical
structures and properties to penetrate the tissues, or of the greater penetration of some substances with higher molecular weights
than others with lower molecular weights, and so on [155, 156,
254, 545, 714]. Most of these facts cannot be reconciled with the
many different hypothetical mechanisms suggested to explain se-
lective passage through tissue barriers.

The rate at which metabolites enter cells, including cells of
the nervous system, is [254, 626, 714, 1249, 1288] determined
both by the ability of the particular compound to pass through the
tissue barriers, and by the intensity of its utilization in biosynthetic and catabolic reactions, i.e., by the rate of its metabolic
conversions in the cell. Thus the rate metabolism can be regarded as a factor determining the selectivity of absorption by

different cells. Some workers [1244, 1584], however, doubt that the intensity of intracellular metabolic processes plays an important role in the actual transport through membranes. Others [952, 953], on the contrary, consider the rate of entry of metabolites into the brain to be determined mainly, if not entirely, by the intensity of metabolism rather than by barrier action.

Some insight was gained by comparing the penetration of natural metabolites into the brain with the penetration of compounds which were chemically related but not metabolized. These findings indicated that penetration into a cell is an active process linked with tissue metabolism, and explained certain aspects of the remarkable selectivity of tissue barriers which had been difficult to interpret. Data confirming a definite connection between the rates of membrane transport and intracellular metabolism will be discussed in more detail later.

The Morphological Substrates of the BBB. Information on the morphological substrates of the BBB is adequately represented in a number of publications [5, 254, 353, 863, 952], and for that reason nothing more than a brief list will be given here of the principal anatomical formations which, in the opinion of most investigators, perform barrier functions in the brain tissues.

Since exogenous metabolites enter the brain tissue chiefly from the blood, it is not by accident that attention was initially directed toward the capillary network as the first and most important barrier along the path of transport from blood to brain and vice versa. There are few who now doubt that the capillary networks of the brain and other organs function as barriers in addition to supporting nutrition. However, the relative importance of these functions in the brain differs from that of other tissues. The chief difference is that the barrier function is more highly developed in the brain and for that reason many substances penetrate with difficulty or not at all whereas in other tissues natural metabolites and occasionally even foreign materials pass comparatively freely through the capillary network.

Some unique structural features of the brain capillaries, which help explain their lower permeability for many substances, have been established [155, 156, 254, 545, 952, 973] with both light and electron microscopy.

Biochemical studies [1465, 1573] have demonstrated the specificity of transport of amino acids and their derivatives through the capillary walls in nerve tissue and the dependence of transport

processes on enzyme systems located in the membrane structures of the brain capillaries. The work of Klosovskii [266, 544],
who studied the blood system of the brain and its ontogenetic development, has shed light on many previously unexplained trophic
and barrier functions of the capillaries.

An important morphological substrate of the BBB is the neuroglia, particularly cells such as astrocytes. These cells are in
close contact with the outer membrane of the brain capillaries
through the "end-plates" of their processes. About 85% of the surface of the brain capillaries is covered by the endings of these
astrocytic processes and only 15% by processes of oligodendro-
cytes, other neuroglia, and neurons [952]. Glial cells surround
the bodies of the neurons densely and have close intercellular con-
tacts with them [8, 215]. The view has actually been expressed
[8] that the glial cells are capable of pinocytosis. On this basis
most workers believe that the astrocytes, with their processes
and end-plates, participate in the nutrition of the nervous tissue
while functioning at the same time as a barrier. In view of the
close contact of the glial elements with the brain capillaries
on the one hand and with the neurons on the other, neurons and
neuroglia together can be regarded as "a single functional system responsible for the integrated function of the brain tissue elements" [8]. Some workers [266], though recognizing the role of astrocytes in cell nutrition, reject their barrier function while others
[952] even deny the role of glia as trophic intermediaries between
the blood and the neurons. We consider the participation of the
glia in the trophic and barrier functions of the brain to be proven.
This view is supported by much experimental evidence [8, 160,
254, 497, 499] obtained by various sophisticated methods including
electron microscopy.

The intercellular ground substance, which is of mucopoly-
saccharide nature and occurs as a component of the connective
tissue, is regarded by many investigators [5, 254, 372, 606, 952,
1124] as a possible anatomical basis of the BBB and also of other
TBBs. The ground substance fills the intercellular tissue spaces
of the brain (as of other tissues) and blood vessels. The experimental evidence [254, 717, 952, 1124] shows that ground substance
is absent in the brain of guinea pig embryos and newborn mice,
i.e., in animals with an undeveloped blood—brain barrier, but is
present in the brain tissue of adult animals in which the BBB is
completely formed.

These anatomical structures do not completely reflect the morphological basis of the BBB. Other structures which doubtless play important roles in regulating the composition of the internal medium of the brain are the vascular plexuses, the leptomeninges, the Virchow—Robin spaces that accompany the blood vessels as they enter the brain tissue and terminate blindly in its depth, and probably more. Their role and the mechanisms by which they participate in regulating amino acids in brain have not been adequately studied although they unquestionably constitute essential morphological elements of the blood ⇌ CSF, CSF ⇌ brain, and blood ⇌ brain barriers.

Variations in the BBB in Different Parts of the Brain. It is clearly established [254, 353, 591, 784, 952, 1778] that the permeability of the BBB varies in different parts of the brain. In some parts of the brain stem, the pineal body, and the adeno- and neurohypophysis, for instance, it is absent altogether or much more permeable than in other parts of the brain. In these structures a higher rate of uptake of ^{32}P [848] and of certain drugs [1778], injected parenterally, is found. Areas of brain tissue with a highly developed barrier function differ morphologically from the tissue structures just mentioned, in which the BBB is absent or incompletely formed. These structural differences occur both in the capillaries, which consist of endothelial cells and of the special cementing substance filling the intercellular spaces, and in the glial elements and the amorphous ground substance lying next to the endothelial tube.

The detailed description of these morphological differences is outside the scope of this book and all that will be said, therefore, is that, together with the characteristic functional and metabolic activity of the various parts of the brain tissue, they probably explain the greater or lesser permeability of the BBB in these areas. Incidentally, there are evidently no absolutely impermeable or completely permeable tissue structures in the living organism. Any organ, at all of its structural levels, must have its own special barrier mechanisms which ensure that metabolites enter the cell and leave it in the necessary amounts.

Differences in the permeability of tissue structures in the various parts of the brain may also be due, to some extent, to differences in the number of capillaries per unit of tissue surface and, consequently, to differences in their total surface area per unit volume of tissue [266, 267, 544, 1031]. According to Grash-

TABLE 1. Dimensions of Bodies of Neurons and Length
of Capillaries in Nuclei of the Cat Brain [46]

Nuclei	Area of cross section of neuron, μ^2		Total surface area of capillaries within a distance of 25 μ from surface of neuron, μ^2	
	large cell	small cell	large cell	small cell
of Deiters'	4002	1800	796	714
interradicular (Klosovskii's)	1254	544	413	268
of Bekhterev's	630	252	307	268
of descending root of vestibular nerve	612	288	369	283
triangular	612	252	332	284

chenkov (cited in [254]), in parts of the brain with the richest capillary network (for example, the paraventricular and supraoptic nuclei) the number of capillaries per unit area of tissue section is several times greater than in some other regions (for example, the cornu Ammonis, the globus pallidus, the 5th layer of the
motor cortex). The investigations of Klosovskii and co-workers
[266, 267, 544] showed (Table 1) that the density of the capillary
network around the nerve cells depends directly on their functional and, consequently, their metabolic activity and is related
to their size.

It must also be remembered that the rate of the blood flow
varies significantly in different structures of the brain [1628].
The quantity of blood (milliliters per minute) flowing through 100 g
of the substance in different parts of the brain is given below:

Cortex:		Superior colliculi	115
somatosensory area	138	Inferior colliculi	180
Auditory	130	Superior olive	117
Visual	125	Reticular formation of the brain	59
Olfactory	77	Cerebellum:	
White matter of cerebral		cortex	69
hemispheres	23	nuclei	79
Medial geniculate body	122	white matter	24
Lateral geniculate body	121	Vestibular nuclei	91
Thalamus	103	Cochlear nuclei	87
Hypothalamus	84	Pyramidal tract	26
Caudate nucleus	110	Spinal cord:	
Hippocampus	61	gray matter	63
Optic tract	27	white matter	14

The velocity of the blood flow is highest in parts of the brain
with the maximal density of the capillary network [1031]. This
may also be of considerable importance in determining the permeability of the BBB in those regions.

The BBB in Different Functional States of the
Organism.   The functional activity of any tissue barrier must
be known in order to interpret correctly metabolic conversions in
the tissues in different physiological and pathological states.
Changes in the intensity of metabolic processes in an organ are
often due to increased or reduced permeability of the tissue barriers for the various metabolites involved rather than due to the
effect of a particular substance or a particular functional state.
On the other hand, the dependence of TBB permeability on various
physiological and pathological states is well known and has been
fully described [144, 155, 156, 254, 353, 545, 550, 715, 952, 973].
Some of the factors significantly altering BBB permeability are
age and seasonal changes, inflammatory processes, disturbance
of the functions of the nervous and endocrine systems, nutrition,
and the influence of various physical factors and pharmacological
agents.

The age factor is one of the most important of the functional
states associated with significant changes in BBB permeability,
and a short review here will not be out of place.

It was concluded from data summarized in the surveys [221,
254, 784, 952, 1778] that the BBB is absent, or present in an incomplete form, in embryos and animals in the early stages of
postembryonic development and that it develops gradually with
age. This conclusion was first reached after the discovery that
certain substances will pass quickly and in high concentrations
into the nerve tissue of embryos and immature animals but will
not penetrate, or penetrate very slowly and in very small quantities, into the brains of adult animals. In the early studies of the
BBB [254, 718, 719, 1731] the substances used in this connection
were mainly not natural metabolites (potassium ferricyanide,
sodium iodide, silver, bromine, various dyes, and other compounds). Most workers regarded the unequal rate of penetration
of these substances into the brain tissue of animals of different
ages as firm evidence of structural obstacles to their free passage into cells in adults. It was accepted that the structures performing barrier functions are gradually formed during ontogeny
of the animals. As mentioned, this view was confirmed by a paral-

lel between changes in the formation of the anatomical structures of the BBB with age and permeability to various substances.

Later investigations showed that the conclusion concerning age differences in the passage of foreign substances through tissue barriers is valid also for compounds that are natural tissue metabolites (phosphates, serotonin, amino acids, carbohydrates, etc.).

The use of radioactive phosphate showed [1032] that it passes through the BBB almost 20 times more rapidly in newborn rats than in adults. Similar but less marked age differences in the penetration of ^{32}P were found by the same workers in other rat tissues (liver, kidneys). They attributed the age difference in the passage of phosphate through tissue barriers more to a decrease in the intensity of phosphorus metabolism with age than to changes in the structural obstacles.

Later investigations [96, 97, 296, 693] also established a link between the rate of passage of phosphate through the BBB and the rate of metabolism in the tissue. In her investigations of the development of BBB with respect to phosphorus in a series of vertebrates at different levels of evolution, Verzhbinskaya [96] concluded that the rate of entry of ^{32}P into the brain is determined chiefly by the rate of phosphate utilization in brain metabolism in general and by the intensity of oxidative phosphorylation in particular, rather than by the permeability of the barrier itself to phosphates.

An age decrease in the permeability of the BBB also has been observed for serotonin [237], histamine [249], cholesterol [912, 1519], and many amino acids [26]. The extensive data on age changes in permeability of the BBB to free amino acids and their structural analogs will be examined later.

Finally, it may be pointed out that the TBBs and the BBB of adult animals are comparatively resistant to the action of ultrasound, ionizing radiation, lowered atmospheric pressure, and various other harmful factors [221-223].

The BBB and Free Amino Acid

Content of Brain Tissue

Under physiological conditions the blood ⇌ brain and blood ⇌ CSF barriers are virtually impermeable to macromolecular substances such as proteins and nucleic acids [981, 1304, 1509]. These compounds are synthesized and renewed at the sites of their function, i.e., in the nerve tissue cells. However, these barriers become permeable to protein molecules in some pathological states such as infectious fevers and inflammations of the meninges of varied etiology [221, 999, 1305, 1663]. The writers are unaware of any data on the passage of peptides through the BBB, including the presumably important peptide glutathione, which accounts for about 95% of the acid-soluble thiol compounds in brain [344]. Brain tissue possesses mechanisms for the biosynthesis of this important tripeptide [1738]. With this in mind and also the fact that the glutathione concentration in the brain is 10 times higher than in the blood plasma [344] there is evidently no reason to suppose any significant exchange of this tripeptide between the plasma and the brain. No other peptides are found in appreciable amounts in the nervous system.

Some Characteristics of the Tissue Distribution of Amino Acids. The concentration of nonprotein amino nitrogen in most tissues of the animal organism is considerably higher than in circulating blood [379, 1029, 1638]. The content of nonprotein amino nitrogen in the brain tissue of rats, cats, and dogs, for instance, is from five to seven times higher than in the blood [379, 929, 1029]. Even higher concentration gradients have been found for individual amino acids and their derivatives: glutamic acid, glutamine, glutathione, GABA, aspartic and N-acetylaspartic acids, taurine, serine. The concentrations of some of these compounds (glutamic and aspartic acids, glutathione, taurine) in the brain are tens or hundreds of times higher than in the blood plasma [227, 344, 826, 1226, 1542, 1732]. The concentrations of methionine, tyrosine, histidine, alanine, glycine, and threonine are only a little higher (not more than twice) than in the blood plasma. On the other hand, the concentrations of proline, lysine, leucine, isoleucine, valine, tryptophan, ornithine, and arginine in the brain are actually lower than in the blood plasma [1542, 1597, 1661].

Depending on their brain/blood concentration gradients the free amino acids can thus be divided into three principal groups: amino acids with a very high brain/blood concentration ratio (5 : 1 or higher); amino acids with a moderate brain/blood plasma concentration gradient (not exceeding 2 : 1), and amino acids with a lower concentration in the brain tissue than in the blood plasma.

Note that the differences between the concentrations of individual free amino acids in the brain tissue are much greater than in

the blood. According to Roberts [1542], in rat blood plasma, alanine has the highest concentration (0.493 μmole/ml) and aspartic
acid the lowest (0.015 μmole/ml). The concentrations of these
amino acids in the blood plasma differ by only 33 times. In the
cerebral cortex glutamic acid has the highest concentration (13.617
μmoles/g tissue) and ornithine (0.014 μmole/g tissue) the lowest;
the ratio of these amino acids in the brain is almost 1000:1. These
great differences in the concentrations of the amino acids in brain
tissue are probably explained by the particularly important role
of some amino acids as substrate for the highly specialized functions of this organ.

It is interesting to note that the relative proportions of the
various amino acids in brain proteins do not correspond to those
of the brain free amino acids. For example, the ratio between
the concentrations of glutamic acid and valine in the total brain
proteins is 10 times less than their ratio in the tissue free pool.
In the proteins of the whole brain of the goldfish the amount of
glutamic acid is only 11 times higher than in the free amino acid
pool, whereas the corresponding figure for valine, leucine, isoleucine, and phenylalanine is approximately 250-300, and that for
aspartic acid as much as 470 [1296]. The lysine concentration in
proteins in various parts of the monkey CNS is also much higher
than the corresponding concentration of free lysine [1036]. Characteristically the cerebral cortex contains less free lysine than
other parts of the brain studied [1036, 1732], whereas the relative
percentage of lysine in the proteins of these parts is, by contrast,
much higher than in the proteins of the cortex [1036].

The tissue/blood concentration gradients also differ for individual amino acids and their derivatives in brain structures with
different functions. This is clearly seen by comparing the concentrations of each amino acid in these parts of the brain [799,
993, 1456, 1615-1617, 1732]. Whereas the concentration of taurine
and serine and the combined concentration of glycine and alanine
in seven parts of the brain (frontal lobe, thalamus, hypothalamus,
medulla, corpus callosum, caudate nucleus, globus pallidus) differ
by no more than a factor of 2, the concentrations of threonine,
leucine, and glutamic acid differ by a factor of 3, that of aspartic acid by a factor of 4, and GABA by a factor of almost 10
[1456]. Consequently, the concentration gradients (tissue/blood)
of these amino acids in these parts of the brain differ by a similar degree.

Most other organs characteristically have high concentration gradients of nonprotein amino nitrogen and free amino acids between the tissue and the blood plasma. The exceptions are tissues with a very low rate of metabolism in general and of protein metabolism in particular (adipose tissue, the lens of the eye) [379, 1029, 1226]. Since considerable differences are found between the concentrations of the various amino acids in different organs [194, 488, 1661], the family of concentration gradients for each organ also differs radically.

It must also be pointed out that many tissues, especially nerve, have a surprising ability to maintain the relative concentrations of free amino acids in various physiological and even in some pathological states [117, 1539], though, of course, with some disturbances (exposure to ionizing radiation, growth of a tumor, poisoning with carbon tetrachloride, hyperthermia, etc.) significant changes in the concentrations of some amino acids are found in the tissue [117, 162, 272, 702, 1542].

The relative constancy of the composition and concentrations of the amino acids in the brain must not, of course, be regarded as the result of a static state. There is no doubt that the free amino acids in brain tissue are constantly being utilized both for biosynthesis and for various metabolic conversion reactions. There is also a constant drain of free amino acids from the brain into the blood stream. The brain is compensated for this constant loss by inflow of amino acids from the blood and by their formation by intracellular metabolism. In the living organism all these processes (inflow into the tissue and outflow from it, formation and loss in the course of biosynthetic reactions and metabolic conversions) are balanced and coordinated by the complex homeostatic mechanisms of the BBB (or of any other TBB) and membrane transport, in consequence of which relative composition of the internal milieu, particularly of the amino acids, is maintained.

These high concentration gradients of free amino acids between the tissue and the blood plasma and the substantial differences between the gradients in different organs and their anatomical formations, though extremely interesting, were until recently completely unstudied. Nothing was known about the mechanisms creating and maintaining these high concentration gradients. The explanation of these mechanisms, and how they are affected in various physiological and pathological states, is ex-

tremely important for understanding the basic activity of these
organs and for discovering ways of restoring their disturbed
metabolism to normal.

The view has been expressed [901] that a disturbance of
amino acid transport between the blood plasma and organs may
play a more important role in disturbances of tissue protein metabolism than changes in the enzyme systems conducting synthesis in the tissues. The experimental data on the possible mechanisms maintaining the amino acid composition of the brain and
those controlling the intensive exchange between blood and brain
even in the presence of high concentration gradients will be discussed below.

Exchange of Amino Acids between the Blood
Plasma and Brain. To all intents and purposes the study
of amino acid transport in the brain began only in 1950 [1180,
1598]. The constant exchange of amino acids between the blood
plasma and the brain could not be demonstrated by the use of
chemical and microbiological methods of quantitative amino acid
analysis [1055, 1180, 1537, 1598] because the BBB effectively
prevents certain amino acids from penetrating into the cells in
excess. For instance, the content of glutamic [1180, 1598] and
aspartic [1180] acids, GABA [363, 1055, 1491, 1537], and proline
[949] in the brain tissues does not rise after a considerable increase in their blood concentration. Some increase in the GABA
level in the brain was observed [363] but only after parenteral injection in very high doses (1-5 g/kg body weight), whereas in
muscle, liver and kidneys the GABA concentration rises very
rapidly after the injection of as little as 500 mg/kg body weight
[1055].

By contrast the concentrations of tyrosine [898], glutamine
[1180, 1598], histidine [1180], methionine [929, 1180], leucine [1543],
and lysine [1180] in the adult brain are substantially increased
after their levels in the blood are raised. But the rapid transport
of all the amino acids from blood to brain in adult animals was
demonstrated only by the technique of radioactive labeling [106,
887, 910, 1066, 1253, 1254, 1258, 1543, 1584], which easily detects
a transfer of metabolites through the cell membranes even
when there is no change in concentrations on the opposite sides
of the membranes.

It is interesting to note that very rapid exchange has been
found [828, 1254] for glutamic acid, the concentration of which

could not be increased in nerve tissue [1180, 1598] even after greatly elevating its level in the blood. The BBB is more permeable for basic than for dicarboxylic amino acids [1066, 1180, 1253].

Despite the comparatively high permeability of the BBB for several amino acids, there is much experimental evidence [365, 929, 1029, 1088, 1180, 1246, 1250, 1253, 1584] that most amino acids are absorbed by the brain much more slowly than by other tissues, i.e., that the BBB is less permeable than the other TBBs to amino acids. Thus amino acids differ in the rate at which they enter the brain compared to various other tissues and, it is important to note, this entry is frequently against quite high concentration gradients.

Though the work described in this section deals mainly with amino acid transport from the blood into the brain, the opposing flow of amino acids into blood is just as important to the study of the nitrogen and protein metabolism. The rapid passage of some amino acids from the brain into the blood plasma has been observed [910, 1044, 1258-1260, 1538] after intracisternal or subarachnoid injection. It is interesting to note that amino acids leave the brain rapidly even when their concentration in the plasma is elevated by parenteral injection of that amino acid [1249, 1259], i.e., when active transport from the cell to the blood stream takes place against a high (perhaps as much as 8 to 15-fold) concentration gradient.

Others [1249, 1259, 1260] have also shown that amino acids differ in their rate of exit from the brain. The rate of outflow of leucine, proline, phenylalanine, and lysine, injected by the subarachnoid route, from the brain of young rats against a high concentration gradient produced by an intraperitoneal injection has been studied. It was found that 40 min after reaching its maximum, brain concentration of leucine had fallen by 94%, that of proline and phenylalanine by 78%, whereas the lysine concentration, 3 h after reaching a maximum, fell by only 60% [1260]. L-Isomers of these amino acids leave the brain faster than the corresponding D-amino acids. Just as with the L-isomers, the rate of exit of D-leucine and D-phenylalanine is always higher than the rate for D-lysine.

Characteristically no direct relationship exists between the rates at which amino acids enter and leave the brain. GABA passes very slowly through the BBB when injected parenterally

[363, 1055, 1537], but passes comparatively quickly out of the brain after intracisternal injection [1538]. Another interesting fact is that the rate of outflow of both L- and D-amino acids is not directly dependent on their concentration gradient between the brain and the blood plasma [1260].

Thus there is a rapid exchange of free amino acids between brain and blood which is brought about by two flows in opposite directions (into and out of the tissue), produced by active membrane transport. The basic components for this transport are probably localized in the structures of the BBB itself and in the membranes of the nerve cells and their organelles. The biochemical systems responsible for the transport of amino acids in opposite directions through the membranous structures of the BBB are so accurately balanced, so well regulated, and in such close harmony with each other and with the metabolism in the cells, that the levels of the free amino acids are maintained in the nerve both under normal conditions and in the presence of substantial changes in the functional state of the organism [117, 1539].

Factors Influencing the Exchange of Amino Acids between Plasma and Brain. The homonymous tissue in animals of different species, different tissues in the same species, functionally different anatomical regions in each organ, and particular types of cells and their organelles all have their own unique compositions of free amino acids, which differ significantly from the composition of blood. Although in the last two decades we have moved a step closer to understanding the most general principles governing the mechanisms of membrane amino acid transport, at our present level of knowledge it is not yet possible to give detailed characteristics of the biochemical systems responsible for maintaining the harmonious operation of these mechanisms. However, we can distinguish a number of factors that play an important role in the regulation of free amino acid levels.

Concentration of the Amino Acids. The permeability of the blood ⇌ brain and blood ⇌ CSF barriers for some amino acids depends on their concentration in the circulating blood [1205, 1249, 1772-1774]. The concentration of some amino acids in the CSF rises after intravenous injection of an amino acid, and the larger the dose injected, the greater the increase. Very high doses of an amino acid (for example, 600 mg alanine, glycine, or serine /kg

body weight) increase the permeability of the barrier not only for
that amino acid but also for others injected simultaneously in
lower doses [1773]. It is notable that amino acids injected in
equally high doses do not increase the permeability of the barrier
for substances not metabolized in the cell, e.g., for dyes [1774].
But even though the brain assimilates more intensively those amino
acids with higher concentrations in the blood stream [1205] and
assimilates more of a given amino acid the higher its concentration in the blood [1249], there is no direct relationship between
the normal blood level and the quantity assimilated by the tissue
[1205, 1249]. That is, if the leucine concentration in mouse blood
plasma is increased by 8 and 52 times, its concentration in the
brain rises by only 2 and 6 times, respectively [1205].

The rate of transport of amino acids into the brain also depends on their brain concentration, as illustrated by the entry of
labeled methionine, lysine, and leucine into areas of the brain
with normally different physiological concentrations of these
amino acids [365, 1249, 1258]. After the parenteral injection
more label was found in those parts of the brain in which the physiological levels of that amino acid were higher. For example, the
lysine concentration in the cerebellum is almost twice as high as
in the anterior lobe of the brain, and the rate of transport of labeled lysine into the cerebellum was also about twice as high [1258].
Unlike lysine, leucine is present in these parts of the brain in
about equal concentrations and the rate of transport of labeled
leucine into these parts was found to be more nearly equal. The
following facts are interesting in this connection. The rate of exchange of intravenously injected labeled lysine between the blood
plasma and the brain is increased after the subarachnoid injection
both of unlabeled lysine and of unlabeled leucine, but in the first
case the rate of exchange is increased more [1258]. In analogous
experiments in which labeled leucine was injected intravenously
the rate of transport of this amino acid into the brain tissue was
increased more by the subarachnoid injection of leucine than of
lysine.

Intensity of the Metabolic Processes. One of the factors believed to regulate the rate of transport of amino acids and other
substrates across cell membranes is the intensity of the intracellular metabolic processes [254, 626, 714, 1249, 1288]. That
active membrane transport of amino acids depends on their utilization in biosynthesis and in metabolic conversions in the cell is

supported by the following important data:

1. Concentrations and rates of active amino acid transport
   are, as a rule, higher in those tissues and in those functional states that are characterized by a higher intensity
   of metabolism, including amino acid and protein metabolism [952, 1090, 1180, 1240, 1444].
2. Amino acids that are metabolized more actively are present in the cells in higher concentrations [488, 1090].
3. Amino acids whose concentrations in the brain tissue are
   higher are assimilated more intensively by tissue slices
   [1261, 1286].
4. L-Amino acids are transported into the cells and from
   them more rapidly than the corresponding metabolically
   inert D-isomers [898, 1249, 1260, 1261].
5. Amino acids enter cells more rapidly than structurally
   closely related compounds (for example, α-aminoiso-
   butyric acid, cycloleucine) that do not participate in intracellular metabolism [1240, 1256, 1584].
6. The concentration gradients of amino acids are extremely
   low in tissues with very low rates of protein metabolism,
   e.g., the crystalline lens and adipose tissue [379, 1029, 1226].

The connection between the transport of amino acids and their
metabolic activity in a particular structure must evidently be regarded as a possible cause determining the rates of absorption and
the levels of the concentration gradients for each amino acid.
These would be expected to differ significantly in cells with different functions and in different metabolic compartments.

As mentioned, there is no unanimity of this hypothesis. The
view that there is not a direct connection [1244] is based chiefly
on the fact that cells can accumulate metabolically inert D-isomers and structural analogs of the amino acids. However, it
must not be forgotten that the rate at which the intracellular concentration of the metabolically inert amino acid analogs reaches
its maximum [1240] and the absolute level of that maximum [1584]
are several times lower than the corresponding indices for amino
acids metabolized in the cell [365, 898, 1090, 1253, 1254].

For the contrary view some experiments with α-aminoiso-
butyric acid [1072, 1240, 1256, 1444, 1584] are of considerable interest. This amino acid does not occur in nature, is not incorporated into tissue proteins [1240], does not undergo metabolic

conversions in the tissues [1240, 1444], but is supposedly [902] transported through cell membranes by the same mechanisms as the natural amino acids. Cycloleucine has also been used for similar purposes [1256]. Comparison of the rates of membrane transport of these nonmetabolized amino acids with the rates for natural amino acids, which actively participate in intracellular metabolism, provides a method of investigating the functions of the BBB, as of any other TBB, separately from the influences of metabolic processes in the cell.

Investigations of this sort have shown [1240] that rat brain takes up labeled α-aminoisobutyric acid from the blood in much smaller amounts and at a much slower rate than do liver and kidney tissue. The active transport of this acid against a concentration gradient is observed in each of these organs. However, the radioactivity of this acid in the brain does not reach its maximum until 20 h after injection, i.e., much later than the peak of radioactivity of some natural amino acids [365, 898, 1090, 1253, 1254]. The maximal concentration of α-aminoisobutyric acid in the liver and kidneys, on the other hand, is observed 80 min after injection. The brain and plasma radioactivity levels become equalized after 6 h; during the next 14 h the nerve cells continue to transport the amino acid actively against the concentration gradient, while its concentration in the blood plasma falls steadily.

When labeled α-aminoisobutyric acid and phenylalanine-^{14}C were injected continuously into the blood stream of a dog (the concentration of the label in the blood plasma was maintained at a constant level throughout the experiment) the concentration of the phenylalanine in the brain tissue was four to six times higher than the concentration of α-aminoisobutyric acid [1584]. However, no such differences were found in the muscles and liver. The level of α-aminoisobutyric acid in the liver, moreover, was 10-50 times higher than in the brain tissue.

These observations can be regarded as evidence of marked functional differences between the BBB and the other TBBs. They also show that active transport mechanisms can carry not only natural metabolites, but also compounds structurally related to them which do not participate in intracellular metabolism. The foreign substances are transported at a slower rate. Hence it can be concluded that there are evidently no grounds for rejecting a connection between the rate of transport of substances (including amino acids) through cell membranes and the intensity of their metabolic conversions in the cell, but by the same token it cannot

be asserted that the only factor regulating the membrane transport of any compound is the rate of its metabolism in the cell.
There is no doubt that the rate of transport of any metabolite
through a cell membrane is determined to some degree by the
rate of its utilization in the cell, which is of primary importance
when appropriate experiments are being designed and their results interpreted.

Supporters of the adsorption theory of cell permeability [379,
626], though recognizing this connection, have claimed that the
rate of metabolic conversion of a compound in the cell determined its transport rate through the membranes only if its concentration in the surrounding medium is higher than the intracellular concentration; if the intracellular concentration of a metabolite is higher, on the other hand, its entry into the cell is determined by the adsorptive properties of the cell colloids, i.e., by
the ability of the intracellular structures to bind metabolites
adsorptively or chemically. At the present stage of development
of the adsorption theory the transport of substances through membranes and the steady-state unequal distribution of metabolites
between the cell and the extracellular medium are regarded [420,
628] as processes that are mutually connected but of different
origin, and which together maintain the specificity of the chemical
composition of the cells and their organelles.

We cannot neglect the possibility that neurohumoral factors
may also play a part in the regulation of amino acid transport
through cell membranes [1072]. Presumably this effect could also
be mediated through intracellular metabolism.

Lipid Solubility. In the case of some amino acids the rate of
their membrane transport has been shown to depend directly on
their lipid solubility. For instance, proline and its derivatives
are ranked in exactly the same order for solubility in lipid solvents (carbon tetrachloride, petroleum ether, arachis oil) as for
ability to penetrate the brain: ethyl ester of N-acetylproline >
ethyl ester of proline > N-acetylproline > proline [949]. However,
no such relationship obtains when the rates of membrane penetration of other amino acids (leucine, lysine, proline, and phenylalanine) were compared with their solubility in a system of chloroform—methyl alcohol—water [1260]. After comparing the lipid
solubility, ionic charges, and membrane permeability of N-acetylproline and the ethyl ester of proline, Dingman et al. [949] concluded that negative charges in amino acids prevent their passage
through cell membranes.

Structure of the Amino Acids. The rate of passage of amino acids from the blood plasma into the brain appears to be independent of the size of the molecules but is largely determined by their chemical structure. The BBB is more permeable to diamino acids than to dicarboxylic acids. As stated, negative charges prevent the penetration of amino acids and their derivatives through cell membranes [949]. This is supported also by the fact that glutamine is accumulated more readily by brain tissue than glutamic acid [1180, 1598]. As a rule, L-amino acids are transported through membranes faster and reach higher concentration gradients sooner than the corresponding D-isomers [898, 1090, 1249, 1260, 1261, 1695]. The effect of stereochemical specificity on transport varies with structurally different amino acids: It is more marked in the case of lysine and tyrosine than for phenylalanine and, especially, leucine.

Mutual Inhibition of Amino Acid Transport. The transport of amino acids from blood plasma into the brain against a concentration gradient is significantly inhibited by other amino acids introduced into the blood stream [639, 898, 1088, 1090, 1256, 1695]. The degree of transport inhibition depends on the structural relations of the amino acids involved [898, 1090, 1695]. As an example the transport of tyrosine into the brain is inhibited by leucine, isoleucine, valine, cysteine, histidine, and tryptophan, but not by alanine, serine, threonine, arginine, lysine, glutamic acid, etc. Stereochemical specificity of competitive inhibition of transport has also been found [1088]. The inhibitory effect of structural analogs of the amino acids is several times weaker than that of the natural amino acids [1256]. Leucine and cycloleucine, for example, mutually inhibit each other's transport into brain tissue [1256], but to obtain the same degree of inhibition, concentrations of cycloleucine from three to eight times higher than those of leucine are required. The carrier evidently has much greater affinity for the natural amino acids. Brain tyrosine uptake is significantly inhibited by many amino acids [639, 1088] but not by acids of similar structure not containing nitrogen [1088]. Amino acids which almost totally inhibit tyrosine uptake by brain tissue have been found inhibitory only to a very slight degree in muscle.

Effect of Age. The permeability of the BBB for different dyes and other histochemically detectable but metabolically inert substances [144, 254, 550, 952] and also for some tissue metabolites [237, 249, 953, 1519, 1731] decreases with age. That age differences in amino acid permeability of the BBB had not been studied

until recently retarded research into the protein metabolism of
the brain and delayed elucidation of the role of proteins in various
functions of the nervous system. It is of historical interest that
the absence of information on the rates at which free amino acids
enter the nerve tissue led to the incorrect conclusion [1081, 1669]
that the turnover of brain proteins in adult animals is very slow.

Recent experimental evidence [364, 1090, 1126, 1240, 1246,
1259] shows that the BBB for amino acids and their derivatives
is formed during ontogenetic development.

The formation of the BBB with age was originally demonstrated in experiments with glutamic acid. These showed that
glutamate is not accumulated in the brains of adult animals after
its blood level is raised by intravenous injection [1598] or prolonged feeding [1126], but in newborn animals its concentration
rises significantly after intraperitoneal injection [1126]. The higher permeability of the BBB of newborn animals than of adult animals was later demonstrated for lysine [1246], leucine [1259],
tyrosine, tryptophan, phenylalanine [1090], GABA [364], and α-
aminoisobutyric acid [1240]. Other recent evidence confirms the
ontogeny of the BBB for amino acids and their derivatives [364,
1090, 1126, 1240, 1246, 1259]. Incidentally the rate of amino
acid transport through the cell membranes in other organs also
varies considerably with age [386].

Attempts have been made [1247, 1253, 1541, 1738] to determine the absolute rates of entry of free amino acids into the brain
at different stages of development. The rate of entry in any particular case is determined by the quantity of amino acid taken up
by the brain from the blood stream per unit time. The results
contradicted the hypothesis that permeability of the BBB for certain amino acids decreases with age. For instance, indices for
the transport of lysine and leucine into the brain were not significantly different in young and adult animals [1247, 1253, 1738].
The calculated rates of passage of glutamic acids from the blood
into the brain were actually lower in newborn animals [1009, 1541]
than in the adult. However, great care must be exercised when
interpreting these results for in calculating the absolute rates of
transport of amino acids into the brain (and other tissues) many
important factors were disregarded and the results are therefore
gross approximations, as the authors themselves pointed out. In
evaluating these virtually identical rates of entry of leucine into
the brain of newborn and adult animals, it must not be forgotten

that labeled amino acids entering the brain tissues of developing animals are incorporated into proteins several times more rapidly than in the adult brain [1247, 1738]. Consequently, allowing for the difference between the rates of incorporation into protein, the actual rate of entry of leucine into the brain is in fact much higher in newborn animals than in adults. Possible differences in the rates of metabolism of the amino acids at different ages must also be taken into account.

We conclude from the experimental evidence now available that the BBB for amino acids and related compounds is formed during development of the organism and that generally the permeability of the BBB decreases with age.

Investigation of Amino Acid Transport through Cell Membranes

The data from the literature examined in the previous section were concerned chiefly with the principles governing amino acid transport between the blood plasma and brain tissue in the intact organism. To understand the mechanisms of amino acid transport through intact brain it is important to study the principles underlying these processes in tissue slices, for under those conditions the permeability of the cell membrane can be studied independently of the permeability barriers of the brain capillaries and of the precapillary anatomical elements.

The Active Transport of Amino Acids through Cell Membranes of Brain Slices. The results of investigations of amino acid transport in slices of nerve tissue agree largely (but not entirely) with results obtained in experiments in vivo. The differences that are found between the basic parameters of membrane transport in vivo and in vitro are predominantly quantitative in character. The in vitro investigations indicate that the process of amino acid transport in tissue slices is enzymic in character and requires the expenditure of energy. This conclusion is based on the following facts. Brain tissue slices assimilate glutamic [1419, 1638, 1689] and aspartic acids [1419], glutamine [1638], GABA [982, 1419, 1638, 1689, 1690], glycine [735], lysine [1244, 1425], leucine [1244], tyrosine [1087, 1419, 1427], histidine, proline, ornithine, methionine, and arginine [1425–1429] against high concentration gradients. This uptake depends on the temperature [1087, 1425, 1689] and the medium [1425], is inhibited under anaerobic conditions [735, 982, 1087, 1425, 1638], stimulated by ATP [735, 1252], glucose, fructose, mannose, and galactose [735, 1087, 1089, 1638, 1689], and is inhibited by respiratory poisons, e.g., 2,4-dinitrophenol, iodoacetate, cyanide, and azide [735, 1089, 1419, 1425, 1427, 1252, 1283, 1689].

These results demonstrate the ability of brain slices to transport amino acids against a concentration gradient, though under physiological conditions these amino acids are present in cells in lower concentrations than in blood plasma (leucine, lysine, proline, etc.). Therefore the mechanisms capable of transporting amino acids into the cellular pool against a concentration gradient must coexist in situ with mechanisms to protect the brain against excessive acceptance of amino acids whose concentration in the blood stream is higher than in the brain tissue. In the intact organism the leading role in maintaining the intracellular concentrations of these amino acids probably belongs to transport processes embodied physically in the BBB itself.

For the establishment of free amino acids levels in the brain, as in any other organ, the rates of exit are just as important as the rates of entry. The mechanism of outflow from the cell has been inadequately studied by experiments in vitro. It has been shown [1284, 1285] that the rates of outflow of lysine, leucine, and D-glutamic and α-aminoisobutyric acids are different and depend on their intracellular concentration and the concentration of other amino acids (and their analogs) in the cell. Glutamic acid exit from the cells is notably tardy. The release of alanine, glycine, glutamic and aspartic acids, glutamine, and GABA into the medium is augmented in the presence of ouabain [922]. It is remarkable that electrical stimulation of the slices increases the rate of liberation of amino acids from nerve tissue cells but does not affect their release from liver and kidney cells [1184].

In different parts of the brain significant differences in the rates of exit have been discovered only for glutamic acid [1285], which may be related to the great disparity in tissue concentrations of this amino acid [827, 1254]. The rates of outward passage of other amino acids are quite similar in different parts of the brain [1285], in contrast to the considerable regional heterogeneity exhibited by some amino acids (lysine, for example) [1287]. The systems of amino acid transport into and out of the cell may not be completely identical (at least for some amino acids).

Amino acid exit from brain slices depends essentially on the
concentrations of other amino acids and their analogs in the
cells. There is some structural specificity displayed in the altera-
tion of rates of outflow by other amino acids [1284]. Despite the
sparse data on this matter it can nevertheless be concluded that
the release of amino acids from the cells in vitro and in vivo do
not take place simply as a result of diffusion but with the aid of carriers,
of which there are evidently several. Even if diffusion processes
play a role in the membrane transport of amino acids, it is by no
means the principal one.

Active Transport of Amino Acids through
Membranes of Structural Components of the Cell.
The heterogeneity of the structural organization, chemical composition, and metabolism of the intracellular organelles suggests
that there are substantial differences in the permeability of different membranes and also in the activity and mechanism of the
biochemical systems responsible for intracellular transport and
for the maintenance of specific concentration gradients of ions
and of various metabolites in these compartments. Differences
of this sort have been discovered for nuclei, mitochondria, and
other particulates [420, 627, 1423, 1448].

Isolated nuclei, mitochondria, and nerve endings from brain
tissue all take up amino acids against a concentration gradient
[1257, 1423, 1448]. Amino acids which are present in vivo in
higher concentrations are assimilated with a greater intensity
and up higher gradients. It is most intriguing that mitochondria
isolated from liver, kidney, or muscle, unlike the corresponding
preparation from brain, do not take up amino acids against a
concentration gradient, which understandably led to proposals
[1423, 1448] that the specificity of transport was embodied in
the mitochondrial membranes of the brain. The transport process
is inhibited at a low temperature and by a variety of substances including sodium cyanide, 2,4-dinitrophenol, ouabain, and numerous
amino acids; it depends on the concentrations of Na^+, K^+, and Ca^{++}
in the medium and it decreases with age. The mechanisms of
amino acid transport through cell (cytoplasmic) and mitochondrial
membranes of nerve tissue may differ, since phenylalanine, for example, inhibits the transport of glycine from the medium into the cell
hardly at all [735], but appreciably inhibits its transport into mitochondria [1448]. Finally, the transfer of activated amino acids

through aminoacyl transfer RNA complexes to the sites of protein
biosynthesis in the cells can be regarded as a special case of intracellular transfer of metabolites by means of intermediate
carriers.

Tissue Differences in the Transfer of Amino
Acids through Cell Membranes. It has already been
stated that under physiological conditions brain tissue differs significantly from other tissues in its membrane transport and its concentration gradients (tissue/blood) for various amino acids. Experiments on tissue slices also revealed substantial differences
in the rates of transfer of free amino acids and their derivatives
through the cell membranes of different tissues [982, 1087, 1426,
1427, 1585, 1638, 1690]. Slices of liver, kidney, and diaphragm,
unlike brain slices, do not transfer tyrosine [1087] or GABA [982,
1690] against a concentration gradient. Spleen slices take up tyrosine from the medium, though less intensively than brain; moreover, unlike brain, the rapidity of this process in the spleen is not
stimulated by glucose [1087]. Lysine, histidine, ornithine, proline, methionine, tyrosine, and 5-hydroxytryptophan are accumulated
by brain slices in higher concentrations than by slices of intestinal mucosa, testis, kidney, spleen, liver, cardiac and skeletal
muscle, and erythrocytes [1426, 1427, 1585]. The only exception
to this rule is slices of small intestine which take up tyrosine
much more rapidly than slices of all other tissues tested, including brain [1427]. Glutamic acid is not concentrated by slices of
liver, small intestine, kidney (medulla), testis, or choroid plexus,
but slices of brain, renal cortex, spleen, and lung do accumulate
this amino acid [1638].

Tissue differences in the effects of acid amino acids on accumulation rates of histine [1429] have been found. The uptake of
histidine by brain slices is inhibited by both L- and D-isomers of
the dicarboxylic amino acids, but in kidney slices is inhibited only
by the D-isomers, and in slices of intestinal mucosa, testis, and
spleen by neither.

Do not conclude from these results that the mechanism of
amino acid transport from the medium into the slice is unique
for nerve tissue. Differences detected between the brain and the
tissues of other organs are as a rule predominantly quantitative
in character. Quantitative differences in membrane transport of
some amino acids are observed when other organs are compared.
Finally, special mechanisms or factors regulating these mecha-

nisms may exist for individual amino acids which are involved in specific functions in the brain or other organ.

If the data on the transport of amino acids between blood plasma and the brain and other organs (experiments in vivo) and between an incubation medium and tissue slices (experiments in vitro) are compared, the following pattern emerges. Some amino acids (histidine, methionine, lysine, tyrosine, proline, etc.) are transported from blood into brain more slowly than into other tissues, whereas these amino acids are assimilated more intensively by brain slices than by the slices of other tissues. Evidently the BBB limits the transport of amino acids into the tissue more strongly than other TBB's. This conclusion is compatible with the observations [1423] of the ability of isolated nuclei and mitochondria of various tissues to accumulate glutamic acid, lysine, and aminoisobutyric acid. Only organelles of brain tissue were found capable of concentrative uptake of these amino acids from the medium.

Factors Influencing the Transport of Amino Acids through Cell Membranes in Tissue Slices. As in the functioning organ, in brain tissue slices the rate of membrane transport (inward and outward) of different amino acids and the concentration gradients attainable are determined by a number of factors, the most important of which are: the structural features of the amino acid to be transported [735, 836, 837, 1425], its initial levels in the cells of tissue slices and the incubation medium [982, 1087, 1244, 1689], the conditions of incubation (pH and ionic composition of the medium, temperature, concentration of other amino acids and sources of energy) of the tissue slices [735, 837, 1088, 1342, 1425, 1428, 1429, 1689], and age and other functional and morphological features distinguishing the tissue investigated [1087, 1090, 1244, 1285, 1287, 1419, 1423]. Let us briefly consider some of the general principles and relationships which have been discovered.

Concentration of Amino Acids. The concentration of amino acids in the incubation medium significantly affects their rate of transport and their ultimate levels in the tissue slices [982, 1244, 1425, 1689]. For example, the concentration of glutamic acid and GABA in slices of the cerebral cortex increases as a linear function of the increase in concentration of these amino acids in the medium [1689]. The uptake of aminobutyric acid and cycloleucine by brain slices also increases with increasing concentration in the incubation medium, but not proportionately: The tissue/medium concentration gradient falls sharply with increasing concentration in the medium [1244]. The rate at which amino acids leave tissue slices also depends on the intracellular concentration of the amino acid concerned [1285].

Structure of the Amino Acids. Structurally different amino acids are concentrated in brain slices to different degrees [735, 836]. For example, slices of rat cerebral cortex, incubated in a salt medium containing glucose at pH 7.5 and at 37°C for 90 min, take up the following amino acids (μmoles/g wet weight of tissue): glycine and serine 14.5-16.4, proline 8.5-10.4, and methionine 1.6-2.0 [735]. It has also been found [836, 837] that the slices take up acid amino acids and GABA to a much higher degree (the tissue/medium concentration gradient is about 30) while basic amino acids such as arginine and lysine reach a concentration gradient of not more than 3-4). Characteristically the relationship between concentrative uptake and the structure of the amino acid transported is opposite to that observed under physiological conditions. Neutral amino acids occupy an intermediate position with the smaller molecules, achieving the greater concentration gradients in the tissue [836].

In short-term (3 min) experiments [837] L-amino acids (glycine, leucine, proline, lysine, glutamic acid, glutamine, etc.) were taken up more intensively by brain slices than the corresponding D-isomers. However, if the incubation were increased to 15 or 60 min the D-isomers of tyrosine [1087], glutamic acid [1689], lysine, and leucine [1244] were taken up to about the same tissue-to-medium ratio as the L-forms. The differences between the results obtained in short- and long-term experiments can be explained by the following circumstances. In investigations of this sort the validity of the conclusions about concentrative uptake as opposed to rate of uptake depends on the duration of incubation of the slices. To obtain reliable measures of the uptake at equilibrium it is necessary to study the indices characterizing the transport process with time, because isomers with higher rates of uptake may also have higher exit rates. Thus varying results will be obtained [1430] depending on the exposure chosen. In experiments in vivo, the process of amino acid uptake by the tissues is also, of course, dynamic and must be studied as such if erroneous conclusions are to be avoided. To find the rates of uptake of amino acids it is only necessary to determine the parameters of transport in short-term experiments. But experiments of longer duration must be undertaken in order to establish the steady-state levels of accumulation of amino acids by the tissue studied, for these are determined by the relative rates of uptake and elimination of the particular amino acid.

The Relation between Amino Acid Transport and Active Transport of Ions through Cell Membranes. There is abundant evidence [735, 1069, 1087, 1244, 1342, 1537, 1689] to show that the process of active transport of amino acids through cell membranes in isolated brain tissue is intimately connected with active ion transport (e.g., tyrosine, glycine, methionine, leucine, glutamic acid, GABA, α-aminoisobutyric acid, and cycloleucine). In a medium deficient in sodium and (or) potassium ions, the uptake of glycine, methionine, leucine [735], GABA [1690, 1689], D-glutamic acid [1342, 1689], and α-aminoisobutyric acid and cycloleucine, which are not metabolized in brain tissue [1244], is appreciably reduced. The absence of calcium and magnesium ions in the medium affects the transport of different amino acids to various degrees: the uptake of tyrosine [1087], α-aminoisobutyric acid, and cycloleucine [1244] is reduced whereas the transport of GABA and D-glutamic acid [1689] is increased. These investigations indicate that high concentrations of ions in the medium also inhibit membrane transport of amino acids. The optimal rates of transport are observed only at certain relative concentrations of sodium, potassium, and other ions. It is interesting to note that ouabain, a substance inhibiting active ion transport through cell membrane, also sharply inhibits the accumulation of some amino acids by nerve tissue slices [1069, 1689] and by isolated brain mitochondria [1448]. The uptake of L-glutamic acid into brain tissue, moreover, is inhibited twice as strongly by ouabain as the transport of its D-isomer [1689].

Mutual Inhibition of Amino Acid Transport. The uptake of some amino acids (glycine, tyrosine, serine, methionine, proline, histidine, glutamic acid, etc.) by tissue slices has been shown to be appreciably inhibited by other amino acids. On the basis of the data available in the literature, it is difficult to deduce any definite rules describing the mutual inhibition of the uptake of each amino acid by brain slices. Only some of the most general facts, about which there is no dispute, and the hypotheses drawn from them can be mentioned. For instance, virtually all amino acids so far studied (neutral, basic, and acidic) inhibit the rate of uptake of

most other amino acids by brain slices [735, 1428, 1429, 1689].
The transport is inhibited by both L - and D-isomers of the amino
acids and by nonmetabolized structural analogs [1428, 1429]
though the L-isomers have a stronger inhibitory effect.

The degree of mutual inhibition of amino acid transport varies
with the pair of amino acids chosen for comparison. β-Alanine
and phenylalanine have virtually no effect on the uptake of glycine
by slices of rat cerebral cortex either in low (2 mM) or in high
(10 mM) concentrations whereas valine and GABA inhibit the uptake of this amino acid by 50% at low concentrations and by 75%
at high concentrations [735]. This mutual inhibition is evidently
explained by direct competition for the common transport mechanism.   The possibility cannot be ruled out that some amino
acids, aspartic and glutamic acids for example, can exhibit their
inhibitory action indirectly by reducing the concentration of ATP
and other sources of energy in the tissue [14, 735, 1320, 1789].
A fall in the ATP level [1252] leads to a corresponding inhibition
of amino acid uptake. On the basis of these observations Lajtha
[1252] postulated an indirect utilization of the energy of ATP for
amino acid transport: The energy was utilized initially to create
ionic gradients which were then used for transporting the amino
acids through the membrane.

Effect of Age. The data for age changes in the absorption of
amino acids by brain slices are still limited in number and concern only a few amino acids. It has been shown that lysine [1244]
and tyrosine [1090] are taken up more rapidly by brain slices
from newborn, whereas leucine [1244] is taken up more intensively
by slices from adult animals. The intensity of uptake of aminoisobutyric acid, which is not metabolized, by brain slices increases with age. It is three times higher in the adult than in the
newborn animals [1244]. It is interesting to note that the ability
of brain slices to accumulate stereoisomers of various amino
acids develops differently with age. In both newborn and adult
animals L-lysine is accumulated to higher concentration gradients
than its D-isomer [1244, 1257]. A different age relationship has
been found for the isomers of leucine wherein L-leucine is more
concentrated in adults but D -leucine in newborn animals. Nuclei
and mitochondria isolated from the newborn brain accumulate
amino acids from the medium to higher gradients than the corresponding subcellular structures from adult brain [1257, 1423].

Under physiological conditions the rates of exchange of amino
acids between the blood plasma and the brain, and the maximal
levels of their accumulation in the cells, are determined by the
functioning of two barrier mechanisms. One is embodied in the
BBB itself and the other in the membranes of nerve cells and their
organelles. One may ask which of these mechanisms is responsible for the decreased amino acid transport with age from the
blood plasma into the brain. To answer this question the transport of tyrosine into the brain of adult and newborn rats was compared in experiments on intact animals and tissue slices [1090].
By conducting experiments with slices it is possible to exclude
the role of transport mechanisms of the capillary structures of
the BBB and thus to investigate the transport systems functioning
in the cell membranes.

These investigations showed that both in vivo and in vitro the
brain of newborns utilizes tyrosine faster and concentrates this
amino acid to a greater extent than the brain tissue of adult animals. On this basis the researchers concluded that the permeability
of the structures of the BBB itself for tyrosine did not change significantly with age. The reduced activity of the cellular mechanisms of membrane amino acid transport can be explained, in their
opinion, by the lower rate of utilization of amino acids by the brain
tissue of adult animals. This problem evidently cannot be regarded as finally solved, for the interpretation of the results
given above is not the only one.

Functional and Morphological Features of the Tissue. In
some investigations [799, 1285, 1287, 1419] important regional differences in amino acid transport have been discovered between
the cell membranes of brain slices. The rates of transport and
the concentration gradients attained in different parts of the nervous system [1287] differ for lysine, a basic acid; glutamic acid,
which is acidic; and the neutral α-aminoisobutyric acid. These
parameters were significantly higher for tissue slices from the
mesencephalon and from functionally and morphologically different areas of the cortex compared to tissue slices from white
matter and from spinal cord. No parallel could be observed between the physiological levels of lysine in the parts of the brain
investigated and the rate of transport of this amino acid into the
corresponding tissue slices. Amino acids such as alanine, histidine, tyrosine, aspartic acid, GABA, etc., were taken up at dif-

ferent rates by slices of different parts of the brain [1133, 1419].
It was shown that at least some of these differences in rates of
transport and concentrative uptake were [1287, 1419] not due
to differences in the volumes of the "inulin space" in the slices of
the regions compared.

An important factor determining the amino acid level in the
tissue slices is the rate at which the amino acids leave the cell
and enter the tissue medium, as mentioned elsewhere [1285]. The
level of any amino acid in the tissue depends on the net rate of
its input. Higher concentration gradients of amino acids are attained in tissue slices from parts of the brain in which the rates
of output do not equal the rates of input until a large difference
in tissue to medium levels is obtained. Regional differences in
the rates of transport and the levels of uptake may also be in part
determined [1419] by the functional and morphological features
distinguishing the different parts of the brain.

Mechanisms of Biological Transport of
Amino Acids through Cell Membranes

The results of the numerous investigations described in this
chapter indicate that an intensive exchange of amino acids takes
place constantly between the brain tissue and blood plasma, between the cells and the extracellular fluid, and between the organelles of the cell and its hyaloplasm. This exchange takes place
while the relative intracellular levels of amino acids are maintained, often against high concentration gradients. What are the
mechanisms that determine the direction and rate of amino acid
transport through a membrane and also maintain a steady and uniform distribution of amino acids between the cells and the medium surrounding them?

Some possible mechanisms for the transport of metabolites,
including amino acids, through the cell membranes have been
discussed [234, 310, 420, 557, 628, 872, 1134]. They include simple diffusion, the passage of dissolved substances within a flow of
solvent, diffusion through the nonaqueous phase, facilitated diffusion, and other complex processes such as active transport, exchange diffusion, and pinocytosis. Considering the particular importance of the last three categories of transport processes, brief
details will be given of the mechanisms by which they function
and of their role in maintaining the intracellular levels of amino

acids in nerve tissue. Two of these mechanisms, exchange diffusion and active transport, unquestionably play a leading role in regulating free amino acid levels in the cells and subcellular structures, a function essential for the intravital renewal of proteins and for the metabolic conversions of amino acids at rates compatible with the functions of the particular tissue or structure.

Active Transport of Amino Acids. Since the vast array of experimental material confirming the active character and enzymic nature of amino acid transport in nerve tissue has already been described at this point, only a brief discussion will be given of some of the problems concerning the mechanism of active membrane transport of amino acids in brain tissue.

The basic evidence in support of the active character and enzymic nature of the processes of membrane amino acid transport is: (1) the transport of amino acids, both inflow and outflow, against high concentration gradients; (2) the dependence of the process on energy and on the temperature and pH of the medium; (3) inhibition by anaerobic conditions and by enzyme poisons; (4) the interconnection between amino acid transport and the active transport of ions; (5) the competitive inhibition of the transport and its stereochemical and structural specificity. On this basis it can be firmly accepted that biochemical systems of active amino acid transport are functioning in vivo in nerve cells and also in the membrane structures of the BBB, which separates the tissue from the blood.

That special carriers participate in the transport of amino acids through cell membranes is no longer disputed, though neither the chemical nature of the intermediate carriers nor their amount, specificity, or mechanism of action has been definitely established. Until recently none of the carriers of amino acids in the tissues of the nervous system had been identified. All that could be said [1135, 1689, 1690] was that certain fractions of phospholipids (phosphatidylcholine, phosphatidic acid) may participate in the transport of amino acids and ions. The hypothesis had been put forward [1331] that phosphopeptides, with a high rate of phosphorus metabolism, may also participate in the active transport of metabolites. The role of phospholipids and phosphoproteins in membrane transport of materials has been discussed elsewhere [258, 328, 618, 1134, 1135]. The implication of phospholipids in membrane transport is of considerable theoretical and practical

importance, for these compounds are important components of
all biological membranes and they are readily able to form lipid-
soluble complexes with extracellular and intracellular substances.

Recent investigations have shown [10] that special transport
proteins, to which the name "transphores" has been given, participate in the transport of ions and various metabolites (including
amino acids) through cell membranes. The transphores of micro-
bial cells have so far been studied in the greatest detail. Some of
these proteins have been isolated and their properties studied.
Proteins performing specialized functions at various stages of
the membrane transport of metabolites are also evidently present
in animal cells [1775]. Research into various aspects of the transport of metabolites through cell membranes has developed rapidly
and it can be hoped that in the near future the transport proteins
will also have been isolated from nerve and other tissues. This
will help to reveal the molecular mechanisms of function of the
transport systems and to explain their role in the regulation of
the biochemical processes and physiological functions of cells, including their role in the specific functions of the nervous system.

Until recently an important problem which had remained un-
solved was the question of whether there is one transport system
common to all amino acids or whether each amino acid has its own specific system. The first enlightening results were obtained by the investigation of the specificity of the mutual inhibition of the transport of different amino acids. The basic assumption for competitive inhibition of membrane amino acid transport, either under
physiological conditions or in tissue slices, is that those amino
acids whose transport is mutually inhibited are transported by the
same carrier. Conversely the absence of mutual inhibition is regarded as evidence that the particular amino acids are transported by different mechanisms. Analysis of the results of investigations in vivo [639, 898, 1088, 1090, 1256, 1695] and on tissue
slices [735, 837, 1087, 1424, 1428, 1689] has shown that one amino
acid may inhibit the transport of several (but not all) amino acids;
on the other hand, the transport of any one amino acid can be inhibited by many other amino acids (but not all). These findings
seem to favor the existence of a special carrier for each group of
amino acids rather than a common carrier for all or specific
carriers for each amino acid.

On the basis of the inhibitory action of 15 amino acids and
their derivatives on the transport of histidine into brain slices it

was first postulated [1428] that at least three types of transport systems exist. These included a stereospecific system for neutral amino acids, one without stereochemical specificity for acidic amino acids, and a third, separate system for aromatic and basic amino acids. A little later, on the basis of similar experiments on the competitive inhibition of the transport of the three classes of amino acids (neutral, acidic, basic) and their analogs, Blasberg and Lajtha [836, 837] concluded that there are six different transport systems: two for neutral amino acids, two for basic, one for acidic, and a special system for GABA and D-glutamic acid.

The role of enzymes in the membrane transport of metabolites, including amino acids, has not been finally explained. For example, we do not know whether enzymes catalyze only the processes of formation and dissociation of complexes between the carriers and the substances transported, or whether the enzymes themselves perform the function of carriers. Kazakova and Neifakh [247] considered the possibility that the actomyosin-like structural proteins of mitochondrial and other membranes are responsible not only for the contractile properties of these membranes, but also for the active transport of metabolites across the membranes ("mechanoenzyme"). There is no general agreement on whether there is one carrier for the transport of a given substrate both into and out of the cell or whether each of these processes has its own carrier. Clearly the actual mechanisms of the processes of active membrane transport of amino acids in brain tissue still remain unexplained.

Exchange Diffusion of Amino Acids. The term "exchange diffusion" implies [628, 1118, 1134, 1174, 1507] a process of exchange of metabolites through cell membranes wherein flow in one direction is counterbalanced by an equal and opposite flow. Exchange diffusion may occur between tissue and blood, between a cell and its intercellular space, or between the cytoplasm and subcellular structures. In contrast to active transport the process of exchange diffusion takes place without the expenditure of energy, independently of the ionic composition of the medium, and undisturbed by uncoupling agents. It balances the inflow and outflow of metabolites in equal quantities, i.e., it does not change the existing concentration for a given metabolite on either side of the membrane. Like the process of active transport, exchange diffusion also involves carriers though these function under the special constraints mentioned.

The work of Quastel [1507] also yielded evidence in support
of two different mechanisms of amino acid transport (active transport and exchange diffusion) in brain tissue. For instance, during
the incubation of brain slices in Ringer's salt medium for 30 min,
a concentration gradient of about 9 was reached for glycine-^{14}C as
determined by uptake of radioactivity. If, however, the slices
were first preincubated with nonradioactive glycine for 90 min,
thus establishing the highest possible concentration gradient of
this amino acid, and the slices were then incubated in medium
containing glycine-^{14}C, a gradient of 9 for labeled glycine was
still attained after 30 min. However, in experiments without preincubation of slices with the nonradioactive amino acid, ouabain
(0.01 mM) almost halved the concentration gradient of the label;
i.e., the process of active transport was inhibited by 47%, whereas in the experiments with preincubation, the concentration gradient of the label was unchanged. Thus ouabain does not inhibit
exchange diffusion. The results of these two series of experiments show that two processes take place in brain slices incubated
under the conditions described: the active transport of the amino
acid from the medium into the cells (experiments without preincubation), which is inhibited by ouabain, and the exchange diffusion process that is insensitive to ouabain (experiments with preincubation).

Pinocytosis. Other mechanisms whereby substances can
enter cells against a concentration gradient have been described
in the literature, one of which is pinocytosis [234, 310, 391, 1134,
1272]. Practically all animal cells in tissue culture and all cells
living freely in the medium (protozoans, microorganisms) possess
this mechanism for the transport of metabolites. It has recently
been shown that many types of cells in the intact living organism
also exhibit pinocytosis. The phenomenon of pinocytosis has been
observed even in the Schwann cells and satellite cells of neurons
of the sympathetic ganglia [1540]. According to Pallade's hypothesis [234], pinocytosis is a feature of all animal cells. The
question of whether pinocytosis is responsible for the transport
of materials through the membrane of cell nuclei is also under
discussion [391, 627].

It is generally suggested that pinocytosis is a mechanism for
transporting into the cell substances of high molecular weight (including proteins), for which the cell membranes are impermeable,
and also substances of low molecular weight (including amino
acids), which pass comparatively easily through cell membranes.

For compounds of high molecular weight this mechanism represents a possible method of transport against a concentration gradient [1272].

Much remains to be explained about the transport of materials into the cell by pinocytosis. In particular, modern views of pinocytosis make it difficult to harmonize the nonspecificity of this system with the high selectivity of transport of metabolites through cell membranes. In this connection the view [310] that pinocytosis is a possible means for the transport of ready-made components of membrane structures into the cell for the construction of organelles, particularly the mitochondria, is most intriguing.

*   *   *

Described in this chapter is the existence of barrier mechanisms and biochemical systems for the transmembrane transport of amino acids both in the physical structures of the BBB itself and within nerve cells. This important role of maintaining the physiological levels of free amino acids in brain tissue and thus in regulating numerous intracellular metabolic processes is carried out by three principal biochemical mechanisms of membrane transport: active intake, active output, and exchange diffusion. The activity of the barrier and the transport systems differs in different parts of the brain and is sensitive to different physiological and pathological factors. Functionally induced changes in the rates of membrane transport of amino acids and other substrates are among the important factors regulating the intensity of intracellular metabolism, including the metabolism of proteins.

Research into the mechanisms for membrane transport of amino acids and other metabolites is just beginning to develop in a coherent manner. Many aspects of this problem remain unexplained and require experimental investigation. Knowledge of these unexplained aspects of normal and pathological metabolism in the tissues of the brain and other organs, as well as the elucidation of the biochemical basis of nervous activity, must largely depend on the successful solution of problems concerning the transport of metabolites through membranes in general and its regulatory functions in particular.

Chapter 4

Some Morphological, Functional, and Biochemical Features Peculiar to Nerve Tissue

Nerve tissue is the most highly specialized tissue in the animal body. It possesses numerous distinctive morphological and functional properties that determine some of the specific features of its biochemical behavior, including its metabolism. A knowledge of these features is important for the correct interpretation of experimental data on the biosynthesis and breakdown of tissue proteins and other biologically important substances and also for an understanding of the role of metabolic processes in the many specific and nonspecific functions of nerve tissue. For that reason, in this section we shall discuss the most important morphological, functional, and biochemical properties of nerve tissue but limit our detailed description to those that are directly concerned with protein metabolism.

The High Degree of Structural and Functional Complexity and of Heterogeneity of Nerve Tissue. The tissues of the nervous system, especially the cerebral cortex, are characterized, as no other tissue of the animal organism is, by the extreme multiplicity of cell types, which differ in structure, shape, size, kinds of intercellular connections, biochemical properties, and functions performed [157, 215, 506, 673, 724]. The most important morphological features distinguishing the neuron, the principal functional unit of the nervous system, from the cells of other tissues are: large size (the largest cell in the body); the presence of a complex and highly developed endoplasmic reticulum and a highly ramified system of dendrites (from a few to several thousands in functionally different neurons), giving a particular neuron interconnections with many other nerve cells (several thousands of neurons can be counted in the sphere of the terminal ramifications of a single nerve fiber); the large mass of the axon (in some neurons it is more than 1000 times heavier than the cell body); and the considerable distance traversed by the axon from the cell body [46, 157, 215, 347, 673, 1759]. It is also important to note that the bodies of the nerve cells in the cerebral cortex occupy only about 5% of the total volume of the tissue, while the dendrites account for another 25%. The myelin, neuroglia, and blood vessels account for 35%, and the extracellular space for the remaining 35% [1493].

The basic functions of nerve tissue are distinguished not only by their extreme heterogeneity and diversity, but also by their high specificity and by their extremely important regulatory role, as well as by the very rapid and rhythmic alternation of the fundamental processes of nervous activity—the states of excitation and inhibition [44, 428, 1759].

The great diversity of the types of cells in nerve tissue and their close structural connections with each other lead to great difficulty in identifying morphologically and functionally homogeneous cells in large numbers and in investigating their biochemical properties. Recent attempts [497, 539, 840, 923, 1004, 1562, 1579] at obtaining preparations of isolated neurons and glial cells must still be regarded as the first stage in the long, arduous, and technically difficult work that neurochemists will have to undertake in the very near future. The main advantages and disadvantages of existing methods of isolating fractions rich in a particular type of cell from nerve tissue have been described in recent surveys by Pevzner [497] and Pomazanskaya [539].

The Proliferation of Nerve Cells. Unlike glial cells and most other types of somatic cells, which remain capable of dividing throughout ontogeny, many neurons evidently function throughout the lifetime of the organism. Neuroblasts lose their ability to undergo mitosis in the early stages of postembryonic development. This view is held by most investigators [127, 170, 214–216, 218, 974] although not by all [158, 493, 554, 762].

Specific Features of the Barrier Mechanisms of the Brain. The BBB, which regulates the exchange of metab-

olites between the brain tissue and the blood stream, differs
from other TBBs in its particularly high structural complexity and
"density," its well-marked regional functional heterogeneity, its
high selectivity, and its exclusive ability to maintain the relative
constancy of the composition of the internal medium of the brain
tissue even during significant changes in the functional state of the
organism. Characteristically this homeostasis exists for some
substances, including many amino acids, despite high concentration
gradients between the brain tissue and blood plasma. The BBB
and its role in brain protein metabolism is described more fully
in Chapter 3.

The High Concentration of Lipids. Nerve tissue
is distinguished by a high concentration of lipids, which vary considerably in their chemical composition and in their functions
[294, 315, 344, 617, 1022]. In adult animals lipids account for
about 50% of the total brain substance. Their predominant constituents are phospholipids.

Much of the protein of nerve tissue exists in the form of
lipoprotein and proteolipid complexes. By contrast with other organs, the tissues of the nervous system and, in particular, of the
white matter of the brain have a high content of lipoproteins of a
special type discovered by Folch-Pi and co-workers in 1951 [1019].
A characteristic feature of the proteolipids is their insolubility in
water, their high solubility in a chloroform—methanol mixture,
the presence of a protein resistant to proteolytic enzymes, and
their metabolic stability [660, 941, 1019, 1022, 1038, 1123, 1201,
1682]. An interesting feature of nerve tissue is that the concentration of proteolipids in the white matter of the brain is 10 times
higher than in other organs and also than in the tissue of the
peripheral nervous system [660, 764, 765, 1123].

Brain tissue contains a special type of gangliosides (belonging
to the group of acid glycolipids), not present in other tissues.
These gangliosides are apparrently localized selectively in the
neurons (chiefly in the membranes of the endoplasmic reticulum)
and in structures (synaptosomes, synaptic vesicles) containing
acetylcholine and acetylcholinesterase [878, 1266, 1308]. It is
interesting to note that gangliosides are probably absent altogether
from the cytoplasm of nerve cells and from myelin [1604]. There
is reason to suppose [617, 878, 879] that gangliosides play an important role in the binding, liberation, and transport of the chemical

mediator of excitation, acetylcholine, and possibly also noradrenalin and others.

The Exceptionally High Intensity of Metabolism and Energy Consumption. Brain cells are characterized by an intense utilization of oxygen and glucose and also by
the "explosive" character of their metabolism. These features of
nerve tissue are in good agreement with the exceptionally high rate
of their physiological actions and the rapid interchange of excitatory and inhibitory processes in it. Brain tissue, accounting for
only 2.0-2.5% of the total human body weight, utilizes 15-20% of
the oxygen assimilated by the body. It receives a similar proportion of the total blood volume pumped by the heart and it utilizes
about the same proportion of total glucose [111, 344, 548, 704,
1188].

The "explosive" character of metabolism in the structures of
the nerve tissue can be attributed to the rapid (milliseconds) interchange of excitatory and inhibitory impulses [44, 428, 724]. Although modern biochemical methods are not yet sufficiently sensitive to record and to measure quantitatively the chemical processes
taking place within such short time intervals, there is no doubt
about the complete correlation in the nerve cells between functional
activity and biochemical conversions.

The Utilization of Amino Acids as a Source of
Energy. Brain tissue differs substantially from other organs
in the very high degree of development of its intracellular biochemical systems for the efficient utilization of amino acids (glycine, alanine, glutamic acid, GABA, etc.) as sources of energy
[274, 1408, 1571, 1738]. This ability of nerve tissue is delineated
more clearly in the case of deficiency of glucose, particularly
during increased functional activity when the expenditure of energy rises dramatically.

Exceptional Sensitivity to Hypoxia. The nerve
tissue of adult animals, especially the cerebral cortex, has an
unusually high sensitivity to hypoxia, which is responsible for the
very short period of survival of its cells during anoxia or circula-
tory insufficiency [128, 704, 1125]. The sensitivity of the brain to
hypoxia results from several factors: the high energy expenditure
coupled with the very rapid response of nerve cells to innumerable
incoming exteroceptive and interoceptive impulses; the absence of

significant reserves of carbohydrates as an alternate source of energy [111, 549]; the comparatively limited reserves of high-energy compounds [704] despite their high concentration [98, 1329]; and the inadequate energy-generating capability of the glycolytic enzyme systems [344, 896].

The Important Role of Glucose as Precursor of Amino Acids. Nerve Tissue, like no other tissue in the animal body, has extremely active mechanisms for the formation of free amino acids from glucose. This important feature of the brain will be described more fully later. Here it will simply be stated that the mechanisms of formation of amino acids from glucose in brain tissue may possibly play an important compensatory role. They may help in stabilizing the level of protein metabolism in nerve tissue in some pathological states connected with a disturbance of the supply of exogenous amino acids, for example, during prolonged and exhausting protein starvation. In this serious pathological condition protein metabolism is known to be disturbed far less in the brain than in other organs [470, 632, 676, 760, 1281, 1344, 1375, 1776]. Of course, other factors, such as active transport, may help ameliorate this condition.

Alternative Pathways of Conversion of Materials. Two pathways exist in the brain for the conversion of certain substances, either of which may predominate in a given metabolic situation [1129, 1687]. Alternative pathways are known, for example, for the conversion of pyruvic acid (oxidation through acetyl-CoA or through CO_2 fixation and oxaloacetic acid) and for α-ketoglutaric acid (conversion into succinic acid through succinyl-CoA and through glutamic and γ-aminobutyric acids). The existence of these and possibly other [1461] alternative pathways for metabolic conversions in the brain is important for maintaining the comparatively stable content and metabolism of various substances, including proteins, should the supply of exogenous metabolites be severely restricted [484, 760, 1281, 1344, 1687].

Compartmentalization of Metabolites with Different Metabolic Activities. Reserves of certain metabolites, including amino acids, spatially separated from each other and differing in their metabolic activity, can be found in the tissue structures of the brain [827, 830, 1039, 1254, 1459, 1733-1735]. Generally one of these stores is quantitatively smaller but

metabolically more active while the other, by contrast, is larger
but less active metabolically and acts as a special kind of reserve
for replenishing the first during special metabolic situations.
This sort of compartmentalization of metabolites is particularly
well marked in the tissues of the nervous system, where it evidently plays an important role in homeostasis in functional states accompanied by sharply intensified metabolism.

The Compensatory Powers of the Tissue. Many
of the morphological, functional, and metabolic features which distinguish the brain demonstrate the very high plasticity of nerve
tissue and its great powers of compensation. This important property of the structures of nerve tissue underlies the stability of its
physiological and biochemical functioning which is of paramount
importance in an organ with the task of regulating activity
under both normal and pathological conditions (for example, in
prolonged starvation), and when metabolism is sharply intensified
in connection with increased physiological functions (for example,
a state of prolonged excitation). Other important compensatory
mechanisms of the brain include: (1) the ability of nerve cells to
undergo hypertrophy at times of active function [544]; (2) the
increased polyploidization of nerve cells with age in response to
natural losses of many of them with age [66]; (3) the possibility of
regulating the intensity of the blood supplied to the brain through
a change in the lumen of the blood vessels, their hypertrophy, and
shifts in the arterial pressure level in response to changes in the
functional activity and metabolic needs of the tissue, especially
in pathological states [121, 134, 265, 392, 734, 1031].

During exposure to mildly injurious factors and at times of
increased functional activity of the cells, the necessary levels of
metabolism in the nervous system are maintained chiefly through
the mobilization of reserve capacities of the tissue structures such
as the use of materials from the "metabolic reserve compartments,"
the "reserve capacity" of enzyme systems, and the switching to
alternative metabolic pathways. In more severe pathological states
e.g., prolonged starvation, when the internal compensatory and
adaptive powers of the nerve tissue are exhausted, additional mechanisms such as the capture of metabolites from other organs
are brought into play.

The Transport of Substances Synthesized in
the Perikaryon of Neurons. In the neuron, as in no other

cell in the animal body, large quantities of proteins and other substances which are synthesized in the perikaryon, at the periphery
of the nerve cell, are transported along the axons and dendrites,
and into the nerve endings. The formation of the axoplasm in the
perikaryon of the neurons and its subsequent movement along the
axon was established in experiments on ligated nerves of animals
[1759], and verified in cells in culture with microfilming techniques and electron microscopy [1760]. Many recent investigations
employing radioactive amino acids (leucine [790-792, 958, 960,
1143, 1144, 1450], lysine [958, 959, 970, 1733], methionine, and
arginine [961]) have proved conclusively that soluble proteins synthesized in the perikaryon of the neuron are transported in a
proximodistal direction. The movement of nucleic acids from the
perikaryon to the periphery of the cell has also been discovered
[860, 888, 1486]. The rate of movement of the axoplasm, and also
of protein and other materials along the axon, is very slight and
differs in neurons with different functions. It varies between 0.1
and 3.0 mm/day [960, 961, 970, 1759] and decreases with age [961].

It was recently established [791] that the specific mitochondrial enzyme monoamine oxidase migrates with the flow of axoplasm.
One must not overlook the possibility that soluble mitochondrial
and other proteins may be transported, not only individually but also
bound with mitochondria, from the body of the nerve cell along the axon
to the nerve endings [791, 792]. Supporting this view is the presence of
clusters of mitochondria near the nodes of Ranvier and in proximal
segments of a crushed nerve trunk [1760], and the data of electron
autoradiography [958]. Data indicating the movement of noradrenalin
from the cell body along the axon toward the nerve endings have also
been obtained [928, 1049, 1050, 1300].

Characteristically the calculated velocity of proximodistal
transport of vesicles containing noradrenalin along the axon is
much higher than the established velocity of movement of the axoplasm. In the sciatic nerve of the rat and cat, noradrenalin-con-
taining vesicles are transported at speeds of 5-6 and 9-10 mm/h,
respectively [928], whereas the rate of movement of the axoplasm
is about 1 mm/day [1759]. If such differences really exist, it can
be assumed that in addition to the passive transport of metabolites
within the flow of axoplasm toward the nerve endings there must
also be mechanisms of active transport. A decrease in the acetylcholinesterase activity of peripheral nerves in the proximodistal
direction has been detected [1310], and this may be evidence of

the transport of this enzyme from the perikaryon toward the nerve endings.

Physiologically active proteinaceous substances, the neurohypophyseal hormones [118, 344, 525, 723, 1587], synthesized in the perikaryon, are transported along the axons of the highly specialized neurosecretory cells, located chiefly in the preoptic, supraoptic, lateral, and paraventricular nuclei of the hypothalamus. The particular features of the structure, function, and phylogenetic and ontogenetic development of these unique neuroglandular cells have been described in detail by Polenov [525]. The movement of various viruses and toxins along the processes of nerve cells has also been described [1031].

Other data [159, 170, 1738, 1760] directly or indirectly confirming the existence of a proximodistal concentration gradient of various substances along the length of the axon have also been obtained.

Since in neurons, in contrast to other somatic cells, physiological regeneration takes place at the subcellular level and not by mitosis, the movement of biologically important compounds to the periphery of the cell, where systems for the biosynthesis of these substances are either absent or very poorly developed, is of absolute importance to the normal functioning of the cell during its long life span.

Other Particular Features. Among the other important distinguishing features of brain tissue are (1) the special anatomical localization of the organ, which greatly complicates long-term studies of experiments requiring repeated measures of the metabolic processes underlying the functions of nerve tissue; (2) the absence of simple and reliable methods of investigating the mechanisms whereby proteins and other compounds participate in such specific brain functions as memory; (3) the higher concentration of slowly metabolized components, including proteins [31, 454, 1678] and nucleoprotein complexes [399], than in other tissues; (4) the high concentrations of compounds such as GABA and N-acetyl-L-aspartic acid, which are found in negligible quantities if at all in other tissues, and the particularly important role of these substances in brain functions [84, 87, 89, 125, 616, 1340, 1659, 1662]; (5) the high activity of the biochemical systems for the formation and fixation of ammonia and the intimate relation of these processes with amination and deamination of the tissue proteins and with brain function [114, 116, 276, 357, 635-637].

Chapter 5

Renewal of the Proteins of Nerve Tissue and Its Subcellular Structures

The Dynamic State of Tissue Proteins

Modern views on the dynamic state, i.e., the constant renewal, of tissue proteins of adult animals have become established as a result of a long conflict between two opposite views [488, 1432]. Some workers postulated that most tissue proteins in the adult organism, which is in a state of nitrogen equilibrium, are metabolically inert. According to others, however, the tissue proteins in adult animals are constantly being reconstructed. The concept of the dynamic state of virtually all tissue proteins in adult animals finally triumphed as a result of much experimental research [1006, 1107, 1395, 1593, 1594, 1635] using compounds labeled with stable and radioactive isotopes. Until quite recently some workers were still inclined to reject the dynamic state of proteins in the mammalian organism [62] on the basis of data obtained during induced synthesis of β-galactosidase by <u>Escherichia</u> <u>coli</u> cells [1132]. However, it no longer seems justifiable to extrapolate to mammalian tissues the results of investigations conducted on microorganisms.

Some possible molecular mechanisms of renewal of tissue proteins in vivo have been discussed in the literature. However, not all have been experimentally confirmed.

For more than two decades the view was held [282, 366, 1541, 1593, 1595] that protein molecules are renewed in the body by partial substitution of individual amino acids or peptides in the polypeptide chains with the aid of enzyme systems capable of hydrolyzing and synthesizing peptide bonds. This hypothesis has not been experimentally confirmed [1302].

The second possible pathway for the renewal of tissue proteins in vivo is by their total breakdown with intracellular proteases, followed by enzymic synthesis de novo from free amino acids [436, 437, 873, 1006, 1407, 1614]. The results of recent studies on the function of intracellular biochemical systems of protein synthesis [178, 366, 703, 1302], together with the presence of enzyme systems capable of hydrolyzing proteins to free amino acids in various tissues and subcellular structures [27, 532, 692, 1250, 1345, 1349], indicate that this cycle is in fact the principal pathway for the renewal of tissue proteins in the living organism.

Another historically interesting view was that isolated protein molecules could incorporate amino acids independently, i.e., without the participation of specific enzyme systems or of template structures. The results of investigations by the few supporters of this hypothesis have been summarized in a recently published work by Konikova and Kritsman [282]. Of course, the nonspecific binding of amino acids with protein structures can take place, especially in experiments in vitro, which must be borne in mind when the results obtained with labeled amino acids are interpreted. However, such an addition of amino acids to protein molecules must not be regarded as a possible pathway of their renewal in vivo, for this process has nothing in common with biological protein synthesis in the intact organism.

Changes in the structures of protein molecules take place during amination and deamination of proteins in the brain [88, 114, 131, 148, 356-358, 635, 1161, 1723, 1724] and other tissues [1586], during the formation of certain proteolytic enzymes from their inactive precursors [196, 1437, 1438], and during the conversion of fibrinogen into fibrin monomer [39, 40, 808, 1264]. Such alterations cannot be regarded as the intravital renewal of protein molecules. What happens in each of these cases is not the synthesis of new protein molecules, but merely the structural conversion of existing protein structures as a result of modifying the amino acid composition. These conversions of protein molecules must be regarded as a final stage of their biological synthesis taking place through the modification of corresponding precursor proteins. The renewal of protein molecules in vivo takes place with the aid of mechanisms that are common to all tissue proteins, viz., their total degradation and subsequent synthesis de novo.

The intensity of renewal of all protein molecules in the body is determined by the rates of enzymic processes in two opposite

directions: synthesis and breakdown. In the various stages of
ontogenetic development, in different tissues and in different functional states, the processes of breakdown and synthesis of protein
molecules take place at different rates. In one case, individual
protein molecules are broken down and synthesized de novo; in
another case, complex organelles undergo destruction and are
re-formed [245, 1006, 1203, 1432, 1433]; in a third case, whole
cells die and are replaced, i.e., regenerated [124, 217, 309,
325]. Protein renewal connected with the physiological regeneration of cells takes place more intensively, for example, in the
hematopoietic organs and in the epithelial cells of the skin and
alimentary tract. The rates of this renewal diminish with age and
are much lower in tissues whose cells have lost their power of
regeneration such as neurons.

Since in most tissues dying cells are constantly being replaced
the question may be asked: Is protein renewal in general unconnected with the physicological wastage? Comparison of the data
on the mitotic activity of various tissues with the intensity
of renewal of their total proteins supplies a positive answer
to this question. For instance, the mitotic activity of the liver
cells is comparatively low [124, 217, 309, 325], whereas the intensity of renewal of its proteins is high, only a little lower than the
intensity of renewal of proteins of the intestinal mucosa [1030,
1301, 1780], a tissue with a very high rate of cell division [309,
1276]. Cells of the basal layer of the cutaneous epithelium are
highly capable of rapid division [217, 309, 1276] although the intensity
of renewal of many of the skin proteins is very low [1104, 1276,
1327]. The intensity of protein renewal of the neurons, cells with
little or no mitotic activity [126, 214-217, 309], is much higher
than the intensity of protein renewal in glial and other cells of
nerve tissue [1579], which are able to divide by mitosis [126, 376,
864].

The rate of intravital renewal of tissue proteins is different
at different stages of ontogenetic development and in functionally
different tissues, cells, cell organelles, and their structural formations. Each individual protein or protein complex has its own
rate of renewal which may depend on its function in the cell. Until
recently there was no experimental support for the view that the
intensity of renewal of proteins, as of other biologically important
substances, varied. The solution to this problem, an important
contribution to the elucidation of the biochemical basis of cell

functions, became possible only after the introduction and extensive use of the isotope method in research. The experimental material obtained by this method, which revealed the considerable metabolic heterogeneity of proteins in the nerve tissue, will be described in the other chapters. Only a few points will be touched upon here.

In studies of the protein-synthesizing activity of the tissue in vivo with the aid of labeled amino acids it is usually impossible to determine separately the rates of renewal of protein due to the replacement of the cells or to the breakdown and synthesis of individual protein molecules in nondividing, functioning cells. The rate of renewal of proteins determined in experiments in vivo, determined from the level of label incorporated in them, is in fact a reflection of all the ways of incorporation of labeled amino acids into proteins, including cell proliferation.

Every protein species arising de novo has its individual, characteristic life span which depends on a complex of functional characteristics of the organ, the type of cells and their organelles, and the function performed by the particular protein (metabolic, transport, regulatory, structural, etc.). Proteins do not have a "life span" in the traditional sense of a timed period of existence but rather suffer a certain probability of degradation. Thus when a particular protein is quickly labeled, the radioactivity begins to be lost immediately and follows a logarithmic decay. Proteins are therefore characterized by "half-lives," that is, the time in which one-half of the protein molecules will be destroyed. A high rate of renewal corresponds to a short half-life. In every tissue, including nerve tissue, the various cells and subcellular structures contain complex sets of proteins which, in accordance with their functional role, are renewed at vastly different rates, i.e., have greatly differing half-lives.

The earlier subdivision of proteins into either a structural (insoluble), metabolically inert class, to which mainly passive structural and protective functions were ascribed, or to a functional (soluble), metabolically active type is nowadays regarded as incorrect [77, 90, 135, 317, 557, 705]. The results of numerous experimental investigations of the physicochemical properties and biological functions of proteins in the membranous structures of the cell indicate that many of these structural proteins either perform enzymic functions (for example, the myosin of the muscles), or are essential components for the functioning of other enzymes (the role of protein—lipid membranes in the coordinated activity of complex

systems of intracellular enzyme reactions) or, finally, participate
in the active membrane transport of metabolites.

It also follows from our own observations, which will be described later, that completely inert protein structures do not in fact
exist in the brain tissue. All the proteins of nerve tissue, both
soluble and structural, undergo structural reorganization to some
extent during life.

Anabolism of Brain Proteins

During the last decade great progress has been made in the
study of one of the most important problems in modern biochemistry, the biological synthesis of protein. This field of biochemistry
has been reviewed in many surveys [17, 45, 228, 367, 556, 609,
610] and monographs [178, 282, 366, 611, 703]. These publications
give a sufficiently full acount both of protein biosynthesis as a
whole and of the various aspects developed most rapidly in recent
years. There is therefore no need to describe the history of the
study of protein synthesis or to expound modern views on the many
reactions of this highly complex and still largely unstudied process.

The biosynthesis of tissue proteins takes place throughout
development. In the developing organism this process provides
chiefly for differentiation and growth of the tissues through the formation of their protein spectra, while in the later stages of ontogeny it functions chiefly for the renewal of tissue proteins. Recent
evidence bearing on the morphogenetic changes in proteins in the
course of differentiation of the various tissues during embryo-
genesis and the early stages of postembryonic development has
been summarized [61, 366, 367, 405]. The history of research
into the biosynthesis of tissue proteins, coupled with their renewal
in the various stages of ontogeny, has been fully described in books
by Nagornyi et al. [399] and Parina [488].

Investigation of the Mechanisms of Protein
Biosynthesis in the Tissue of the Nervous System.
Although the basic mechanisms of protein synthesis are largely
the same in the various tissues of the animal body [178, 282, 366,
671, 703], the specific functions of the neurons and their unique
structural organization introduce a number of special features into
the renewal of their protein structures during life. These will be
described more fully below.

Research into the role of intracellular structures in the many
reactions of protein synthesis was tradiationally conducted chiefly

on liver, pancreas, and other organs [178, 282, 366, 671, 703].
The intensity of renewal of the proteins of the subcellular structures of brain tissue was investigated simultaneously in the laboratories of Palladin [29, 450, 460], Waelsch [1036], and Richter
[910] in experiments in vivo with the aid of labeled amino acids.
The results of these and of other investigations [1143, 1387, 1737]
showed that, as in other organs, labeled amino acids are incorporated most intensively in the brain tissue into proteins of the
microsomal fraction, containing membranous structures of the
endoplasmic reticulum and ribosomal granules, and also into the
soluble cytoplasmic proteins. It was later shown [739, 740, 1250]
that in brain tissues the proteins of the ribosomal granules incorporated labeled amino acids faster than did the membranous
proteins of the original microsomal fraction. It was concluded
from these results that protein biosynthesis in the cells of nerve
tissue is performed by the same biochemical systems found previously in other mammalian tissues and that there were no substantial differences between the mechanisms of protein synthesis
in different tissues. The few special features of protein metabolism in nerve cells can be explained by the unique structural organization of these cells and are concerned more with the mechanisms of transport of metabolites to the sites of synthesis and of
synthesized protein molecules to the sites of their function than
of the actual processes of biosynthesis and degradation.

More recent studies to determine the rates of incorporation
of labeled amino acids into brain proteins in vitro have confirmed
the important role of microsomal structures and soluble cytoplasmic fractions in the processes of intracellular protein biosynthesis.
These investigations were carried out initially on nerve tissue
homogenates [1250, 1588] and later on tissue slices [1012, 1045,
1367, 1464, 1654, 1655, 1658], subcellular organelles [748, 781,
882, 909, 1054, 1122, 1200, 1201, 1566, 1642, 1793], preparations
of myelin [1625], isolated ganglia [1379], and isolated nerve fibers
[971].

An important contribution to successfully investigating the
mechanisms of protein synthesis and the discovery of the intracellular localization of the protein-synthesizing systems in the
structures of nerve tissue was the development of various procedures of differential centrifugation for homogenates of brain [460,
463, 866, 1732, 1737, 1738, 1770] and peripheral nerve [1481].
With these methods purified preparations of nuclei [460, 463, 477,

532, 756, 1093], mitochondria [567, 866, 900, 1167, 1306, 1664, 1665], ribosomes [740, 933, 934, 1160, 1575, 1657, 1804], nerve endings [736, 1159, 1239, 1338, 1532], and synaptic vesicles [1533, 1534] may be obtained. In the development of these methods for fractionating nerve tissue homogenates it was found that the conditions of homogenization and the composition of the medium used [986, 1771] are very important factors. Manual homogenization in a blender of the Dounce type is recognized as the most sparing method for brain tissue [1771].

The ability of isolated structures of nerve tissue to incorporate labeled amino acids into their proteins was first demonstrated for brain homogenates 20 years ago by Greenberg et al. [1080]. Almost ten years later Lipmann [1298] first found enzymes in the cell juice of nerve tissue that activated amino acids. These observations were confirmed and expanded by other workers [1412, 1653, 1764]. It was found that brain homogenates can activate all of the amino acids from which their protein molecules are formed. The rate of biosynthesis of a protein in the tissues of the nervous system is evidently determined not only by the tissue concentration of enzymes activating amino acids, but also by the other components of the protein-synthesizing systems of the cell. This conclusion was based on the following facts: (1) the absence of correlation between the content of individual amino acids in nerve tissue proteins and the intensity of activation of these amino acids by the enzyme systems of the brain [1764]; (2) the lack of a direct relation between the intensity of incorporation of labeled amino acids into proteins in different parts of the nervous system and the concentration of the activating enzymes in these parts [1653]; (3) the similar direction and degree of age changes in the intensity of incorporation of leucine-^{14}C by microsomal brain preparations when pH 5 fractions from the brain of young animals, in which the enzyme systems are more active, and of adult animals were used [1412].

The content of activating enzymes in nerve tissue is probably not the limiting factor in protein biosynthesis. The activity of the systems transporting the activated amino acids and the quantity of ribosomes in the cells are evidently the limiting factors.

Brain slices, like those of other organs, have been found to incorporate labeled amino acids into proteins and the peptide glutathione [1012, 1014, 1015, 1017, 1367, 1654, 1655]. The rate of this incorporation, moreover, is higher in brain slices than in

liver slices, though data to the contrary have also been obtained
[1805]. The intensity of incorporation of amino acids into brain
proteins is reduced by 2,4-dinitrophenol, methionine sulfoxide,
chlorpromazine, certain amino acids, and also under anaerobic
conditions. It depends on the concentrations of potassium, calcium,
and magnesium ions in the medium. The incorporation of amino
acids into proteins of liver slices, on the other hand, is independent
of the presence of these ions.

Evidence was later obtained of the ability of cell-free systems
of brain tissue to synthesize protein actively. To achieve the maximal level of incorporation of labeled amino acids into brain protein,
the biochemical system was found to require the following
components: a microsomal fraction (or a preparation of ribosomal
granules), cell juice (or the pH 5 fraction), sources of energy
(ATP, GTP, or biochemical systems generating nucleotide triphosphates), and also Na^+, K^+, and Mg^{++} ions [740, 845, 1122, 1412,
1413, 1566, 1577, 1580, 1637, 1656, 1738, 1792, 1805]. The incorporation of amino acids into protein by this biochemical system
is inhibited by specific inhibitors of protein and nucleic acid biosynthesis such as ribonuclease, chloramphenicol, and puromycin
[1409, 1566, 1577, 1580, 1637, 1656, 1792]. Many investigations
[740, 845, 1413, 1656, 1805] have shown that cell-free brain systems containing ribosomal granules incorporate amino acids into
proteins more intensively than the analogous systems containing
the original microsomal fraction.

It is another very interesting fact that in analogous experiments with subcellular fractions of the liver, no difference was
found in the intensity of incorporation of labeled amino acids between the microsomal and ribosomal fractions [1656]. The more
active incorporation of labeled amino acids into proteins by the
ribosomal particles of the brain can be attributed to the fact that
the activity of the microsomal system is inhibited by lipid components, in which the microsomal fraction is very rich. This explanation is based on investigations showing the inhibitory effect of
free fatty acids on the incorporation of amino acids into proteins
by a cell-free system containing purified ribosomes [739].

The character of the effect of a complete set of amino acids
on the rate of protein biosynthesis by cell-free brain systems is
not yet clear. Some workers [1413, 1566, 1637] observed a stimulant effect of an amino acid mixture on the incorporation of labeled
amino acids into protein, others [1656, 1805] found an inhibitory

effect, while a third group [1580] observed no effect at all. It has recently been shown [1657] that the addition of glutamic acid alone to the system can provide for 90-95% of the total volume of protein synthesis taking place when the complete set of amino acids is added.

Experiments in vitro and in vivo showed [1120, 1440] that labeled amino acids are incorporated more intensively into polysomes than into ribosomes. The greater part of the label, moreover, is found in polysomes and ribosomes bound to membranes, and much less in free ribosomal granules. The latter findings, however, were not confirmed by other investigators [1599].

Investigations comparing the intensity of incorporation of labeled amino acids into proteins by cell-free systems of brain and liver [1412, 1577, 1656, 1805] showed that the ribosomal and microsomal brain fractions incorporated amino acids into protein at equal or greater rates than the corresponding preparations from liver. Only in one investigation [740] was slower incorporation of amino acids by microsomal preparations of the brain found compared to the corresponding preparations of the liver. These divergent results may be partly explained by differences in the composition of the microsomal preparations obtained due to inadequate preparative methods.

Characteristically the investigations described indicated only the overall ability of cell-free brain systems to synthesize tissue proteins. From such data it is impossible to judge simply on the rate of incorporation of labeled amino acids into total protein whether all the intermediate stages of protein biosynthesis in brain tissue were the same as those in other organs or not. Protein synthesis in cell-free systems of the brain had to be studied step by step. The first such investigations were carried out in Satake's laboratory [1578, 1656]. They showed that all the known stages of biological protein synthesis, discovered previously in analogous cell-free systems from liver and other organs, were present in nerve tissue.

The Role of Subcellular Structures of Nerve Tissue in Protein Biosynthesis. The experimental investigations cited showed that labeled amino acids are incorporated most rapidly into proteins of the microsomal (ribosomal) fraction and less rapidly into proteins of other cell organelles. The same pattern was discovered in experiments in vivo and in vitro. It was concluded that primary protein biosynthesis in the

cell takes place only on ribosomes: Hence the newly formed protein must be transported into other subcellular structures.

It was difficult to harmonize this view with the fact that a
considerable portion of the cell proteins with specific compositions,
for example, the basic and acidic proteins of nuclei, is selectively
concentrated in certain cell structures. Also, according to the
concept of the dynamic state of the tissue proteins, all protein
structures of the cell, including the soluble proteins of the cytoplasm and the organelles and structural proteins, must undergo
renewal.

Whereas the primary synthesis of the soluble proteins could
take place elsewhere, followed by their transport into the organelles, it was difficult to postulate a similar mechanism for proteins
forming insoluble structural components of the cell. One expected
that such protein structures would be renewed at the sites where
they actually functioned, i.e., in the organelles themselves. Consequently, independent mechanisms for the synthesis, and probably
the breakdown, of protein molecules should exist in the principal
cell organelles.

As a result of the further research into the biochemical functions of cell structures, evidence was in fact obtained from which
it could be concluded that organelles such as the nuclei and mitochondria in tissues of the liver and some other organs can synthesize protein independently. Extensive information on this problem
is presented in a series of surveys [178, 257, 282, 366, 565, 703].
The data on nerve tissue are inadequately represented in these
surveys and for that reason a brief account will be given here of
the facts confirming independent protein biosynthesis in cell organelles of the nervous system.

On the basis of observed functional changes in the size and
shape of the nuclei and on the results of many histochemical investigations, it has long been held that the nucleus participates
directly in synthetic processes in the cell, including protein biosynthesis [67, 257, 366], though until recently there was no direct
experimental proof of any independent protein biosynthesis in the
nucleus [60, 137, 187]. During the last decade several investigations [90, 145, 257, 282, 431, 671, 876] demonstrated that the cell
nucleus not only supplies the templates for cytoplasmic protein
synthesis as the "storehouse of most of the stock of inherited information" [366], but also possesses its own independent mechanisms of protein biosynthesis. These mechanisms are basically

identical with those found in the cell cytoplasm. They include the activation of amino acids, the formation of amino acyl-tRNA, and transport to the nuclear ribosomes, followed by biosynthesis of a polypeptide chain [430].

Practically all the components necessary for protein synthesis, including ribosomes [45, 205, 209, 232, 430, 579, 1026] and enzymes activating amino acids [139, 140, 430, 432, 672, 1138, 1754], have been found in the cell nuclei of the various tissues. This nuclear protein synthesis is dependent on sources of energy [139, 430-432], and is otherwise similar to the processes of protein synthesis in the cytoplasm. Protein biosynthesis in isolated nuclei does have some distinguishing features. The intensity of protein synthesis in the nuclei depends on sodium ions, but in the cytoplasm, on potassium ions. This difference, however, results not from the mechanism of protein synthesis itself, but from the character of transport of the protein precursors and other biochemical substrates through the cytoplasmic and nuclear membranes [282, 380, 430, 431, 1512].

In the case of mitochondria the historic development was different. The existence of independent mechanisms of protein synthesis in the cell nucleus was postulated on the basis of cytomorphological observations made long before labeled amino acids were available for the study of subcellular protein biosynthesis. Until recently there was no proof, direct or indirect, of protein biosynthesis in the mitochondria. It was generally accepted that the mitochondria were only suppliers of energy for intracellular syntheses taking place outside the mitochondria. Only in the last decade was conclusive experimental evidence obtained [510, 565] of autonomous protein biosynthesis in the mitochondria.

The possibility of protein biosynthesis within the mitochondria was first demonstrated by autoradiographic, histochemical, electron-microscopic, and biological investigations showing that the mitochondria of different tissues contain RNA [206-208, 589, 696, 697, 1231, 1557, 1558, 1790] and DNA [207, 246, 589, 849, 880, 1179, 1231, 1622]. But the discovery of nucleic acids in the mitochondria suggested, but did not prove, the existence of independent biochemical systems of protein synthesis in these structures. Direct experimental proof was first obtained in experiments in vitro using mitochondrial preparations from various tissues [501, 1101, 1230, 1555-1558, 1641], including tissues of the nervous system [781, 1073, 1200, 1448], and also fractions of submitochondrial structures isolated from fragmented mitochondria [565, 803, 1229, 1231, 1555, 1557, 1558]. Mitochondrial preparations from the brain incorporated amino acids as quickly as [781], or more quickly than [1073], liver mitochondria.

A necessary condition for the incorporation of labeled amino acids into mitochondrial proteins of any tissue is the maintenance of a minimum level of oxidative phosphorylation [781, 1556, 1558]. Incorporation of amino acids into proteins of intact mitochondria takes place quite rapidly without the addition of sources of energy [781, 1073, 1393], amino acids [781], ribosomes [1557], or pH 5 enzymes [1231, 1557, 1558], for all these components are present in the mitochondria themselves. The incorporation of labeled amino acids into mitochondrial proteins discovered in experiments in vitro, as further investigations showed, is due neither to bacterial contamination of the preparations used [804, 1229, 1558], nor to the presence of cytoplasmic ribosomes and pH 5 enzymes as impurities [781, 1025]. The mitochondria contain enzymes activating amino acids [1101, 1229, 1557, 1688] and ribosomes [206, 208, 767, 1101, 1229, 1688]. The latter are concentrated chiefly on the inner surface of the mitochondrial membranes, in good agreement with the observation that the incorporation of leucine-^{14}C into protein is most intensive in the inner membranes of the mitochondria [803]. Another fascinating discovery was that the mitochondria of tissues with a high mitotic index are richer in ribosomes than the mitochondria from tissues with a low mitotic index [767]. In summary the mitochondria of various cells, including nerve cells, contain all the components necessary for protein biosynthesis: ribosomal, transfer, and messenger RNA, DNA, enzymes activating amino acids (aminoacyl tRNA-synthetases), DNA- and RNA-polymerases, sources of energy, free amino acids, etc. Generally the mechanism of protein biosynthesis in the mitochondria in vivo is identical with that in the cell nucleus and cytoplasm and includes all the known stages from activation of the amino acids to the formation of the polypeptide chain.

Protein synthesis in isolated mitochondria possesses certain special features distinguishing it from other cell-free tissue systems. For instance, neither ribonuclease nor deoxyribonuclease inhibits the incorporation of amino acids into the proteins of intact mitochondria [781, 1025, 1179, 1231, 1393, 1557, 1558], although they inhibit this process in isolated submitochondrial fragments [565, 1179]. Mitochondrial membranes are evidently not readily

permeable to substances such as ribonuclease or actinomycin
[820, 1434]. Thus it can be postulated that the specific features
of protein biosynthesis observed in the mitochondria in vitro, as
well as the effects of various ions [1229, 1558, 1655] and inhibitors
of protein synthesis and oxidative phosphorylation (puromycin,
penicillin, actinomycin D, chloramphenicol, 2,4-dinitrophenol, etc.)
[501, 781, 1073, 1229, 1231, 1557], are due basically not to differences in the mechanisms of protein synthesis, but to differences
in the supply of energy and access of metabolites, metallic ions,
or inhibitors to the sites of synthesis.

Protein biosynthesis in the brain mitochondria evidently differs
somewhat from that in the mitochondria of other organs [1073],
for inhibitors such as acetoxycycloheximide depress protein
synthesis considerably in the mitochondria and synaptosomes of
the brain but do not affect the intensity of this process in liver
mitochondria.

It would appear that not all mitochondrial proteins are synthesized in the mitochondria, just as not every protein synthesized
in the mitochondria is utilized at the site of synthesis. Some proteins functioning in the mitochondria, water-soluble enzymes, for
instance, are evidently synthesized on the cytoplasmic ribosomes
and then transported into the mitochondria [707, 802, 1101, 1177,
1178]. Transport in the reverse direction probably can also take
place, i.e., the passage of protein synthesized in the mitochondria
out through the membrane into the cell cytoplasm [1558]. And, of
course, there are proteins, evidently of the membranous structures
of the mitochondria, which are insoluble in water, that are synthesized in and function in the mitochondria [117, 1555-1558].

The mitochondria of the brain may participate in protein biosynthesis for the formation of proteolipids, specific protein—lipid
complexes of nerve tissue utilized in the formation of myelin and
other membranes [1201, 1682]. Another hypothesis is that mitochondrial systems synthesize only the proteins of the inner membranes (structural proteins and enzyme proteins firmly bound with
the membranes), while the proteins of other membranes, including
the outer membrane of the mitochondria itself, are synthesized by
the cytoplasmic system [510, 707].

Two possible enzymatic mechanisms for protein biosynthesis
have been postulated in the mitochondria: one dependent on oxidative phosphorylation, going to completion during incubation for
1.0-1.5 h, and inhibited by chloramphenicol, puromycin, and peni-

cillin; and another independent of oxidative phosphorylation, functioning during incubation for several hours, and not sensitive to the above-mentioned inhibitors [1641]. It was suggested that the second pathway of protein biosynthesis in the mitochondria might act by a transpeptidase process, catalyzed by intramitochondrial cathepsins. Others have postulated [1682] that the mechanisms of biosynthesis of proteolipid proteins for the nerve cell membranes may differ from those of the ribosomes. Haldar [1100] claims there are two protein-synthesizing systems in the mitochondria of the brain which could be localized either in the same or in different organelles. One synthesizes insoluble proteins and is inhibited by chloramphenicol, whereas the second synthesizes both soluble and insoluble proteins and is inhibited by cycloheximide. These conclusions were drawn from studies of purified mitochondria.

We conclude from these in vitro investigations that the nuclei and mitochondria of brain tissue possess independent systems for protein biosynthesis which are not significantly different from the systems synthesizing proteins on cytoplasmic ribosomes and are similar to preparations from other organs. The differences between the incorporation of labeled amino acids into proteins in cell-free systems of the brain and other organs are quantitative rather than qualitative and are probably attributable not to the protein-synthesizing systems themselves but to differences between the barrier and transport mechanisms.

Attempts to detect ribosomal granules with the electron microscope [958, 1403], and primary protein synthesis by an autoradiographic method [170], in the axons proved unsuccessful. Lately evidence was obtained indicating local systems of protein biosynthesis were present in axons and nerve endings (synaptosomes). These investigation have been devoted chiefly to the study of acetylcholinesterase, an enzyme protein specific for nerve structures. The ability of axons and synaptic structures to synthesize some proteins is confirmed by the following facts: (1) a distoproximal descending gradient of recovery of acetylcholinesterase activity in nerve trunks when irreversibly inactivated by special inhibitors [911]; (2) the higher acetylcholinesterase activity in the distal (fivefold) segments of the divided hypoglossal nerve than in the intact control nerve [1208]; (3) the RNA content in axons [969, 1208] and synaptosomes [775, 1403]; (4) the intensive incorporation of labeled amino acids into proteins by preparations of isolated

synaptosomes [775, 778, 1403, 1404, 1266], and (5) the fact that
the activity of irreversibly inhibited acetylcholinesterase in the
peripheral nerves is restored much sooner than in the CNS [1210].
It would be possible [778, 1073] that the biosynthesis of protein
in preparations of synaptosomes is carried out by the mitochondria
contained in these morphologically complex preparations of nerve
tissue. However, it has recently been shown by experiments in
vitro by the method of electron-microscopic autoradiography [959]
that intensive nonmitochondrial protein biosynthesis from free
amino acids takes place in the nerve endings of rat brain tissue.

The intensity of renewal of the brain tissue proteins in vitro
is approximately equal to that of the liver proteins, whereas in
vivo liver proteins are metabolized more rapidly than brain proteins. These differences might be the result of the influence of
the BBB, which in the intact organism restricts the supply of metabolites to the brain tissue cells. Another possibility is that in
vivo the potential capacity of the brain cells to carry out protein
biosynthesis is used only partially, i.e., the cell possesses a definite reserve of synthetic activity which is not utilized under
ordinary conditions.

Sources of Free Amino Acids. The biosynthesis of
intracellular proteins can take place only if free amino acids are
present. The extent to which the potential capacity of the biosynthetic systems of the cell can be utilized largely depends on the
concentration of protein precursors at the sites of protein synthesis. Accordingly, before the results of investigations of the metabolic activity of the tissue proteins can be evaluated it is necessary
to obtain information about the levels of free amino acids in the
different parts of the tissue, the individual cells, and their compartments. The importance of this index of the intracellular metabolic situation for the correct interpretation of results obtained
during determinations of incorporation rates into brain proteins
is discussed in other sections. Here the main sources from which
the intracellular amino acids are derived will merely be enumerated.

Under normal conditions the basic requirements of the brain
are satisfied by amino acids split off by the proteolytic enzymes
of the gastrointestinal tract from exogenous (food) proteins. These
pass from the blood vessels through the BBB to the sites of protein synthesis in the nerve cells. In special metabolic situations
important additional sources of free amino acids for protein bio-

synthesis in such vitally important organs as the brain, heart, and
liver are the proteins of the skeletal muscles [522, 1610].

An important source of free amino acids in brain tissue is
the endogenous proteins. In adult animals during prolonged protein
starvation, i.e., when no dietary free amino acids are supplied to
the brain, the protein content and the intensity of renewal in nerve
tissue are maintained at virtually the same level as in animals
feeding normally [461, 470, 1375]. Amino acids liberated during
the hydrolysis of endogenous proteins are utilized more efficiently
for the biosynthesis of new protein molecules than exogenous amino
acids [1432, 1482]. The intensive utilization of endogenous tissue
proteins for maintaining the levels of free amino acids and the
reutilization of these free amino acids in the biosynthesis of new
protein molecules is confirmed by the continuing formation of milk
proteins in the isolated mammary gland, i.e., when the supply of
exogenous amino acids is discontinued [242].

No less important as a source of free amino acids in the brain
tissue is their metabolic formation de novo [106, 575, 826, 1044,
1045, 1098, 1543], primarily from glucose, and by interconversion
of different amino acids.

Convincing evidence of the ability of nerve tissue to form non-
essential amino acids from glucose is given by the results of
numerous investigations conducted in vivo [817, 819, 1281, 1459,
1460, 1725, 1728, 1796], on brain homogenates [1508, 1621, 1781],
tissue slices [889, 921, 1092, 1158, 1506], and also on brain per-
fused in situ [789, 1051, 1343]. Compared with other organs, the
brain was found to have the highest rate of formation of amino
acids from glucose [1040, 1041, 1522, 1541].

In newborn and adult mice, labeled carbon is incorporated in
vivo into the glutamic acid of the brain from the glucose of the
blood 10 times more rapidly than into the blood glutamic acid
[1009]. As early as 22 min after the subcutaneous injection of uniformly labeled glucose more than 70% of the total ^{14}C content of
its acid-soluble fraction was found in the free amino acids in various parts of the cerebral cortex of the cat, whereas the proportions found in the blood plasma, liver, kidneys, muscles, spleen,
and lungs were only 3, 8, 15, 12, 16, and 13%, respectively [1041].
The specific radioactivity values of the free amino acids in the
cortex and other parts of the brain were 3-5 times higher than in
the blood and kidneys, 4-7 times higher than in the liver, 7-10
times higher than in spleen and lungs, and 15-20 times higher than

in the muscles [1041]. Of the total radioactivity in the free amino acids of the brain 70-80% belonged to dicarboxylic, 20-30% to neutral, and not more than 1-2% to the basic amino acids [1041, [1281]. It was also shown [889, 1009, 1039, 1040, 1460, 1506, 1541] that ^{14}C from labeled glucose is incorporated most intensively into amino acids formed from components of the tricarboxylic acid cycle, i.e., glutamic and aspartic acids, glutamine, alanine, and GABA.

The formation of amino acids from glucose takes place at different rates in different parts of the brain [889, 1041]. The rate of their formation declines in the following order: cerebral cortex > cerebellum > pons > medulla > spinal cord.

The free amino acids formed in the brain tissue from glucose are utilized intensively for the biosynthesis of the tissue proteins [1508, 1621, 1726, 1728, 1781, 1797, 1800]. Much of the label passes comparatively quickly from the glucose and into the other components of the brain tissue such as lipids and nucleic acids [1726, 1728, 1800].

Glucose in thus not only the chief energy-yielding substrate in nerve tissue, but also an important source for the formation of free amino acids. Characteristically brain tissue has a greater ability to form amino acids from glucose than the tissues of other organs. Note that while the BBB significantly restricts the access of certain amino acids to the brain, it transports glucose without hindrance. Hence the biochemical mechanisms for the active conversion of glucose in the brain into amino acids may be particularly important for maintaining protein and other types of metabolism at a definite level, especially if amino acids are deficient in the diet. It may be relevant in this connection that the rate of formation of amino acids from glucose is considerably lower in the brain of newborn animals, whose BBB is more permeable to amino acids, than that of adult animals [1541].

Protein Metabolism in Structurally and Functionally Different Parts of the Nervous System

The tissue of each part of the nervous system and of each of its individual morphofunctional formations is characterized by a particular structural organization and a particular chemical composition. Since all the metabolic processes lying at the basis of these functions take place with the participation of proteins, their metabolism must differ in different parts of the CNS.

Biochemical Investigations. In the preisotopic period of research into the protein metabolism of nerve tissue it was found [443, 447, 452] that morphologically and functionally different parts of the brain differ from one another in their protein content, with the higher concentrations occurring in those parts and structures that perform the more complex and more active physiological functions. Furthermore the protein composition of the main parts of the nervous system undergo substantial changes during ontogeny. From the results obtained by these methods it could be postulated that the intensity of protein metabolism varies in different parts of the CNS and the PNS. This hypothesis was supported by the determination of the activity of the enzymes concerned with nitrogen and other types of metabolism and the intensity of proteolysis in different structures of the nervous system under normal and some pathological conditions. However, experimental data demonstrating differences in the intensity of renewal of the tissue proteins in different parts of the nervous system were obtained only after the introduction of the labeled materials [451, 447, 448, 1468, 1732, 1738]. With labeled amino acids and other compounds substantial differences in the rates of renewal of proteins were found not only in different parts of the nervous system, but also in the cellular layers of the cortex, with their different morphological composition, in different types of cells and their organelles, in isolated protein fractions, and even in the individual proteins of nerve tissue.

Research with labeled amino acids in the laboratories of Palladin [99, 461, 473, 587, 1467, 1468], Richter [914], Waelsch [1036, 1255, 1736], and others [360, 481, 483] showed for the first time that the proteins of structurally and functionally different parts of the nervous system are renewed at different rates. The various parts of the cat's nervous system were shown to be arranged in the following order of diminishing intensity of renewal of their tissue proteins (Table 2): cerebellum > gray matter of the cerebrum > thalamus > mesencephalon and diencephalon > medulla > white matter of the cerebrum > spinal cord > sciatic nerve. Characteristically the ability of the tissues of these parts of the nervous system to form amino acids from glucose decreases in roughly the same order [889, 1041].

TABLE 2. Intensity of Incorporation of Labeled Amino
Acids into Total Proteins of Different Parts
of the Nervous System of the Cat and Monkey

Parts of the nervous system	Cats				Monkeys
	[99, 461]	[360]	[483]	[587]	[1036]
Cerebral hemispheres					
gray matter	0.23	21.2	431	34.5	43—61*
white matter	0.10	—	146	22.6	39.0
Cerebellum	0.28	30.1	530	—	41.5
Mesencephalon	0.17	14.8	—	—	—
Diencephalon	—	16.3	—	—	—
Medulla	0.14	11.3	307	—	32.4
Thalamus	0.19	—	—	—	27.5
Spinal čord	0.08	7.9	170	—	15.6
Sciatic nerve	—	—	—	5.6	—
Hypothalamus	—	—	353	—	—

* The specific activity of proteins from functionally different parts of
of the cerebral cortex varies within these limits.

As the data in Table 2 show, proteins of the cat sciatic nerve
incorporate methionine-^{35}S four and six times less intensively than
proteins of the white and gray matter of the cerebral hemispheres,
respectively. The intensity of renewal of spinal cord proteins is
also very low. Similar differences in the intensity of renewal of
proteins in different parts of the CNS were found in other investigations [113, 174, 287, 288, 826, 1223, 1301, 1733, 1800] undertaken
on animals of other species and with the use of other labeled amino
acids, as well as of labeled sodium sulfate and glucose-^{14}C.
Waelsch and co-workers [1036] found that proteins of morphologically and functionally different parts of the cerebral cortex of the
monkey differ significantly in the intensity of their renewal. The
highest rate of renewal is shown by proteins of the sensory cortex
(0.66 pulse/μg lysine/min), and the lowest by proteins of the temporal cortex (0.43 pulse/μg lysine/min).

It could thus be concluded from the results given in Table 2
and those of other investigations that the intensity of protein renewal as a rule is higher in phylogenetically younger, structurally
more complex, and functionally more active parts of the nervous
system.

There are, however, some exceptions, difficult to explain at
the present time, that call for experimental verification. We refer

to the very high intensity of incorporation of lysine-^{14}C into proteins of the corpus callosum. Although this structure is composed of the metabolically more stable and functionally less active white matter of the brain, its proteins are renewed more intensively; i.e., they have a shorter half-life than the proteins of the cerebral cortex and cerebellum [1036, 1255]. No satisfactory explanation of this fact has yet been found. According to the authors cited, the apparently short half-life of proteins in the corpus callosum may in fact be an artifact due to compartmentation of the lysine precursor. In this case a small pool of relatively high specific activity would be the precursor for protein synthesis. Thus the measured specific activity would be the average for the two pools and would be lower than the actual precursor specific activity, resulting in an erroneously high calculated rate of synthesis. On the other hand, there may be a high intensity of protein renewal in the corpus callosum based on mechanisms connected with the specific structure and functions of this part of the brain. The corpus callosum consists of a well-developed bundle of axons which forms multiple connections between the neurons of the right and left hemispheres and provides for functional interaction between them.

It is noteworthy that the content of nucleic acids and the intensity of their renewal as a rule are also higher in parts of the brain that are metabolically and functionally more active [321, 350, 592, 1341, 1372]. These findings are in full accord with the renewal rates of proteins in different parts of the nervous system and with modern views of the role of nucleic acids in protein biosynthesis.

Lipoproteins, the major component of the white matter of the brain, are also characterized by different rates of renewal (estimated as phosphorus) in different parts of the nervous system [109]. Surprisingly the lipoproteins of the white matter of the cerebral hemispheres have the highest rate of renewal, whereas lipoproteins of the spinal cord have the lowest, though these parts of the brain are similar in anatomical structure and are equally rich in structures containing myelin. An intermediate position as regards the rate of renewal of lipoprotein phosphorus is occupied by the cerebral gray matter, the cerebellum, the mesencephalon, and the diencephalon. The rate of incorporation of labeled amino acids into lipoproteins of the brain is very low, an order of magnitude below those for other protein structures of the brain [1736]. Another somewhat unexpected finding is that the renewal of phosphoproteins (as phosphorus) is a little lower in the tissues of the

cerebral cortex and cerebellum of the rabbit than in the brain
stem — the diencephalon, mesencephalon, and medulla [297].

Individual proteins perform different functions in the nervous
system: structural, metabolic, transport, etc. It is therefore natural to suppose that the half-lives of these will vary. This hypothesis was confirmed in a series of investigations [287, 288, 473,
587, 1223, 1224], in which labeling was used to study the intensity
of renewal of protein fractions consisting of individual proteins
with similar physicochemical properties.

Work in Palladin's laboratory [288, 473, 587, 1468] showed
that four protein fractions isolated from tissue homogenates of
the cerebral gray and white matter and the spinal cord of the cat
by the method of Mirsky and Pollister [1390] differ significantly
in their rate of incorporation of methionine-^{35}S (Fig. 4). A fraction
obtained by extraction with 0.14 M NaCl, and consisting chiefly
of ribonucleoproteins, had the highest intensity of protein renewal
for all parts of the nervous system studied. Proteins of a fraction
subsequently extracted with 1 M NaCl solution, containing most
of the deoxyribonucleoprotein, had a rather lower intensity of
renewal. Next in order of decreasing intensity of protein renewal

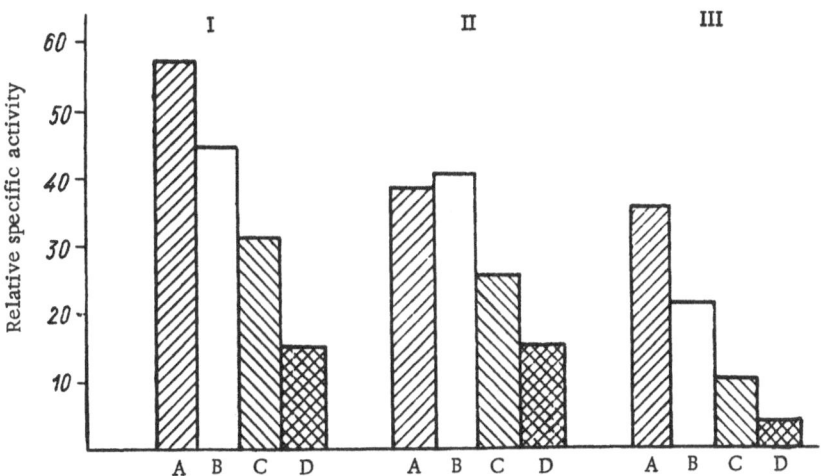

Fig. 4. Intensity of incorporation of methionine-^{35}S into protein fractions of the
gray and white matter of the cerebral hemispheres and sciatic nerve of the cat:
I) cerebral gray matter, II) cerebral white matter, III) sciatic nerve; A) protein
fraction extracted with 0.14 M NaCl, B) with 1 M NaCl, C) with 1 M NaOH; D)
protein insoluble in 1 M NaOH.

was the fraction soluble in 1 M NaOH solution, and lastly the alkali-insoluble residue.

The original conclusions that were drawn from these investigations were that the metabolically most active proteins in the gray and white matter of the cerebral hemispheres (the ribonucleoprotein fraction) are renewed 3.5 and 2.5 times more rapidly, respectively, than the relatively inert proteins of the alkali-insoluble residue of these tissues. In peripheral nerve tissue, the intensity of renewal of the rapidly metabolized proteins is almost eight times higher than that of the alkali-insoluble residue.

The metabolically most active ribonucleoprotein fraction of peripheral nerve is renewed at a rate slightly less (by 40-30%) than the proteins of the same fraction of the cerebral gray and white matter. More substantial regional differences are found for the proteins of metabolically more stable tissue structures. For instance, the alkali-insoluble proteins of the sciatic nerve are metabolized 3.3 times less intensively than the proteins of the corresponding fraction of the cerebral gray and white matter.

It can be concluded from these observations that the rapidly metabolized proteins of the tissue structures of the CNS and PNS evidently perform similar or identical functions, mainly metabolic, whereas the slowly renewed protein structures of peripheral nerve tissue differ functionally from the analogous proteins of the cerebral gray and white matter.

Clouet and Richter [910] studied the metabolic activity of the same four protein fractions mentioned above but isolated from subcellular structures of the rat brain. They found that the metabolically most stable of all the subcellular fractions are proteins of the alkali-insoluble residue. The highest metabolic activity in the nuclear and mitochondrial fractions is possessed by proteins extractable with 1 M NaCl solution, but in the microsomal fraction the alkali-soluble proteins are the most highly labeled. This protein fraction consists of liponucleoprotein components of Nissl's granules.

The high rate of renewal of brain proteins which are soluble in water and salt solutions and the very low rate of incorporation into proteins insoluble in solutions of neutral salts or in alkali were confirmed by investigations [1224, 1340, 1800] in which methionine-^{35}S, other labeled amino acids, glucose-^{14}C, and sodium sulfate-^{35}S were used.

Finally, the activity of enzyme systems concerned with cellular nitrogen and protein metabolism also varies in different parts

of the nervous system and corresponds to the activity of the biochemical systems responsible for the appropriate stages of intracellular metabolism. This applies in particular to intracellular
peptide hydrolases [514, 532, 745, 769, 1248, 1250, 1251, 1495],
enzymes activating amino acids [1653], enzymes of intermediate
amino acid metabolism [471, 726, 766, 1031, 1536, 1567], deaminat-
ing enzymes [402, 537], ATPase (3.6.1.3) [259, 292, 721], inorganic
pyrophosphatase (3.6.1.1) [645, 647], alkaline (3.1.3.1) [1031] and
acid (3.1.3.2) [485, 1031] phosphatases, enzymes of nucleic acid
metabolism [409, 553, 593, 846, 1466], and some oxidative enzymes
[292, 504, 506, 507, 1683].

A u t o r a d i o g r a p h i c   I n v e s t i g a t i o n s . The unequal
intensity of protein renewal in structurally and functionally different parts of the nervous system has also been confirmed by
using autoradiographic methods [170, 346, 543, 914, 1001, 1002,
1301, 1375, 1524] which, unlike existing biochemical methods, can
determine the localization of compounds and changes in their concentration in the tissue microstructures even at the level of the
individual cell, its organelles, and their microstructures. Despite
some shortcomings [170, 562, 962], this method has established
a correlation between the functional activity of the microstructures
of nerve tissue and the intensity of metabolism, especially protein
metabolism, in them.

It was shown for the first time by autoradiography in Richter's
laboratory [914, 1524] that methionine-^{35}S is incorporated most
intensively into proteins of those parts of the rat brain in which
the bodies of nerve cells are concentrated (the supraoptic and
paraventricular nuclei of the hypothalamus, the pyramidal layer
of the cornu Ammonis, and the granular layer of the cerebellum).
A lower level of labeling was found in the white matter of the brain,
which is rich in fibers. Other autoradiographic investigations
[346, 543, 1001, 1002, 1301, 1375] also showed that the gray matter
of various parts of the brain takes up methionine-^{35}S much more
intensively than the white matter. In the dog's brain the highest
level of labeled methionine was found in the granular layer of the
cerebellum [1001, 1301].

The results of autoradiographic investigations showing a high
rate of incorporation of labeled amino acid into the tissue structures of the cerebral cortex and cerebellum and a low level of
labeling in structures of the white matter are in good agreement
with the following facts.

1. The degree of vascularization of the cerebral cortex is much higher than that of the white matter [265, 1031]. According to Campbell [1031], the length of the capillary network in 1 mm^3 of tissue in cell layer IV of the parietal cortex is 882 mm, whereas its length in 1 mm^3 of tissue in the corresponding area of white matter is only 374 mm.

2. Of all parts of the brain, the tissues of the cerebellum have the highest packing density of the cells and the highest concentrations of DNA and RNA. According to May and Grenell [1372], the cell density (the number of cells in 1 g wet tissue) in the rat cerebellum is 5.88×10^8, whereas in the cerebral cortex and white matter, the thalamus, hypothalamus, and medulla the figure varies between 1.08×10^8 and 1.58×10^8.

3. The content of proteolipids, whose intensity of renewal is the lowest [1038] in the cell layers, gradually rises from the surface of the cerebral cortex to the white matter [797]; calculated per cell and per unit of residual tissue protein, their content rises sixfold from the surface layer of cortical cells to the white matter, and calculated per dry weight of tissue it rises from 2% in cortical cell layers I, II, and III to 5.5% in the white matter (Fig. 5). Considerable differences in the relative content of proteolipid and nonproteolipid proteins are observed between the gray and white matter of the brain. In certain parts of the human cerebral cortex the content of nonproteolipid protein is 15-20 times higher than that of proteolipid protein, whereas in the corpus callosum it is only four times higher.

4. Changes are observed in the successive layers of the cerebral hemispheres and cerebellum [20, 1031, 1129, 1548, 1549], at increasing depths from the surface of the cortex in the content

Fig. 5. Distribution of proteolipid proteins in cell layers of
the rat brain [797].

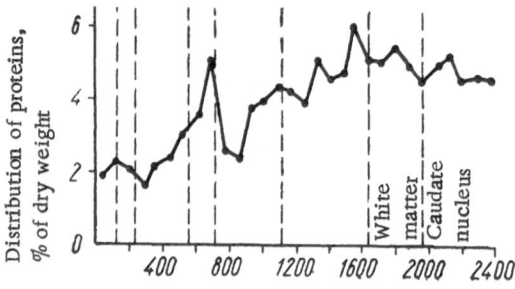

Distance from surface of cortex, μm

of proteins, lipids, nucleic acids, and some low-molecular-weight substances. The activity of certain glycolytic, oxidative, hydrolytic, and other enzymes, directly or indirectly connected with tissue protein metabolism, also changes. The pattern observed, moreover, is in basic agreement with the variations in the types of cells and in the levels of their functional activity in the different layers. These observations are fully described and discussed in surveys by Baranov and Pevzner [20] and in Friede's monograph [1031].

The autoradiographic investigations of Manina [346] showed that structurally and functionally different parts of the rat brain can be arranged in the following order of descending intensity of methionine-^{35}S incorporation: cerebellum, medulla, cornu Ammonis, cerebral cortex, and white matter. Other workers [1375] found that the parts of the adult rat brain can be arranged in the following order of decreasing intensity of labeling: gray matter of the cerebellar vermis, cerebellar cortex, pyramidal layer of the hippocampus, vestibular nucleus, olfactory area, supraoptic nucleus, olive, dentate nucleus, cerebral cortex, caudate nucleus, white matter of the cerebral hemispheres, and cerebellum. Portugalov and co-workers [543] found the intensity of incorporation of methionine-^{35}S highest in the Purkinje cells of the cat's cerebellum, somewhat lower in the motoneurons of the anterior horns of the spinal cord, and lowest in neurons of the cerebral cortex. Gracheva [170] also found that neurons of various types synthesize proteins at different rates. The highest rate was found in nerve cells of peripheral ganglia, spinal motoneurons, and Purkinje cells of the cerebellum, a much lower rate in neurons of the posterior horns of the spinal cord and the granular layer of the cerebellum, and a very low incorporation of labeled glycine in nerve fibers and glial cells of the CNS and PNS.

Different levels of labeling have also been found by the autoradiographic method in different types of cells in the gray matter of the spinal cord. The level of radioactivity is higher in motoneurons than in other nerve cells [1001, 1301]. Even among cells of the same type in the same tissue formation, it is possible to find [170, 1001] cells with either high or low concentration of labeled amino acid. This is presumably evidence of differences in their functional activity resulting from the periodic alternation of states of excitation and inhibition.

Variations in the intensity of incorporation of methionine-^{35}S have also been found [543] in different structural parts of the neuron. The structural elements of motoneurons in the anterior horns of the spinal cord can be arranged in the following order of decreasing intensity of incorporation of label: cytoplasm of the cell body, dendrites, axon hillock, nucleus, axon, and nucleolus. Glycine-^{14}C is, however, incorporated more intensively into the nucleus than into the cytoplasm of nerve cells of the peripheral ganglia [170].

Neurons with predominantly nuclear and others with predominantly cytoplasmic types of protein synthesis have been discovered by the autoradiographic method. The presence of nerve cells with protein biosynthesis localized in either the nucleus or cytoplasm has also been confirmed by the cytospectrophotometric method [64, 65]. When the results of autoradiographic investigations with labeled amino acids are assessed, it must be remembered that different amino acids may be incorporated preferentially into different cell organelles. Experiments on frog oocytes showed [343] that some amino acids (phenylalanine, glycine, aspartic acid) are incorporated more intensively into the nucleolus and others (glutamic acid) into the nucleoplasm.

This survey of the literature shows that the autoradiographic data on levels of protein metabolism in functionally different parts of the CNS and PNS are basically in agreement with the results of biochemical investigations although there are some points of disagreement. For instance, autoradiographic studies showed [346, 1001, 1301] that some phylogenetically older parts of the CNS incorporate labeled amino acids more intensively than the younger structures. This conclusion is opposite to that drawn from the biochemical observations described above.

These conflicting observations can be explained to some extent if the following technical problems are evaluated when interpreting the results of autoradiographic investigations [346, 1001, 1301] of the intensity of incorporation of labeled amino acids into the tissues of different parts of the brain. In most of the investigations cited* the total label was determined in the tissues, that

---

* Exceptions to this rule are the investigations of Gracheva [170], who combined quantitative autoradiography with the consecutive extraction of lipids, nucleic acids, and proteins from tissue slices and in that way was able to determine the metabolic activity of each of these substances.

is, that bound with protein and other morphological structures.
Some of these phylogenetically older parts of the nervous system,
characterized by a lower rate of protein metabolism, have higher
concentrations of some free amino acids [1036, 1504, 1598, 1732].
More importantly, different structures of nerve tissue differ significantly in the permeability of their membranes to amino acids
(see also Chapter 3). Factors such as these may account for the
higher content of labeled amino acids found autoradiographically
in the tissue of the phylogenetically older parts of the brain. In
these cases, the amount of label would not be an indicator of the
rate of intravital protein renewal. In fact the intensity of incorporation of labeled amino acids into proteins, i.e., the rate of
renewal of the protein structures in these parts of the brain, is
evidently lower than in the phylogenetically younger formations of
the nervous system.

Comparison of the Metabolic Activity and Composition of Protein Fractions in Different Parts
of the Nervous System. Proteins and protein complexes
with different physicochemical properties have been found electrophoretically in the tissues of different parts of the nervous
system. For instance, substantial differences in the relative percentages of protein in the individual electrophoretic fractions
have been found between the tissues of the gray and white matter
of the cerebral hemispheres and the cerebellum [302, 466, 467,
527, 1199, 1547]. Other workers [110, 1546], however, found only
extremely small quantitative differences or no difference whatsoever in the composition of the proteins of these tissues. Differences in the protein content in certain electrophoretic fractions
also exist between other parts of the brain (cerebellum, medulla,
pons, hypothalamus, corpus callosum, etc.) [302, 467, 521, 1183,
1199]. Slight differences between the protein content of the electrophoretic fractions are even found in the various parts of the
cerebral hemispheres (in the frontal, temporal, parietal, and
occipital lobes) [847]. On this basis the hypothesis of chemical
individuality (referring to protein composition) of functionally different parts of the brain has been put forward.

Characteristically parts of the nervous system that differ
significantly in structure and function also differ considerably not
only quantitatively (an unequal percentage distribution of protein
between the electrophoretic fractions), but also qualitatively, for
protein fractions may be present in one part but absent in the

others. These differences have been found by comparing the electrophoretic spectra of soluble proteins of the cerebral hemispheres and spinal cord [302], the brain and the peripheral nerves [527, 528], the spinal cord and peripheral nerves [302, 527], the spinal cord and its roots [469, 526, 527, 555], the peripheral nerves and roots of the spinal cord [526], myelinated and nonmyelinated nerves [527, 528], different parts of the auditory areas of the brain [940], and other structures of nerve tissue [9, 528, 555, 940].

By means of an ultramicromethod of radial electrophoresis in polyacrylamide gel Aleksidze [9] recently showed that even functionally different nuclei of the brain stem may differ in the composition of their protein fractions. He suggested that these differences may be due to differences in the relative number of neurons and glial cells in the nuclei studied. This conclusion is based on the fact that significant differences in the number of postalbumin fractions were found in the soluble proteins of isolated neurons and glial cells of the lateral vestibular nucleus.

It can thus be concluded from the data described above that variations in the metabolic activity of the total proteins of different parts of the nervous system may be due not only to variations in the quantitative proportions of the electrophoretic protein fractions in the tissues of each part studied, but also to the presence of proteins specific for a particular part, or for a particular structure, in them. The second conclusion that can be drawn from a comparison of the biochemical and autoradiographic data on the metabolic activity of proteins with the results of electrophoretic investigations of the composition of their fractions in different parts of the nervous system is that the more these parts differ in their structural organization, the functions they perform, and the biochemical processes on which these functions are based, the more clearly the differences in the intensity of protein renewal and in the composition of the protein fractions in the various structures of the nerve tissue are manifested.

The results of the biochemical and autoradiographic investigations into the metabolic activity of proteins in different parts of the nervous system are basically in agreement with the observed content of sulfhydryl and other functionally active protein groups [52, 541, 543, 572] and free radicals [7] in these structures. As a rule, the intensity of metabolism of the proteins is higher in parts and microstructures of nerve tissue that are richer in functionally active groups and free radicals.

When the metabolic activity of the total proteins in different parts of the brain and their microstructural areas is compared, allowance must always be made for their cellular composition, since morphologically and functionally different cells differ significantly in the rates of their intravital protein renewal. Experiments in vivo [839, 874] and in vitro [1680] have shown that proteins of fractions rich in neurons incorporate labeled amino acids more intensively than proteins of fractions rich in glial cells. The RNA of a fraction rich in neurons also incorporates labeled precursors in vivo faster than the RNA of a fraction rich in glial cells [1005].

Reserves of Free Amino Acids in Different Parts of the Nervous System. The intensity of protein metabolism in structurally and functionally connected areas, in cell layers, in groups of similar cells, and even in individual cells and their microstructural formations is determined primarily by the degree of development of the biochemical systems of protein synthesis and breakdown and by the level of functional activity of the structure concerned. However, the intensity of labeling of intracellular proteins also depends on the content of precursors of protein synthesis (free amino acids) in the tissue. Unless the levels of intracellular amino acid reserves are determined in the various parts of the nervous system it is virtually impossible to assess correctly the results obtained in investigating the metabolic activity of proteins by labeling with radioactive precursors.

When the appropriate data are analyzed it must be remembered that the concentration of free amino acids in the metabolic reserves of nerve tissue varies in animals of different species [402, 1455, 1535]. The concentration of free amino acids also varies substantially depending on the way the experimental animals are killed, the methods used to extract and determine the amino acids, and other technical factors. These remarks apply in particular to amino acids that undergo very active intracellular metabolic conversion, and especially to those whose metabolism is linked with the tricarboxylic acid cycle. This explains why the experimental data so far obtained on the amino acid reserves of the tissues of various structures of the nervous system is so heterogeneous and contradictory and not always suitable for comparison. At this stage of the book we shall therefore describe only the most general principles, the authenticity of which are above question, and supplement this with some details of the special character of the composition of the free amino acid reserves in phylogenetically, structurally, and functionally different parts of the nervous system. The analysis of these special features can shed light not only on differences in the rates and character of protein metabolism in functionally different structures of nerve tissue, but also on the role of other important aspects of intracellular metabolism, including the exchange of some metabolites specific for nerve tissue (GABA, serotonin, etc.).

All amino acids are found in each part of the CNS and PNS. About four-fifths of the amino acid pool of nerve tissue are accounted for by glutamic and aspartic acids, glutamine, GABA, and alanine, amino acids whose metabolism is linked with the tricarboxylic acid cycle and with ammonia metabolism.

Different parts of the nervous system differ significantly from each other in their content of these and other amino acids. The greatest differences in concentrations are observed for amino acids that are not components of protein molecules, viz., GABA, N-acetylaspartic acid, and cystathionine. (Because of their preferential localization in nerve tissue these amino acids are considered to be specific compounds for it.)

The GABA concentration is higher in the gray matter of the brain than in the white [403, 1233, 1659]. The GABA concentration in the thalamus (4.16 μmoles/g tissue) is twice as high as in the cerebellum, cerebral hemispheres, mesencephalon and diencephalon [521]. High concentrations of GABA are also found in the globus pallidus, thalamus, and hypothalamus [831, 1233, 1456, 1615, 1617], but its concentration in the corpus callosum is very low [831, 1456]. Significant differences in GABA concentration have been found [1129] in the cellular layers of the monkey's cerebral cortex. In the motor cortex, for instance, the GABA concentration falls steadily from the outer to the inner cell layers, to reach a minimum in the white matter (from 20-21 to 3 mmoles/kg dry weight).

The concentration of N-acetylaspartic acid in nerve tissue is almost two orders of magnitude higher than in the liver, kidneys, and other mammalian organs [1662]. Its highest concentration is found in the cerebral cortex [1659].

The concentration of the sulfur-containing amino acid cystathionine is very high in the tissue of the pineal gland [1243]; there

its concentration (128 mg %) is 180 and 130 times higher, respectively, than in the cerebral cortex and the anterior pituitary. It is
interesting to note that the levels of other amino acids in these
tissues are either almost equal or differ by only a few times. In
the corpus callosum the concentration of cystathionine is several
times higher than in other parts of the brain [1456].

Significant differences in the concentration of amino acids
that are components of protein molecules — glutamic acid, glutamine, alanine, glycine, and tryptophan — have also been found in
different parts of the nervous system. The concentration of glutamic acid in the gray matter of the brain is much higher than in
the white [403, 1686]. The concentrations of glutamic acid and
glutamine in the tissue reserves of the medulla and pons are lower
than in the cerebral cortex, cerebellum, thalamus, hypothalamus,
and other parts of the nervous system [826, 831, 1233, 1686]. The
highest levels of tryptophan are found [1504] in the pons and
hypothalamus, whereas the content of this amino acid in the tissues
of the cerebral hemispheres and spinal cord is extremely small.
The highest concentrations of serotonin are also found in the hypothalamus and other parts of the brain connected with the central
autonomic sytems [173, 387, 1031]. The alanine content in the
frontal lobe is almost twice as high as in the tissues of the mesencephalon; elsewhere its concentration is rather lower than in the
frontal lobe [1617]. Significant differences in amino acid concentrations have also been found in certain parts of the spinal ganglia
and peripheral nerves [963].

It is important to note that aspartic acid is distributed more
uniformly in different parts of the brain than other amino acids
linked metabolically with the tricarboxylic acid cycle [826, 1456,
1539, 1615]. The concentration of most amino acids in the mesencephalon [1615, 1617], the corpus callosum [1456], and peripheral
nerves [1535], incidentally, is lower than in other parts of the
nervous system.

These experimental data concerning the distribution of free
amino acids in different parts of the nervous system and changes
in the concentration of amino acids in the brain in various functional states (see Chapter 6) and in ontogeny (see Chapter 7) are
evidence that amino acids specific for nerve tissue or playing an
active part in metabolic conversions are less uniformly distributed
between the structural formations of nerve tissue and are functionally more variable. Examples of this type of metabolism are the forma-

tion and fixation of ammonia, reactions of the tricarboxylic acid cycle, and the formation of biologically active substances such as GABA and serotonin. The rest of the amino acids are uniformly distributed in the different parts of the nervous system and their concentrations are more stable during changes in the functional state of the nervous system.

Because of the limited sensitivity of the methods used and because of technical difficulties, no information is available on the intravital concentration of free amino acids in the microstructures of nerve tissue – in the different types of cells and their organoids, in spatially identifiable groups of cells and their organelles. However, there is no doubt that the metabolic reserves of each of these structural elements possess a specific complement of free amino acids and other metabolites [1734]. There is much indirect evidence in support of this view, notably the existence of reserves (compartments) of amino acids and other metabolites, differing in their volume and metabolic activity, in the tissues of the brain and other organs (see Chapter 4).

Analysis of the experimental data on the concentration of free amino acids in structures of nerve tissue performing different functions also indicates that the intensity of renewal of the tissue proteins of some parts of the nervous system does not always correlate with the concentration of the free amino acids or, at least, of some of them, in these parts. For example, the concentration of individual amino acids [1036, 1732] and the rate at which they enter the tissue from the blood [106] are sometimes smaller in tissue structures with higher rates of protein renewal.

It can be concluded from these findings that although the content of free amino acids in the tissue pools is of great importance for protein synthesis, the rates of protein metabolism in some parts of the nervous system, when functioning normally, are determined chiefly by the overall activity of the intracellular biosynthetic systems. The concentration of the free amino acids, on the other hand, probably becomes a limiting factor under conditions accompanied either by a sudden intensification of protein synthesis or by a substantial restriction of the supply of protein precursors to the reserves of the cell.

Another important conclusion follows from these facts. Evidently in parts of the nervous system characterized by a low intensity of protein metabolism and high levels of free amino acids, the latter are utilized in intracellular metabolic processes not

related to protein biosynthesis more intensively than in parts with
a high rate of renewal of protein structures and low concentrations
of amino acids.

Protein Renewal in Subcellular Fractions from Various Regions of the Brain

Appraisal of the Methods. Almost half a century ago,
in his "Lectures on Physiology," Pavlov formulated the problems
facing cell physiology in the following words: "... we shall have
to divide the cell into microscopic parts, to find out how they work
separately, how they interact with each other, and how all this
adds up to the work of the cell as a whole ...." Yet for several
decades after these words were spoken science had neither methods
nor suitable equipment for dividing cells into their separate
structural components or for investigating their chemical composition and metabolic functions. The histochemical and cytochemi-
cal methods that existed at that time, as well as the method of
electron microscopy introduced somewhat later, still gave only
very limited opportunities for studying processes taking place in
single cell organelles and for examining their role in the activity
of the living cell. The dynamics of the biochemical functions of
the intracellular ultrastructures still remained almost impossible
to study.

The situation changed radically after the development of the
technique of differential centrifugation of tissue homogenates and
its introduction into biochemical research. The basic principles
of this method, the ways in which the homogenates are prepared,
and the conditions of their fractionation have been adequately reviewed in many publications [188, 202, 311, 324, 671, 1376]. We
shall therefore simply list the most important advantages and disadvantages of this method.

First, this method is at present virtually the only one by which
sufficiently homogeneous preparations of the principal structural
components of the cell (nuclei, nucleoli, mitochondria, lysosomes,
ribosomes, membranous formations performing various functions,
and so on) can be obtained in isolated form and in sufficient
quantities for reliable results with most forms of biochemical
analysis. This is also true for the isolation of cells of different
types. Differential centrifugation combined with electron-microscopic and biochemical methods provides a way not only of deter-

mining the intracellular localization of particular chemical components, enzyme systems, biochemical processes, and so on, but
also of investigating the metabolic conversion of substances in
each subcellular structure and the functional interaction between these
substances in the course of metabolism. A wonderful illustration of
the advantages and extensive possibilities of this method was the discovery of the lysosome. Through differential centrifugation of
homogenates, in conjunction with biochemical determinations of
the activity of certain hydrolytic enzymes, the existence of lysosomes, hitherto unknown intracellular organoids, was first postulated and later proved [202, 966, 771, 1713].

Naturally the method of differential centrifugation is not
without its disadvantages. The structural components of the cell
isolated by this method from different tissues differ in some characteristics (chemical composition, structure, activity of their
enzyme systems, etc.) from the corresponding structures in the
intact cell. Also, this method cannot yet provide preparations of
cell organoids uncontaminated by other morphologically and functionally different subcellular structures closely resembling the
organelle sought in their sedimentation properties. It cannot always be taken for granted that compounds contained in the soluble
cytoplasmic fraction were in fact located in the cytoplasm of the
intact cell. At least some substances pass into the soluble fraction as a result of the disintegration of subcellular structures and
others are extracted from them during preparation and fractionation
of the homogenate.

However, the disadvantages of the method of differential centrifugation noted above do not justify the doubts sometimes expressed
during the interpretation of results obtained with isolated preparations of subcellular structures. Every other method of investigation of the structural-chemical systems maintaining the activity
of the living cell has no less serious, and sometimes more serious,
disadvantages. For instance, cells in tissue culture differ both
physiologically and biochemically from cells functioning in the
multicellular organism. The electron-micrographs of the ultrastructures of the cell by no means reflect accurately the situation
as it exists in situ. Enzyme proteins isolated in the purified form
differ significantly in their functional characteristics and physicochemical properties from those functioning in the intact cell.
Many other similar examples could be given. Nevertheless, these
and many other methods have brought about the tremendous pro-

gress that has taken place in the last two decades in the study of the chemical basis of living processes and they have enabled the first steps to be taken into what was regarded as an impenetrable territory of living phenomena.

Because of the tendency to underestimate the value of differential centrifugation as a research technique it is worthwhile pointing out some of its recent results. Special comparative investigations of the electron-microscopic characteristics of subcellular structures in situ and in isolated fractions obtained by differential centrifugation from homogenates of various parts of the brain have shown [502] that isolated mitochondria, membranes of the endoplasmic reticulum, nerve endings and, in particular, nuclei remain in a perfectly satisfactory state of preservation. Other workers [57] also found that the cell organelles are comparatively resistant to harmful factors and that some changes taking place in them may actually be reversible. In different species of animals the resistance of some subcellular structures (for example, nerve endings and synaptic vesicles) to injury evidently varies [298].

During the development and use of variations of the method of differential centrifugation of brain homogenates efforts were made to obtain fractions containing more homogeneous subcellular structures. For example, to prevent contamination of the mitochondrial fraction with nuclei and of the microsomal fraction with mitochondria, rather stringent conditions of fractionation were used during precipitation of the nuclei and mitochondria [31, 34–36, 532]. The original nuclear fraction, which contained mitochondria and other nonnuclear structures as impurities, was repeatedly washed in sucrose solutions of increased density [31, 34–36, 460, 532]. In other investigations [2, 29] the nuclear fraction was isolated by methods [4, 477] yielding preparations with a minimal quantity of other subcellular structures as impurities. Mitochondrial fractions were washed in an isotonic sucrose solution to remove lighter cytoplasmic structures precipitated along with them [38, 460, 532]. During the separation of fractions of unpurified mitochondria by the gradient centrifugation method, fractions of more homogeneous structures (myelin, nerve endings, and mitochondria) were obtained by separation of intermediate zones [32, 33]. The comparative homogeneity of the morphological composition of the subcellular fractions studied was verified by microscopic [29, 31, 34, 35, 38] and biochemical methods [34, 35, 38, 621].

The radioactive indicator method has played an important role in the development of research into protein metabolism at the level of the cell and its microstructures. The scope, advantages, and disadvantages of the various forms of this method are well known [170, 1432]. Therefore only a problem directly related to the use of this method and the interpretation of results will be briefly discussed.

Labeled amino acids can be attached to proteins with the aid of nonpeptide bonds [233, 283, 916]. This binding of labeled amino acids to proteins can significantly distort the results of investigations if the preliminary treatment of the protein preparations to determine their radioactivity is carried out under conditions that do not ensure complete removal of amino acids adsorbed on the proteins or linked to them by nonpeptide (e.g., disulfide) bonds. In our own investigations all protein preparations intended for radiometric measurements were treated by a method [29, 460] leading to the virtually complete removal of lipids, nucleic acids, free amino acids, and other low-molecular-weight compounds. Control determinations showed that when the incorporation of methionine-^{35}S into brain proteins is investigated in vivo, using protein preparations treated by this method, there is no risk that any significant part of the labeled amino acid is bound to the protein by nonpeptide bonds. This conclusion is in good agreement with results obtained by other workers [170, 694, 869] who showed that the adsorption of labeled amino acids in vitro is reduced to a minimum in experiments on animals.

Protein Renewal in Tissue Subcellular Fractions of the Cerebral Hemispheres and Cerebellum. The numerous papers [288, 408, 410, 461, 473, 517, 587, 1043, 1044] published before 1957 contained data mainly on the intensity of renewal of total brain proteins. At that time nothing was known about the metabolic activity of the proteins of the cell organelles. Observations of differences in protein renewal in morphologically and functionally different cellular microstructures are essential to the elucidation of their role in the functions of the cell. Therefore research was undertaken to examine the characteristics of protein metabolism, at first in the nuclear and cytoplasmic fractions [29], and later in the various cytoplasmic structural components of the brain as a whole and of its principal parts — the cerebral hemispheres and cerebellum [460]. The experimental animals, adult cats, received a subcutaneous injec-

tion of methionine-^{35}S, in a dose of 0.09 μCi/g body weight 2 h
before sacrifice. The purified nuclear fraction was isolated from
the brain homogenate by a slightly modified [29] method of Palladin et al. [477]. The anuclear residue of the homogenate formed
the cytoplasmic fraction.

The following conclusions can be drawn from the facts illustrated in Fig. 6 [29]. Proteins of the purified nuclear fraction
of brain tissue incorporate the radioactive label more intensively
than proteins of the cytoplasmic fraction. The specific radioactivity of the nuclear proteins in tissues of the cerebral hemispheres was more than 70% higher than that of the cytoplasmic
proteins. The corresponding subcellular fractions in the tissues
of the cerebellum are virtually indistinguishable in the level of
radioactivity of their proteins. Proteins of the nuclear fraction
of the cerebellum and cerebral hemispheres are renewed at equal
rates, whereas the metabolic activity of the proteins of the cytoplasmic fraction of the cerebellum is considerably higher than
that of the cerebral hemispheres. These differences in the rate of
intravital renewal of nuclear and cytoplasmic proteins are evidently due to differences in the functional activity of the protein
structures compared. The different rate of renewal of proteins

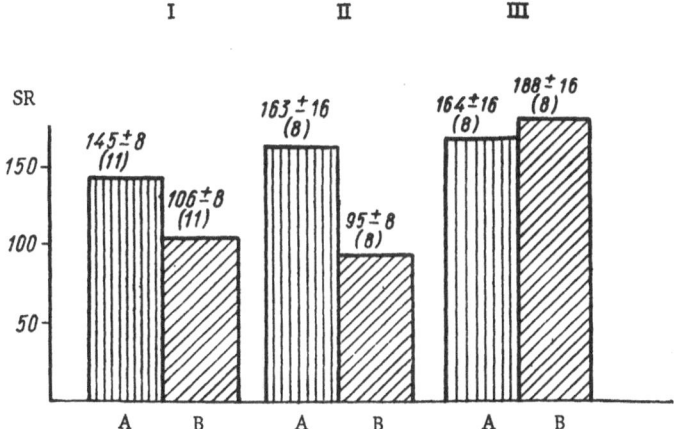

Fig. 6. Specific radioactivity (pulses/min/mg nitrogen) of proteins of
nuclear and cytoplasmic fractions of cat brain tissue: I) whole brain,
II) cerebral hemispheres, III) cerebellum; A) nuclear fraction; B) cytoplasmic fraction. (Here and in other figures the numbers in parentheses
denote the number of experiments).

in the fractions of the cerebral hemispheres and cerebellum is dependent on the functional characteristics of these parts of the brain. Of course it is possible that the specific nature of the structural organization of these parts of the brain may have some influence on the character of distribution of their intracellular components between the nuclear and cytoplasmic fractions.

Having determined significant differences between the rates of renewal of the total cytoplasmic proteins in the cerebral hemispheres and cerebellum, the next step was to determine which structural components of the cytoplasm are responsible for the higher activity of the cerebellar cytoplasmic proteins. To study this problem the rates of incorporation of labeled methionine into proteins of the mitochondrial (12,000g, 20 min), microsomal (50,000g, 90 min), soluble cytoplasmic (final supernatant), and the nuclear (600g, 25 min) fractions of the tissues of the cerebral hemispheres and cerebellum of the cat were determined [460]. The brain homogenates contained 18 mM calcium chloride to increase the resistance of the nuclei to harmful factors and to ensure better sedimentation of the light cytoplasmic structures during exposure to relatively low centrifugal forces [140, 671].

The results (Fig. 7) showed that the specific radioactivity of proteins of the microsomal fraction in the cerebellum was on the average 42% higher, and that of the mitochondrial fraction 80% higher, than the radioactivity of the corresponding fractions of the cerebral hemispheres. Proteins of the nuclear and soluble cyto-

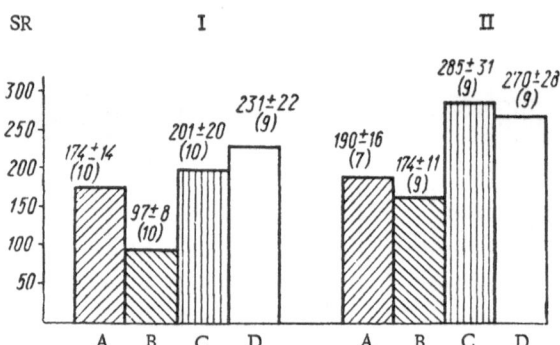

Fig. 7. Specific radioactivity (pulses/min/mg nitrogen) of proteins of subcellular fractions of tissue from the cerebral hemispheres and cerebellum of the cat: I) cerebral hemispheres, II) cerebellum. Subcellular fractions: A) nuclear, B) mitochondrial, C) microsomal, D) soluble.

plasmic fractions of these two structures are renewed at practically the same intensity, although in this case also the specific
radioactivity of the proteins of these fractions was a little higher
in the cerebellum than in the cerebral hemispheres. Proteins of
the different subcellular fractions in the cerebellum and cerebral
hemispheres incorporate the labeled amino acid at different rates.
The highest rate of incorporation was found in the microsomal
and soluble cytoplasmic fractions and the lowest in the mitochondrial. The level of radioactive label was particularly low in proteins of the mitochondrial fraction from the cerebral hemispheres.
In the cerebellum the specific radioactivity of the nuclear proteins
was 33 and 30% lower, respectively, than the radioactivity of the
microsomal and soluble proteins. The specific radioactivity of
the nuclear proteins in tissue from the cerebral hemispheres did
not differ significantly from that of the microsomal proteins but
was 25% lower than the radioactivity of proteins of the soluble
cytoplasmic fraction. The intensity of renewal of the proteins of
the microsomal and soluble fractions was practically the same.

When the data on the metabolic activity of proteins of the soluble cytoplasmic and microsomal fractions are analyzed, it must
be borne in mind that under the conditions used to obtain the microsomal fraction the lighter components of the endoplasmic reticulum were not completely sedimented. According to the observations of Lajtha et al. [1255], it is the proteins of these light cytoplasmic structures that incorporate labeled amino acids most
intensively. Furthermore, experiments in vitro have shown [1410]
that light structures of the cytoplasm isolated from the brain tissue
of two-week-old rats by consecutive centrifugation (78,500g, for
5 h, followed by 78,500g, for 16 h) of the supernatant after sedimentation of the microsomal fraction increase the rate of incorporation of labeled phenylalanine into proteins of the microsomal and ribosomal fractions by three to five times. The mechanisms
of this activation of protein synthesis have not been explained. It
has been postulated that the postmicrosomal fractions can activate
the process of incorporation of labeled amino acids into proteins
of the microsomes and ribosomes in vitro in two ways: by inhibiting the splitting of polysomes and by activating the binding of
ribosome monomers with messenger RNA.

Metabolic Activity of Proteins of the Cytoplasmic Fractions with Different Sedimentation
Characteristics. Brain tissue is distinguished by con-

siderable morphological and functional heterogeneity and nerve cells by their very large size and extremely complex structure and physiological and biochemical organization [44, 46, 67, 170, 215, 347, 529, 541, 673, 1759]. The functional and morphological diversity of the cells of nerve tissue (the various types of nerve, glial, neurosecretory, and other cells) and the fact that they consist of functionally different parts (the cell body, its many processes, synapses, and other microstructures) led to the suggestion that metabolically and structurally different mitochondria may exist. These cell organoids possess morphological polymorphism, functional diversity, and considerable functionally determined variability [505, 508, 696, 697, 699]. In view of the absence of data on the metabolic activity of the proteins of morphologically and functionally different mitochondria in the literature, it was decided to investigate the intensity of intravital protein renewal in a series of cytoplasmic fractions with different sedimentation characteristics.

The following fractions were isolated consecutively from a tissue homogenate of the adult rabbit brain by differential centrifugation: nuclear fraction ($1300g$, 10 min), six fractions of cytoplasmic structures (3F10, 5F10, 8F15, 12F20, and 20F20),* and the soluble cytoplasmic fraction. The sedimentation characteristics of these fractions, the results of phase-contrast microscopy and staining with Janus Green B, and some biochemical parameters (content of protein, RNA, and DNA; succinate dehydrogenase activity) indicate that the subcellular fractions obtained differ significantly from each other not only in size, shape, and density of the mitochondrial granules that they contain, but also in their chemical composition [34, 621]. These differences, moreover, are found both in the original unwashed mitochondrial fractions and in fractions washed in the appropriate way in order to remove the lighter cytoplasmic structures and coprecipitated soluble proteins. The 20F20 fraction was contaminated the most with membranes of the endoplasmic reticulum and ribosomal granules.

The results (Fig. 8) showed that the rates of protein renewal vary considerably in the cytoplasmic fractions isolated. Proteins of fraction 5F10, containing the heaviest mitochondrial granules, have a very low intensity of renewal. With an increase in the

---

*The number in front of the letter F indicates the centrifugal force used in thousands of g, the number after it the duration of centrifugation in minutes.

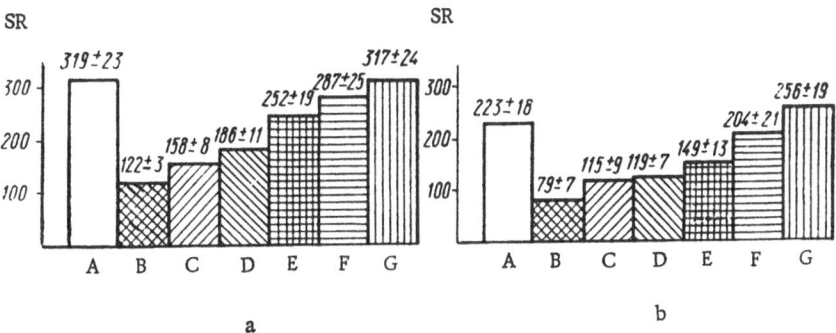

Fig. 8. Specific radioactivity (pulses/min/mg nitrogen) of proteins of the nuclear and cytoplasmic fractions of rabbit brain tissue: a) unwashed cytoplasmic fractions, experimental animals received a subcutaneous injection of methionine-^{35}S, 0.13 μCi/g body weight; b) washed cytoplasmic fractions, experimental animals received a subcutaneous injection of methionine-^{35}S, 0.09 μCi/g body weight. Subcellular fractions: A) nuclear, B) 3F10, C) 5F10, D) 8F15, E) 12F20, F) 20F20, G) 75F120 (microsomal). Sacrificed 2.5 h after injection.

centrifugal force at which the fractions are sedimented, the specific radioactivity of the protein rises. For instance, in a series of experiments with washed granules (Fig. 8b) the proteins of fraction 3F10 are metabolized 1.5 times more slowly than the proteins of fractions 5F10 and 8F15, and 1.9 and 2.6 times more slowly than the proteins of fractions 12F20 and 20F20, respectively. The pattern of changes in the rates of intravital protein renewal in morphologically and functionally different mitochondria was generally the same in fractions of unwashed granules (Fig. 8a). It is also important to note that the intensity of protein renewal in the heaviest mitochondrial fraction was about 2.6 (in the unwashed granules) or 2.8 times (in the washed granules) lower than the intensity of renewal of the nuclear and microsomal proteins.

The results of the experiments with washed cytoplasmic granules suggested that brain mitochondria contain soluble proteins with a high rate of renewal, about equal to that of the proteins of the nuclear and microsomal fractions. This hypothesis was confirmed by experiments which showed [34] that proteins with a specific radioactivity two or three times higher than that of proteins in the residue of the mitochondrial structures can be extracted by means of 0.25 M sucrose solution. The original mitochondrial granules were sedimented by centrifugation of the anuclear fraction of brain homogenate at 12,000g for 20 min. Soluble proteins of the mitochondrial matrix are renewed at practically the same rate as proteins of the microsomal and soluble cytoplasmic fractions of the brain and blood serum proteins.

Renewal of Proteins of Fractions of Purified Mitochondria, Nerve Endings, and Myelin. The comparatively low intensity of renewal of the mitochondrial protein fractions discovered by our own investigations and those of other workers [910, 1036, 1255, 1520] was difficult to reconcile with the active role of the mitochondria in cell metabolism [318, 324]. This is especially true since it has become accepted in recent years that autonomous mechanisms of protein biosynthesis exist in these organelles [510, 565, 781, 1073, 1682]. It might be supposed from the available experimental evidence that the low rate of protein renewal of the unpurified mitochondrial fraction was due to the metabolically inert protein structures contained in it. One such structure is myelin, which contains very slowly metabolized proteolipid proteins [941, 1036, 1038].

To test this hypothesis the rate of intravital renewal of proteins of morphologically and functionally different structures was determined. These fractions were isolated from unpurified mitochondria of the rabbit brain by centrifugation in a sucrose density gradient by our modifications [33, 621] of Whittaker's method [1769]. Besides the principal fractions A, B, and C, another four zones (I-IV) were identified. This considerably reduced the degree of mutual contamination of the isolated fractions, and the main mass of soluble cytoplasmic proteins contained in the original fraction of unpurified mitochondria was separated. The scheme of fractionation of the unpurified mitochondria is illustrated in Fig. 9. Depending on the predominant localization of the tissue structures, established by electron-microscopic and biochemical investigations [33], the principal fractions A, B, and C were named "myelin," "nerve endings," and "purified mitochondria," respectively.

As the results of radiometric measurements have shown [33], the proteins of the isolated structural components differ substantially in the intensity of their renewal during life. The proteins of the myelin fraction have the lowest intensity of renewal (SR = 39 pulses/min/mg protein). The anabolic activity of the proteins of the purified mitochondria is nearly twice that of the protein structures of the myelin fraction. Proteins of the nerve-ending fraction and the intermediate zones are renewed at virtually the same intensity, 1.5 times higher than the renewal rate

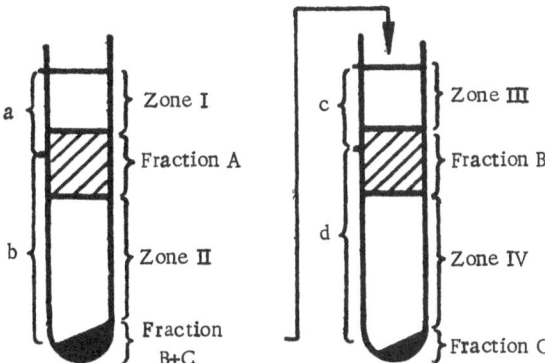

Fig. 9. Scheme of fractionation of unpurified brain
mitochondria in a sucrose density gradient by Whittaker's method as modified by the authors: a, b, c,
d) layers of gradient before centrifugation.

of the myelin fraction. Proteins of intermediate zone I are renewed more intensively than those of the other zones. As con-
firmatory experiments showed, the high metabolic activity of the
proteins of this zone is due chiefly to the soluble proteins.

In summary, the results described above confirm the considerable heterogeneity of the mitochondrial fraction isolated
from brain homogenates by differential centrifugation. The morphologically and biochemically heterogeneous structural components of this fraction were separated by gradient centrifugation
into subfractions containing more homogeneous structures —
myelin, nerve endings, and mitochondria. A very low intensity
of protein renewal was found in the myelin fraction, and a comparatively high intensity in the pure mitochondrial fraction.

The results described in this section are evidence that in
adult animals, characterized by a state of nitrogen equilibrium,
the proteins of all intracellular structural components of the brain
tissues are in a constant state of intensive intravital reconstruction. Of the six subcellular fractions studied (nuclear, mitochondrial, microsomal, soluble cytoplasmic, nerve endings, and
myelin) only the protein structures of myelin have relatively low
intensity of renewal.

Proteins of the microsomal and soluble fractions, in which
the main biochemical systems of cytoplasmic protein biosynthesis
are concentrated, have the highest rate of renewal; proteins of the

nuclear fraction are also renewed rapidly. This last fact correlates closely with the functions of the microstructures of the nucleus, which not only play an important role in the transmission
of inherited information, but also play an active part in processes
of intracellular (nuclear and cytoplasmic) metabolism, including
protein biosynthesis [120, 229, 366, 381, 673, 685]. Another important factor determining the intensive renewal of nuclear proteins is that the nucleus functions actively during the long period
of activity of nerve cells.

By contrast with these observations showing a high metabolic
activity of nuclear proteins, other workers [910, 1255, 1520] observed a comparatively low rate of incorporation of labeled amino
acids into proteins of the nuclear fraction of brain tissue. The
explanation is that with the methods of differential centrifugation
of homogenates used in the cases cited, the nuclear fraction was
contaminated by considerable quantities of metabolically more
stable anuclear structures (mitochondria, fragments of myelin,
etc.). The method we used to fractionate the homogenate, however,
yields a nuclear fraction with only very small quantities of other
structures as impurities; there is therefore good reason to assert
that the high rate of renewal of the proteins of this fraction is determined mainly by the protein structures.

The findings showing a comparatively high rate of renewal of
proteins in the fraction of purified mitochondria isolated by centrifugation in a sucrose density gradient are very interesting. Together with the data in the literature on the existence of independent biochemical systems of protein synthesis in the mitochondria,
these results are evidence of the active participation of these organelles in the renewal of the protein structures of the cell.

Tests on cytoplasmic granules belonging to five different sedimentation classes showed that the mitochondria of brain tissue
are highly heterogeneous both morphologically and functionally.
In the heterogeneous population of mitochondria there are granules that differ substantially in size, shape, density, chemical
composition, and metabolic activity of their protein structures.
In connection with the structural and functional heterogeneity of
nerve tissue and also with the extreme morphological complexity
of nerve cells, whose individual microstructural regions differ in
their chemical composition and in their levels of metabolism [46,
170, 540, 541, 543], it can even be postulated that brain tissue
contains an even larger assortment of morphologically and functionally heterogeneous mitochondrial granules than that which we

investigated. This conclusion is supported by results obtained by Pigareva and Shabadash [505, 508, 696, 697, 699, 701] and also by data [177, 1186, 1203, 1431, 1683] pointing to the marked polymorphism of the mitochondria. This polymorphism may arise either because the mitochondria belonged to cells of different anatomical formations in the brain and to different regions of the cells or because of differences in the stage of their development or the level of functional activity of the corresponding cells.

Some soluble proteins synthesized in the perikaryon of the neutron are known to migrate along the axons and dendrites to the periphery of the cell, toward the nerve endings [525, 790, 791, 960, 961, 1450, 1760]. Most probably the transported proteins consist chiefly of soluble enzymes and other metabolically active specific proteins of the peripheral systems of the cell. The structural proteins of the cell processes and nerve endings, on the other hand, are evidently renewed at the sites where they function, with the aid of peripheral biochemical systems of protein synthesis and proteolysis. This conclusion is supported by our observations of the comparatively high rate of intravital protein renewal in the fraction of nerve endings and the presence of active proteolytic systems in these structures (see Chapter 8) and also by data in the literature, mentioned above, on the ability of isolated axons [970, 1084, 1209, 1658] and nerve endings [775, 778, 1403, 1404] to incorporate labeled amino acids intensively into proteins.

The differences found in the rates of renewal of nuclear and cytoplasmic proteins in the tissues of the cerebellum and cerebral hemispheres of adult animals [99, 113, 408, 461] can be explained by the higher functional activity of the cytoplasmic systems of protein synthesis in the cell structures of the cerebellum.

To sum up, the comparatively high rate of intravital protein renewal in most subcellular structures of brain tissue so far studied is a factor of particularly great importance in ensuring the very long duration of function of the neurons — cells which lost their ability to divide by mitosis in the early stages of ontogenetic development.

Metabolic Heterogeneity of the Proteins of Subcellular Fractions

Intracellular organelles perform different functions and are characterized by the wide diversity of biochemical processes taking place within them. These processes are represented structurally by a complex and highly heterogeneous system of chemical components, including many individual proteins. These function chiefly by forming complexes with each other and with other biologically important compounds. These proteins and their complexes differ in their chemical composition, structure, and physicochemical properties. Each individual protein and protein complex functioning in a cell must evidently possess its own characteristic and unique rate of metabolism. This assumption has been confirmed experimentally [2, 3, 33, 35, 36, 621] by the study of the metabolic activity of protein structures with different physicochemical characteristics isolated from subcellular fractions of brain tissue.

In Palladin's laboratory, Belik, Smerchinskaya, Terletskaya, and their collaborators have investigated the intensity of renewal of various proteins isolated from the following subcellular fractions of brain tissue: soluble cytoplasmic, consisting mainly of proteins of the cell hyaloplasm; heavy and light mitochondria, isolated by differential centrifugation of a homogenate (Fig. 10); purified mitochondria, nerve endings (synaptosomes), and myelin, obtained from a suspension of the unpurified mitochondrial fraction by centrifugation in a sucrose density gradient (Fig. 9), and also from the nuclear fraction isolated by the method of Avdeev and Palladin [4]. Unlike others, this method provides a rapid way of obtaining comparatively pure preparations of nuclei from brain tissue by centrifuging a 25% tissue homogenate once only ($40,000g$, 20 min) in a sucrose (d = 1.273) or sucrose–glycerophosphate (d = 1.27) medium. This method gives a 30% yield of nuclei as DNA.

The protein of the initial homogenate is distributed among the various subcellular fractions (Fig. 10) as follows (mean values of six experiments expressed as a percentage of the total protein content in the homogenate): nuclear 6.8 ± 0.4; mitochondrial (heavy) 31.4 ± 2.3; mitochondrial (light) 4.2 ± 0.2; microsomal 5.5 ± 0.2; soluble 19.6 ± 0.6; and washings 29.3 ± 1.6. The fractions of heavy and light mitochondria stain deeply with Janus Green B, with maximal color intensity occurring after 30 and 120 min incubation, respectively (pH 7.4, 37°C, final concentration of dye 1:100,000). Under these conditions the nuclear and microsomal fractions do not reduce the dye even after 7 h. The highest concentrations of RNA and DNA are found in the microsomal and nuclear fractions, respectively. Succinate dehydrogenase is concentrated chiefly in the mitochondrial fractions (96% of the ac-

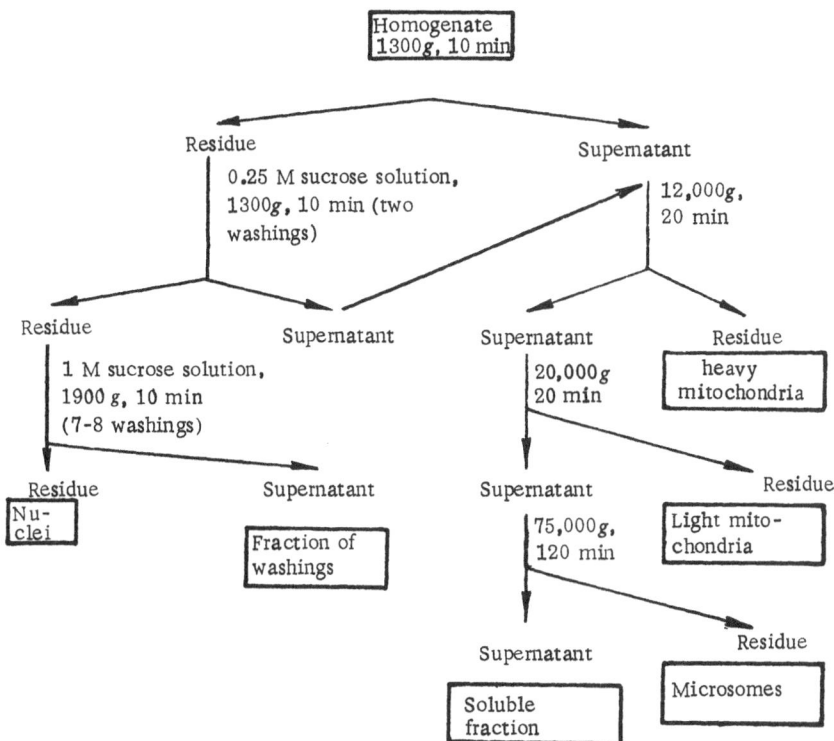

Fig. 10. Scheme of differential centrifugation of brain tissue homogenate made up in 0.25 M sucrose solution [36].

tivity in the heavy mitochondrial fraction and 4% in the light mito-chondrial fraction).

In a preliminary series of experiments in which the period from the administration of labeled amino acid to sacrifice varied (from 15 min to 96 h), the character of intravital renewal of total (unfractionated) proteins of the above-mentioned subcellular fractions of the brain (Fig. 10) was investigated. The following con-clusions were drawn:

1. Differences between the levels of radioactive label in the proteins of the subcellular fractions as reflected in the SR and RSR values (Table 3) were most marked in the experiments with durations of 15 and 30 min; later the differences gradually dis-appeared and in the experiment with a duration of 12 h a virtually constant value was reached.

TABLE 3. Changes in SR and RSR of Proteins of
Homogenate and Subcellular Fractions of Rabbit Brain
Tissue Versus Duration of Labeling

Duration of experiments	Homo-genate		Subcellular fractions										
			nuclear		heavy mitochondria		light mitochondria		microsomal		soluble		washings	
	SR	RSR	SR	RSR	SR	RSR	SR	RSR	SR	RSR	SR	RSR	SR	RSR
15 min	37	12	34	11	21	7	54	18	76	26	41	14	18	6
30 min	64	17	73	19	27	12	154	40	160	42	86	22	23	6
2 h	296	66	266	59	181	40	473	105	482	107	417	92	175	39
4 h	340	140	244	101	275	114	541	224	488	202	—	—	213	88
12 h	442	181	362	148	337	138	489	200	579	237	613	251	250	102
48 h	402	490	319	389	300	366	485	592	466	568	516	629	264	322
96 h	366	416	309	351	286	325	457	519	443	503	465	528	219	249

Each rabbit received a subcutaneous injection of methionine-^{35}S, 0.09 μCi/g body weight.

2. After the time intervals investigated the highest levels of
SR and RSR of the proteins of all subcellular brain fractions were
found in the experiments in which the duration of stay of the labeled
amino acid in the body was 12 and 48 h, respectively. The exception was the washings fraction, in which these two parameters
reached their maximum in the experiments with a duration of 48 h.

3. The radioactivity of the proteins of all subcellular brain
fractions declined very slowly after reaching the maximum; even
in the rapidly metabolized proteins of the microsomal and soluble
cytoplasmic fractions the value of SR 84 h after reaching the maximum had fallen by only 24%. These observations show that in the
brain tissue of adult animals the free amino acids of the cell, including those removed from proteins by intracellular proteases,
are repeatedly utilized for intravital renewal of protein molecules.

On the basis of the results of this series of experiments an
exposure of 2-2.5 h was taken as the optimum for such investigations. With experiments of this duration the levels of radioactivity required to obtain reliable results can be reached in proteins
of the metabolically most stable tissue structures in the brain of
adult animals with their well developed BBB. Characteristically
in experiments with a duration of 2-2.5 h substantial differences
still remain between the levels of radioactive label in the proteins of different intracellular structures; this is important when

studying the metabolic heterogeneity of tissue proteins at different stages of postnatal development of the brain.

The first step in the investigation of the metabolic heterogeneity of intracellular proteins was to determine the intensity of renewal of the various proteins of the soluble cytoplasmic fraction of adult rabbit brain [36]. Proteins of the cell hyaloplasm, labeled in vivo, were separated into subfractions by salting out with ammonium sulfate. This process yielded successively proteins salted out between saturations of 0 and 0.2 (subfraction I) and 0.2 and 0.5 (subfraction II) of ammonium sulfate and proteins not salted out in a semisaturated solution of the salt (subfraction III). The specific radioactivity (pulses/min/mg nitrogen) of the proteins of subfraction I was taken as 100 and the corresponding values for proteins of subfractions II and III were expressed as percentages of this figure. The duration of the experiments was 2.5 h.

The results (Fig. 11) showed that the level of radioactive label in the proteins of subfractions II and III was 11 (P < 0.001) and 30% (P < 0.001) below that in the metabolically most active proteins of subfraction I. The specific radioactivity of the proteins of subfraction III was 21% lower (P < 0.01) than that of the proteins of subfraction II. The proteins of subfraction III, which by their salting-out properties with ammonium sulfate belong to the group of albumin-like proteins, are thus metabolically the most stable.

After separation of the proteins of the soluble cytoplasmic fraction by dialysis against distilled water into subfractions of "albumins" and "globulins" the intensity of renewal of the albumin-like brain proteins, i.e., proteins soluble in water, is 22% less (P < 0.001) than the intensity of renewal of the globulin-like proteins precipitated during dialysis [36].

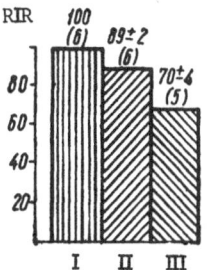

Fig. 11. Relative intensity of renewal of proteins isolated from the soluble cytoplasmic fraction of the rabbit brain by fractional salting out with ammonium sulfate: I, II, III) subfractions of proteins salted out with ammonium sulfate between saturations of 0-0.2, 0.2-0.5, and 0.5-1.0, respectively.

However, proteins salted out by ammonium sulfate between saturations of 0 and 0.5, like proteins not soluble in distilled water, can be called globulin-like proteins ("globulins") and proteins salted out between saturations of 0.5-1.0 of ammonium sulfate or soluble in water can be called albumin-like proteins ("albumins") only conventionally, by analogy with blood serum proteins. It would be wrong to identify these fractions with the corresponding fractions of blood serum proteins, for investigations by Polyakova and Lishko [535], in Palladin's laboratory, have shown that the proteins remaining in solution after prolonged dialysis against distilled water, i.e., those with the salting-out properties of albumin-like proteins, could be separated by electrophoresis on agar into 16 fractions. Only one of these fractions, accounting for only 8% of the total content of the original proteins, migrated with the speed of blood serum albumin. The remaining proteins migrated either faster or much slower than albumin. On the other hand, the fraction of brain proteins salted out by ammonium sulfate between saturations of 0.5 and 1.0 also contains many proteins migrating with the speed of serum globulins.

Dingman and co-workers [950] also found significant differences in the intensity of renewal of protein subfractions isolated from a rat brain extract by fractional salting out with ammonium sulfate. The extract was obtained by centrifugation ($11,900g$, 30 min) of a homogenate prepared in glycine buffer, pH 9.5. Although the original composition of the protein solutions before fractionation differed in our experiments and in those of these workers, the results are in good agreement. They found that the intensity of renewal of protein subfractions of a brain extract decreases with each consecutive increase in the degree of saturation of the salt.

The higher rate of renewal of the globulin-like brain proteins is in agreement with experimental results obtained by other workers who used an isotope method to determine the intensity of intravital renewal of various protein subfractions from blood plasma, muscles, and other tissues [253, 306, 571, 1386]. Some workers [1594, 1670] found no difference in the incorporation rate of protein fractions of blood plasma which they tested. However, their results are more likely to be explained by the particular conditions of their experiments rather than by truly equal intensities of protein renewal in the fractions. Conditions crucial to the outcome of these experiments include the period of labeling, the age of the animals, and so on.

For example, experiments in which the labeled amino acid was given long before sacrifice of the animals [306] showed that the differences in the rates of renewal of plasma protein fractions disappeared partly or completely, whereas in short-term experiments differences were clearly seen. In long-term (18 h) experiments we found no difference in the intensity of renewal of subfractions of soluble brain proteins [36]. This is perfectly understandable because the level of the label in tissue proteins is determined by the rates of two simultaneous processes – the incorporation of the labeled amino acid into protein and its removal from the protein. In the proteins of the fractions studied the rates of these processes differ; in rapidly renewed proteins these rates are higher than in slowly renewed proteins. In the early period after a single injection of labeled amino acid the level of the label in the tissue proteins is determined chiefly by the rate of incorporation, whereas in experiments of increasing duration the process of removal of the label gradually becomes predominant. As a result, after a certain time the levels of radioactivity of the protein fractions renewed at different rates become equal.

The age factor also plays an important role in the production of differences in the rates of renewal of individual protein fractions, for the rate of renewal of some proteins (for example, blood serum γ-globulin) increases with age whereas that of others (albumin) decreases [1164].

It can thus be concluded from the results of these investigations that the globulin-like proteins of the soluble cytoplasmic brain fraction possess higher metabolic activity than proteins of the albumin type. The heterogeneity of the rates of intravital renewal of the protein fractions studied is undoubtedly based on the unequal role of their constituent proteins in the cell functions, i.e., the unequal functional activity of these proteins.

Smerchinskaya and Belik, working in Palladin's laboratory, found considerable metabolic heterogeneity of the soluble brain proteins by means of electrophoresis in agar gel, autoradiography of the proteinograms, and direct radiometry of proteins eluted from the corresponding electrophoretic zones. Proteins of the soluble fraction of rabbit and cat brain in these investigations were labeled in vivo with methionine-^{35}S or lysine-^{14}C. Distinct authoradiographs (using RM-1 and "Supervidox" x-ray film) of the protein zones were obtained only in experiments with high doses of labeled amino acids (up to 3 μCi/g body weight) and with long (up to 60 days) exposure of the proteinograms. The intensity of blackening of most zones corresponded to their protein content. In some zones the protein content was not directly proportional to the optical density of the autoradiographs. This applies chiefly to protein zones (Fig. 1) migrating rapidly toward the cathode (zones 15 and, in particular, 16) and toward the anode (zones 1a, 1b, and 1c). The proteins of zone 15 give darker autoradiographs than would be expected from the protein content of this zone. Zone 16 is hardly visible on the proteinogram but gives a dark shadow on the x-ray film. Protein zones with a high rate of migration toward the anode either give disproportionately weak autoradiographs (zones 1a and 1b) or no darkening whatever (zone 1c). Autoradiographs of protein zones 1a and 1b were so weak that densitometric recording was virtually impossible. It could be concluded from the visual observations that the rate of renewal of the proteins contained in zones 1a, 1b, and 1c is comparatively low. This conclusion was confirmed in experiments with direct radiometry of the protein extracts of the six electrophoretic zones studied (1c, 1b, 1a, 1, 2, and 3) using proteins salted out between 0.6 and 1.0 from the water-soluble proteins of the rabbit and cat brain. The extracted proteins were prepared for determination of radioactivity by the method suggested by Mans and Novelli [1335]. The radioactivity of the protein samples was measured with a scintillation counter using the FÉU-42 photomultiplier and the PS-5M apparatus. The specific radioactivity of the proteins of zones 1a, 1b, and 1c was found to be much lower than that of the proteins of the other zones investigated (1, 2, 3) and of the original fraction salted. It can thus be concluded from the concurrent results of the autoradiographic investigations and direct measurement of the radioactivity of the proteins from the six electrophoretic zones studied that the strongly acidic proteins of brain tissue (zones 1a, 1b, and 1c) undergo comparatively low rates of metabolic activity.

Bondy and Perry [845] also observed considerable differences in the rates of intravital renewal of the fractions of the soluble brain proteins obtained by DEAE-cellulose chromatography and starch gel electrophoresis.

Considerable differences have also been found between the rates of renewal of the mitochondrial proteins. Mitochondria are the most complex and heterogeneous structural components of the cell both morphologically and functionally [176, 177, 318, 324, 388, 389, 707]. They are composed of proteins and protein complexes performing various functions, including structural, enzymic, transport, contractile, etc. It ought therefore to be expected that the metabolic activity of these proteins would vary. To obtain data confirming the presence of metabolically heterogeneous proteins in the mitochondria of brain tissue and also to determine the rates of renewal of mitochondrial protein fractions differing in their physicochemical functional properties, the experiments described below were carried out [35].

The protein subfractions isolated successively from heavy and light mitochondria labeled in vivo and obtained by the scheme described in Fig. 10 were: (A) proteins extracted with 0.25 M sucrose solution; (B) proteins extracted with 0.5% sodium deoxycholate solution; and (C) proteins insoluble in sodium deoxycholate. The proteins soluble in sodium deoxycholate (subfraction B) were further separated into three subfractions by salting out with ammonium sulfate at 10°C. In this way protein subfractions D, E, and F were salted out by ammonium sulfate between saturations of 0-0.2, 0.2-0.5, and 0.5-1.0.

In their protein composition the fractions of heavy and light mitochondria differed significantly: The proteins of each of these subcellular fractions were extracted to different degrees by the solvents and were salted out in a different pattern by ammonium sulfate (Table 4).

The proteins of these mitochondrial subfractions also differed substantially in the intensity of their renewal. The difference in the levels of radioactive label in the individual mitochondrial subfractions was considerable. Proteins extracted with 0.25 M sucrose solution (subfraction A) had the highest intensity of renewal in the heavy mitochondria. Their radioactivity was virtually on the same level as that of the proteins of the microsomal fraction of the brain and blood serum proteins. Proteins extracted with 0.5% sodium deoxycholate solution (subfraction B) incorporated the radioactive label only half as intensively (P < 0.001) as the proteins of the sucrose extract. Mitochondrial proteins not solubilized by sodium deoxycholate (subfraction C) are renewed more slowly. Their specific radioactivity was 3.4 and 1.6 times less (P < 0.001) than that of the proteins of the sucrose and deoxycholate extracts, respectively.

The proteins of the light mitochondrial fraction also differed in their metabolic activity, but the differences between the rates of renewal of the protein fractions in the light mitochondria were

TABLE 4. Specific Radioactivity (M ± m) of
Proteins of Subcellular and Submitochondrial
Fractions of Rabbit Brain and Blood Serum
(mean of 7 experiments)

Fractions	Specific radioactivity, pulses/min/ mg protein nitrogen	Distribution of protein, %	
		exp. 1	exp. 2
Homogenate	178± 6	—	—
Heavy mitochondria	125± 7	100	100
A	263± 8	6	7
B	128± 5	58	59
C	78± 3	36	34
D	141± 7	63	64
E	123± 7	22	21
F	99± 7	15	15
Light mitochondria	191± 8	100	100
A	285±11	12	11
B	175±15	71	71
C	163± 9	17	18
D	209±12	76	72
E	159±11	13	15
F	126± 8	11	13
Microsomes	261±13	—	—
Blood serum	275±18	—	—

The figures given for subfractions D, E, and F represent the percentage distribution of proteins of the deoxycholate extract (subfraction B) from which these subfractions were obtained (by salting out).

less marked than in the heavy. Proteins of the sucrose extract had the highest metabolic activity in the fraction of light mitochondria. The intensity of their intravital renewal was practically the same as that of the microsomal proteins and the blood serum proteins. By contrast with the fraction of heavy mitochondria, the proteins soluble and insoluble in sodium deoxycholate in the light mitochondria were renewed at an almost identical rate, 1.6-1.7 times slower than the proteins of the sucrose extract.

The protein fractions of the deoxycholate extracts of the heavy and light mitochondria salted out between different degrees of saturation of ammonium sulfate also differed significantly in their rates of intravital renewal. Experiments showed (Table 4) that the higher the salt concentration at which the protein fraction of the deoxycholate extract of mitochondria was salted out, the lower the intensity of renewal of its proteins, i.e., the same pattern is found as for the corresponding fractions of the soluble cytoplasmic proteins of brain tissue (Fig. 11).

As shown in Table 4, proteins of sucrose extracts of heavy and light mitochondria were metabolized at virtually the same rate, whereas the proteins of the other subfractions of the light mitochondria incorporated labeled amino acid more intensively than proteins of the corresponding subfractions of the heavy mitochondria. It is interesting to note that the detergent-insoluble proteins, which are the metabolically most stable proteins of the heavy and light mitochondria, differ the most (by twice) in their rate of intravital renewal.

It might be supposed that the high rate of renewal of the soluble mitochondrial proteins was due to the presence of membranous structures of the microsomal fraction in the sucrose extracts of the mitochondria. However, this suggestion was not confirmed by experiments in which proteins of the sucrose extract of the heavy mitochondrial fraction were separated by centrifugation ($75,000g$, 60 min) of the mitochondrial suspension into soluble proteins and proteins bound with the structures. Although under these conditions the residue contains not only mitochondria but also traces of membranes of the endoplasmic reticulum as impurities, the specific radioactivity of the proteins of the soluble fraction was several times higher than that of the proteins of the precipitated material.

Some interesting results in connection with the unequal rate of renewal of proteins performing different functions were obtained by Marks et al. [1346] who, in experiments in vivo and in vitro, found a higher rate of renewal of the proteins of the inner mitochondrial membranes than that of the outer membranes.

Nerve endings, structures specific for nerve tissue, are also distinguished by their morphological, chemical, and functional complexity and heterogeneity [673, 724, 1532]. The molecular organization and the chemical (including protein) composition of the structures of the myelin sheath are also highly complex [660, 673]. It therefore might be expected that the functionally heterogeneous proteins of these structures, like the functionally heterogeneous proteins of the mitochondria, would be metabolized at different rates, some of them faster, others slower. To detect any such differences, the intensity of incorporation into proteins of purified mitochondria, myelin, and nerve endings, soluble and insoluble in sodium deoxycholate, was investigated [33]. These structural components were isolated from the unpurified mitochondrial fraction by centrifugation in a sucrose density gradient

TABLE 5. Specific Radioactivity (M ± m) of
Proteins Soluble and Insoluble in Sodium
Deoxycholate from the Fractions of
Myelin, Nerve Endings, and Purified
Mitochondria of the Cat Brain
(mean of 5 experiments)

Fractions	Specific radioactivity, pulses/min/mg protein		A/B
	soluble protein (A)	insoluble protein (B)	
Myelin	46.0±4.1	13.0±1.8	3.8±0.3
Nerve endings	55.0±6.0	29.0±1.8	1.9±0.2
Purified mitochondria	71.0±6.2	54.0±0.4	1.3±0.1

Methionine-^{35}S was injected subcutaneously into the experimental animals in a dose of 0.3 μCi/g body weight. Animals were sacrified 2.5 h after injection.

by a modified Whittaker's method (Fig. 9). The results (Table 5) showed that these fractions contain proteins with high and low metabolic activity. Proteins in purified mitochondria and in the fractions of myelin and nerve endings that are insoluble in detergents are renewed at a slower rate than soluble proteins. Characteristically the differences between the rates of intravital renewal of the proteins, soluble and insoluble in detergent, differ in degree in the fractions of purified mitochondria, myelin, and nerve endings. For instance, whereas the intensity of renewal of detergent-soluble proteins in the mitochondrial fraction is only 1.3 times higher than that of the insoluble mitochondrial proteins, in the fraction of nerve endings the difference is 1.9 times, and in the myelin fraction almost four times (Table 5).

It is most interesting that the proteins of purified mitochondria, myelin, and nerve endings that are soluble in sodium deoxycholate differed from each other much less than the proteins of the same fractions insoluble in detergent. For instance, the specific radioactivity of the mitochondrial proteins soluble in detergent was 54 and 30% higher than that of proteins of myelin and nerve endings, respectively. The corresponding figures for proteins insoluble in detergent were 315 and 86%. The rates of renewal of proteins soluble in detergent from the fractions of heavy and light mitochondria also differed less than those of the insoluble proteins of these fractions (Table 4).

Since the myelin fraction obtained by this modification [33] of Whittaker's method [1769] is morphologically very heterogeneous, it was impossible to be sure that the differences found in the metabolic activity of the proteins of this fraction soluble and insoluble in sodium deoxycholate are attributable entirely to the myelin protein structures. The question arose whether the myelin of the CNS contained proteins with such widely differing metabolic activities (Table 5). To investigate this problem we obtained preparations of purified myelin free from contamination by soluble cytoplasmic proteins or membranes of endoplasmic reticulum, by the method of Autilio et al. [779], and investigated the renewal rate of the myelin protein fractions.

Purified myelin preparations were obtained from the brain stem, the white matter of the cerebral hemispheres, and the cerebellum of cats into which methionine-^{35}S had been injected in a dose of 2 μCi/g body weight 5 h before sacrifice. Unlike in Autilio's method, in our investigations the myelin preparations were not separated into subfractions of "light" and "heavy" myelin. Electron-microscopic investigation showed [478] that the purified myelin preparations were morphologically homogeneous (Fig. 2). In contrast to the myelin fraction isolated by Whittaker's method [1769], these preparations contained practically no membranous structures of endoplasmic reticulum or ribosomal granules. Furthermore, washing the myelin preparations three times with distilled water removed practically all the soluble proteins from them. The protein of the resulting myelin preparations constituted on the average 5-6% of the total protein content of the original homogenate and about one-quarter (23 ± 0.9%, mean of six experiments) of the dry weight of the lyophilized myelin preparations.

The acetone-treated myelin preparations were extracted three times, first with Triton X-100, and then with sodium dodecylsulfate in concentrations of 0.5 and 0.1%, respectively, the residue being removed by centrifuging the suspension at 75,000g for 90 min. Triton X-100 solubilizes a little more than one-fifth (22 ± 3%, mean of five experiments) of the protein in the original myelin preparation, whereas sodium dodecylsulfate solubilizes about four-fifths (79 ± 3%, mean of six experiments) of protein insoluble in Triton X-100 [478]. The radioactivity of proteins of the Triton and dodecylsulfate extracts of purified myelin was determined by liquid scintillation spectrophotometry on the PS-5M apparatus with the FÉU-42 photomultiplier. Specimens for counting were

prepared by the method of Mans and Novelli [1335], except that
No. 4 nitrocellulose ultrafilters, giving a higher counting efficiency, were used instead of the Whatman 3MM paper discs used
in the original procedure. Proteins of the dodecylsulfate extract
of myelin were found to be synthesized in vivo several times more
slowly than proteins in the Triton extract. The SR of the proteins
of these fractions was 35 ± 2 and 188 ± 5 pulses/min/mg protein
(mean of three experiments), respectively. The considerable metabolic heterogeneity of proteins of the myelin structures of brain
tissue was thus demonstrated in purified preparations of myelin.

Since the rates of renewal of the structural proteins of different cell components are determined chiefly by the intensity of
their function, it can be concluded that the levels of metabolic
processes in the different subcellular structures compared differ
less than the rates for protein functions within the structures themse

The results of these and other investigations [289, 1267] of
differences in the protein–lipid composition of membranes of different functional specialization indicate that though the molecular
organization of biological membranes and their protein elements
is universal in character [1078] there are still differences between the structural proteins in membranes of functionally different microstructures. This is easily understood if one re-
members that structural proteins play a role in organizing the
activity of the numerous enzymes attached to membranes, in the
membrane transport of metabolites, the conduction of nervous
impulses, and in other biochemical processes and physiological
functions. They do not serve only a simple plastic function. Other
specific functions of the brain, such as memory, are probably also
connected more with the proteins of the membranous structures
than with the soluble proteins of the cytoplasm.

The different renewal rates of proteins of submitochondrial
structures of brain are determined by their role in mitochondrial
functions. Proteins with the highest rate of renewal, which are
easily extracted by isotonic sucrose solution, are evidently the
soluble proteins of the mitochondrial matrix, i.e., they are not
fixed to its membranous structures. These proteins evidently
perform active enzymic and other metabolic functions in the mitochondria. Although the intensity of renewal of mitochondrial proteins soluble in detergent is much lower than that of the proteins
of the sucrose extract, it is higher than that of the proteins insoluble in detergent. Despite their structural functions, it will be

noted that the latter possess a comparatively high rate of renewal. For instance, the proteins of purified mitochondria which are insoluble in detergent are renewed two and four times more rapidly, respectively, than the analogous proteins of nerve endings and myelin (Table 5). Proteins solubilized by detergent are probably not in the free state in the intact mitochondria, but fixed to the membranes. A large proportion of the proteins of this fraction, like the rapidly renewed soluble mitochondrial proteins, consists of enzymes.

This conclusion was confirmed in Palladin's laboratory by the solubilization with detergents of a certain proportion of the inorganic pyrophosphatase (3.6.1.1) [632], acid proteinase [33], acid phosphatase (3.1.3.2) [33], neutral proteinase, and ATPase (3.6.1.3) [464, 465], fixed to tissue structures, and also by the data of others on the partial conversion of membrane-bound enzymes into the soluble form with the aid of detergents or other agents. The enzymes studied included NAD-nucleosidase (3.2.2.5) [1779], acetylcholinesterase (3.1.1.7) [1003], glutamine synthetase (6.3.1.2) [1603], hexokinase (2.7.1.1) [834], and several others [855, 870, 1003]. In this connection it should be noted that the ability of structural proteins of brain mitochondrial and microsomal membranes to form complexes with enzymes in deoxycholate extracts of these organelles has been established experimentally [1074]. The work of Green et al. [924, 1078, 1517] demonstrated that isolated mitochondrial structural protein can form complexes with lipids and with proteins, including enzymes.

The slowly metabolized proteins of the heavy mitochondrial fraction, i.e., those insoluble in sodium deoxycholate, are evidently structural proteins of the mitochondrial membranes. It is interesting that our observations (Table 4) of their content in this fraction (34-36%) is about the same as that found by Green et al. [1078] for structural protein of the mitochondria of the heart. The comparatively low rate of renewal of these proteins is in full agreement with their plastic functions. Characteristically, however, the sodium deoxycholate-insoluble proteins of the light mitochondrial fraction are more active than the analogous fraction of heavy mitochondria (Table 4). This could be due to any of several circumstances. The light mitochondrial fraction may contain mitochondria in a metabolically more active state of maturation or it may be richer in mitochondrial granules belonging to more active cells and cell regions. A possibility which cannot be ex-

cluded is the presence of rapidly metabolized ribosomal and endoplasmic reticular proteins in this fraction as impurities.

The metabolic activity of the detergent-insoluble proteins in
microstructures of nerve tissue call for more detailed discussion.
That these proteins are renewed less intensively than those extracted by sucrose and sodium deoxycholate solutions is not unexpected, for the insoluble protein fraction consists chiefly of the
metabolically most stable structural proteins of membranous
formations. The metabolic activity of proteins insoluble in detergent is also interesting from another point of view. First, these
proteins, despite the low level of activity of their structural functions, have a comparatively high rate of renewal, as confirmed by
the specific radioactivity of the corresponding protein fraction
from mitochondria (Tables 4 and 5). This is in harmony with
views that intensive function may result in a reorganization of the
mitochondria [699, 701] and a periodic interchange of their popula-
tions in the functioning cell [318, 406, 1006, 1203, 1433]. Accordingly it is important to note that the intensive renewal of mitochondrial structures is evidently of particularly great importance
to the activity of nerve cells, most of which function throughout
the life of the organism.

Second, membrane proteins of the various microstructures
of nerve tissue which are insoluble in sodium deoxycholate differ
in the intensity of their renewal much more than the soluble proteins of the corresponding structures. This is clearly seen when
the radioactivities of structural and soluble (in solutions of su c-
rose and sodium deoxycholate) proteins from heavy and light mitochondria (Table 4) and also from purified mitochondria, nerve
endings, and myelin (Table 5) are compared. Whereas the deoxycholate-insoluble proteins of these microstructures in fact
perform only plastic functions in the cell, the rate of their renewal is presumably determined mainly by the duration of function of the particular microstructure in the cell, and to a much
lesser degree by the level of metabolism in it. Naturally, therefore, the structural proteins of myelin, the most stable of the
membranous structures, are metabolized more slowly than the
insoluble proteins of nerve endings and mitochondria.

Third, the relatively high level of label in both the water-
insoluble proteins of the subcellular structures and even in the
proteins insoluble in detergents must be regarded as evidence
that practically all the brain proteins — soluble, membrane fixed,
or structural — are in a dynamic state.

The membranes of nerve structural components which possess biochemical specialization thus differ not only in their protein—lipid composition [289, 1267] but also in the metabolic activity of their structural proteins and possess an important role in the biochemical organization of the membrane components. Differences in the rates of protein renewal in the mitochondria, nerve endings, and myelin are probably due to differences in the functional role of these structures and their individual protein components.

Proteins of the nuclear fraction were found to be no less heterogeneous from the metabolic point of view. The nucleus is a complex intracellular structure whose activity is connected not only with the transmission of genetic information but also with the metabolism of the whole cell [67, 229, 230, 366, 381, 673, 685]. The main components of the nucleus are the double nuclear membrane, the nucleolus (of which there are more often several), the chromatin, and the nuclear sap, also called nucleoplasm or karyolymph. These components perform different functions and they differ essentially from each other in their chemical composition and ultrastructural organization. Because of the differences in the physiological role of these components in the functions of the nucleus and the whole cell, the intensity of their biochemical processes, including protein renewal, must also differ. There are many data in the literature [90, 211-213, 578] characterizing the intensity of protein renewal in the various ultrastructures of the nuclei in the liver and other organs, but the metabolic activity of functionally different nuclear proteins has not been investigated.

The metabolic activity of the proteins and protein complexes composing the karyoplasm and nuclear ultrastructures was investigated in Palladin's laboratory [1-3] by determining the intensity of incoporation of labeled amino acids into the various protein components of brain cell nuclei. Adult cats received a subcutaneous injection of labeled amino acids (lysine-1-^{14}C or glycine-2-^{14}C) in a dose of 0.16 μCi/g body weight 2.5 h before sacrifice. Sufficiently pure preparations of well preserved nuclei were obtained by the method of Avdeev and Palladin [4]. These preparations were separated into fractions containing predominantly morphologically and functionally heterogeneous nuclear components. Fractionation of the nuclear proteins was carried out by our modification [2] of the scheme proposed by Zbarskii and Georgiev [231], the five principal stages of which are illustrated below.

Scheme of Fractionation of Proteins of the Nuclear Fraction

Isolated nuclei

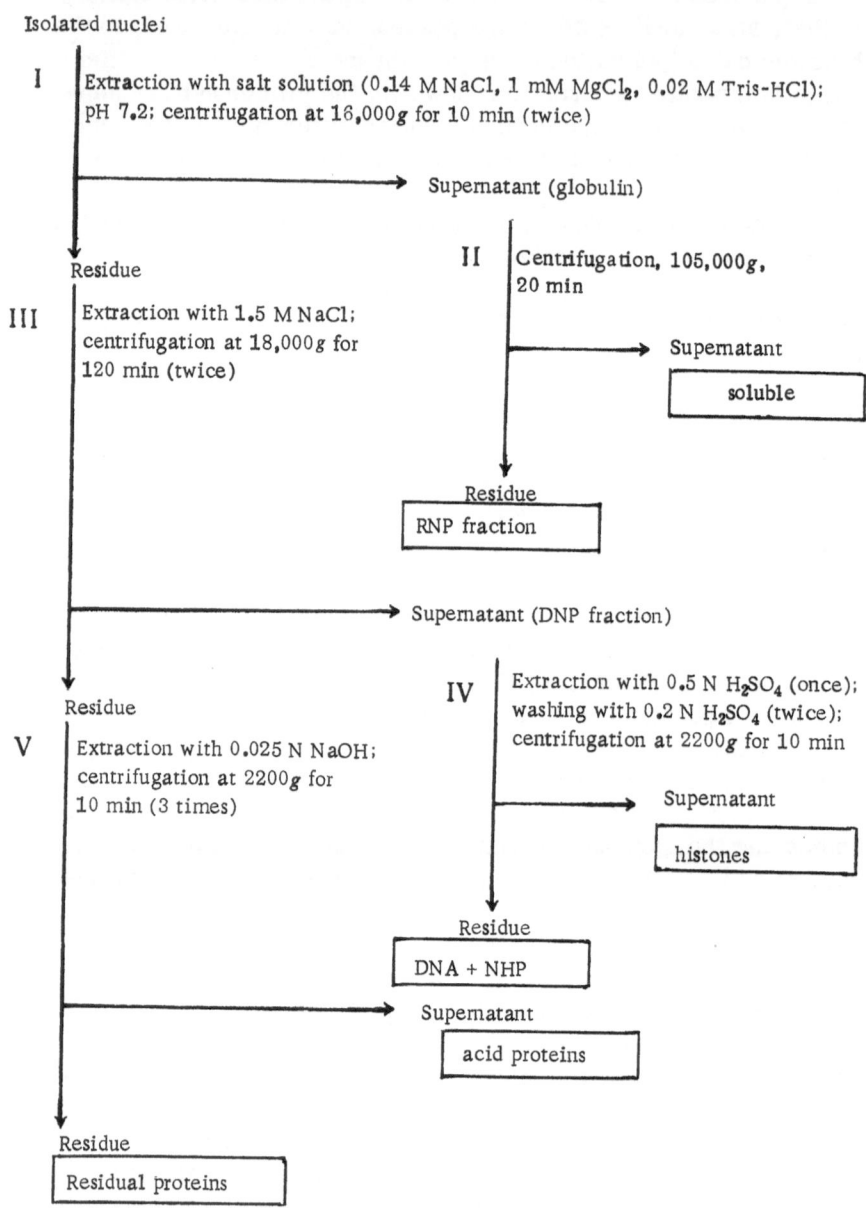

I   Extraction with salt solution (0.14 M NaCl, 1 mM MgCl$_2$, 0.02 M Tris-HCl);
     pH 7.2; centrifugation at 16,000g for 10 min (twice)

Supernatant (globulin)

Residue

III   Extraction with 1.5 M NaCl;
      centrifugation at 18,000g for
      120 min (twice)

II   Centrifugation, 105,000g,
     20 min

Supernatant

soluble

Residue

RNP fraction

Supernatant (DNP fraction)

Residue

V   Extraction with 0.025 N NaOH;
     centrifugation at 2200g for
     10 min (3 times)

IV   Extraction with 0.5 N H$_2$SO$_4$ (once);
     washing with 0.2 N H$_2$SO$_4$ (twice);
     centrifugation at 2200g for 10 min

Supernatant

histones

Residue

DNA + NHP

Supernatant

acid proteins

Residue

Residual proteins

Six subnuclear fractions were obtained: (1) soluble, containing most of the soluble proteins and other components of the karyoplasm; (2) an RNP fraction rich in nuclear ribonucleoproteins; (3) a DNA + NHP fraction, containing nearly all the nuclear DNA and the nonhistone proteins (NHP) of the nuclei; (4) histones — consisting chiefly of basic proteins; (5) acidic proteins, containing material of the nucleoli; and (6) residual proteins — corresponding to the material of the nuclear membranes.

The protein content in the isolated subnuclear fractions was determined by Lowry's method [937, 1309] and nucleic acids were estimated by color reactions: RNA by Meibaum's orcine reaction [371], and DNA by Burton's diphenylamine reaction [877]. The data on the chemical composition of the isolated subnuclear fractions given in Table 6, together with results described in [231, 381], indicate that fractions containing predominantly particular subnuclear structures, differing substantially in their content of chemical components, can be isolated from cell nuclei by this method.

Proteins of analogous fractions of other tissues are known to differ considerably in their amino acid composition [90]. The values obtained for the rate of incorporation of a given amino acid into the proteins of the subnuclear fractions tested would therefore not reflect correctly the true intensity of their renewal. Accordingly, in order to determine the metabolic activity of subnuclear proteins differing in their amino acid composition, two labeled amino acids were used: lysine-1-^{14}C and glycine-2-^{14}C [1, 2].

The data for the metabolic activity of proteins of the subnuclear fractions are given in Fig. 12 as two parameters: specific radioactivity (SR; pulses/min/mg protein) and the relative intensity of renewal (RIR) of the proteins of these fractions. The pa-

TABLE 6. Percentage Distribution of Protein, DNA,
and RNA (M ± m) between Subnuclear Fractions
of Cat Brain Tissue (mean of 4 experiments)

Substance	Soluble fraction	RNP	DNA + NHP	Histones	Acidic proteins	Residual proteins
Protein	43.6±4.0	5.9±0.6	12.2±0.7	26.4±2.8	9.7±1.3	2.1±0.7
DNA	—	—	94.2±1.0	—	5.7±1.0	—
RNA	8.8±1.4	15.4±1.5	38.0±1.6	—	33.6±2.8	3.6±0.5

Total content of the compound in the subnuclear fractions taken as 100%.

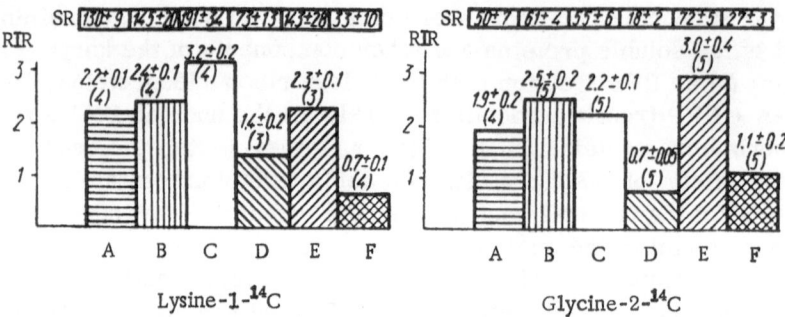

Fig. 12. Specific radioactivity (pulses/min/mg protein) and relative intensity of renewal of proteins of subnuclear fractions of cat brain tissue: A) soluble fraction, B) RNP, C) DNA + NHP, D) histones, E) acidic proteins, F) residual proteins.

rameter RIR is the ratio between the SR of the proteins of each fraction tested and the SR of the proteins of the original homogenate.

The specific radioactivity of the total tissue proteins and the proteins of the separate subcellular fractions depends on the amount of free amino acid in the brain (within certain concentration limits) corresponding to the labeled amino acid used. That is, an important factor besides the amount of labeled amino acid injected is the degree of its dilution by the corresponding non-radioactive amino acid. By contrast with SR of the proteins, the value of RIR for the proteins of a particular fraction is independent of the level of the label in the pool of free amino acids and is determined purely by the rate of incorporation of that particular amino acid into the proteins of the fractions tested. This distinguishing feature of RIR is of great importance in the interpretation of results obtained when the labeled amino acids used differ substantially in their ability to penetrate the BBB. The RIR value of the proteins shows directly how much the rate of incorporation of a given amino acid into the protein structures of a subnuclear fraction is higher (or lower) than the rate of its incorporation into the unfractionated proteins of the original brain homogenate. The value of RIR points clearly to differences in the rates of protein renewal in the subnuclear fractions that are determined by the role of the proteins concerned in these structures (Fig. 12).

It follows from the data shown in Fig. 12 that the proteins of the isolated subnuclear fractions are metabolized at different intensities. A very high rate of renewal is a feature of the proteins of three fractions rich in ribonucleoproteins (DNA + NHP,

RNP, and acidic proteins) while a lower rate is characteristic of the histones and residual proteins. The renewal rate of the proteins in the last two fractions is practically the same, but it is several times lower (P < 0.001) than that of the other four subnuclear fractions. The proteins of the soluble fraction occupy an intermediate position with slightly lower RIR and SR values than the proteins of the three most active fractions, but much higher values (P < 0.02) than the histones and the residual proteins. The renewal rate of the four metabolically most active subnuclear fractions (DNA + NHP, RNP, acidic proteins, and soluble) is two to three times as high as the rate for total brain protein and two to four times as high as the rate of the nuclear histones and residual proteins. These four subnuclear fractions account for over 70% of the total nuclear protein. It is the proteins of these fractions, containing nuclear chromatin (DNA + NHP fraction), the ribonucleoprotein components of the nucleus (RNP fraction and acid proteins), and the enzyme proteins of the nucleoplasm (soluble fraction) that are responsible for the most important and most active functions of the nucleus.

The SR of the proteins of most subnuclear fractions in the experiments with lysine-1-^{14}C was two to four times as high as in the experiments with glycine-2-^{14}C. The excess was a little less only in the fraction of residual protein. Such substantial differences in the SR values of the proteins in experiments studying the incorporation of different labeled amino acids are attributable not to the specific character of incorporation of the particular amino acids into the proteins but to differences in the intensity of the accumulation of the amino acids in the free amino acid pools of the brain. This conclusion is confirmed by the following facts.

If the ratios

and

$$K_1 = \frac{\text{SR of protein for lysine-1-}^{14}\text{C}}{\text{SR of protein for glycine-2-}^{14}\text{C}}$$

$$K_2 = \frac{\text{RIR of protein for lysine-1-}^{14}\text{C}}{\text{RIR of protein for glycine-2-}^{14}\text{C}}$$

are calculated for the proteins of the subnuclear fractions studied, the value of K_1 in each fraction is found to be approximately twice the value of K_2:

	K_1	K_2
Soluble fraction	2.60	1.18
RNP	2.38	0.96
DNA + NHP	3.47	1.45
Histones	4.06	2.00
Acidic proteins	1.99	0.77
Residual proteins	1.22	0.64

As was pointed out above, SR of the proteins (and, consequently, the value of K_1) depends on the rate of incorporation of
the labeled amino acid into protein and on the level of that amino
acid in the tissue, whereas the RIR of proteins is determined
purely by the rate of incorporation of the label into the protein.
Hence it can be concluded that the twofold excess of K_1 over K_2
observed is due less to a difference in the rates of incorporation
of the label from lysine-1-^{14}C and glycine-2-^{14}C into the proteins,
than to the unequal levels of the labeled amino acids in the
brain tissue, which depends primarily on the ability of each amino
acid to pass through the TBB (into the blood stream from the site
of injection) and the BBB (from the blood stream into the brain,
and vice versa).

It can be concluded from the values of K_2 and RIR (Fig. 12)
that the lysine-1-^{14}C label is incorporated into the proteins of
two fractions — histones ($P < 0.02$) and DNA + NHP ($P < 0.01$) —
more intensively than the label of glycine-2-^{14}C. However, it
must be remembered that the histone fraction contains a high
level of lysine. This could account, at least partly, for the higher
level of radioactive label in the proteins of the histone fraction
when lysine-1-^{14}C was used. The two amino acids are incorporated at practically identical rates into proteins of the soluble
and RNP fractions. A lower rate of incorporation of labeled ly-
sine-1-^{14}C was observed in the fractions of acidic and residual
proteins, but the differences were not statistically significant.

The method used to fractionate the nuclei yielded protein
fractions which differ substantially from each other in their functions and in their rates of renewal. Nuclear proteins undoubtedly
differ from each other even more in their metabolic behavior, as
the data given above demonstrate. Each protein fraction isolated
consists in turn of a series of individual proteins, each with its
own rate of synthesis. Although the protein fractions studied are
highly complex in their composition, the results can be regarded
as evidence of the metabolic heterogeneity of the nuclear proteins.

How can the levels of metabolic activity of the isolated subnuclear protein fractions be reconciled with the functions of the
intranuclear structures which they contain? Taking the mechanisms of cytoplasmic protein synthesis as the starting point, the
high metabolic activity of the subnuclear fractions containing soluble proteins of the nucleoplasm (soluble fraction) and nuclear ribosomes (RNP fraction) is not in dispute. This is in accord with

modern views of the existence of independent systems of protein biosynthesis in the nucleus. The proteins of these subnuclear fractions are known to be renewed rapidly in other tissues also [381, 578, 1745].

The high metabolic activity of the nonhistone proteins in the brain nuclei is in good agreement with the intensity of synthesis of proteins of the analogous fractions of liver [578, 624] and spleen [624]. Nonhistone proteins are components of the DNP complex of the nuclei. Some proteins of nonhistone nature can occur in the form of a complex with nuclear RNA [381]. It must also be remembered that in addition to proteins of the chromatin structures, the metabolically active proteins of the nuclear ribonucleoprotein reticulum and of the nucleoli [90] as well as some unextracted soluble proteins of the nuclear sap are found in the subnuclear DNP fraction.

There is abundant cytological and biochemical evidence [90, 257, 381, 557] of the important role of the nucleolus in the close functional interconnection between the nucleus and cytoplasm and, in particular, between nuclear and cytoplasmic protein synthesis. For example, a linear relationship has been found [972] between the volume of the nucleus and the RNA content in the body of the neurons of the supraoptic nucleus. The highest concentration of RNA and protein in the cell is found in the nucleolus [90, 381, 720]. RNA localized in the nucleolus incorporates label several times more intensively than cytoplasmic RNA [381, 1165]. These, like many other observations, point to the important role of the nucleolus in the active biosynthesis of intracellular proteins. This role of the nucleoli in the cells of nerve tissue is supported by the results of our own and other investigations [212] described above, showing the very high rate of renewal of the acid proteins of the nucleoli.

Although the structural proteins of the nuclear membrane are renewed several times less intensively than the other nuclear proteins, their metabolic activity is quite high and is almost equal to that of the total protein of brain tissue (Fig. 12). This is consistent with the functions of the nuclear membrane. This membrane is regarded [381] as a specialized part of the general cell membrane system, with an active part in nucleo-cytoplasmic interaction, including protein biosynthesis and the transport of metabolites from nucleus to cytoplasma and vice versa.

The biological role of the nuclear histones has not yet been precisely established. However, there is no question about the

multiplicity of their functions and their close connection with the
mitotic activity of the tissues and with the function of the genetic
system of the cell [90, 189, 229, 264, 1489]. The rates of renewal of histones and DNA in brain and liver tissue are characteristically almost identical and on the whole match the mitotic
activity of these organs [1489]. The metabolic activity of the
histones varies in cells with different specializations. This is
confirmed by data in the literature [381, 578, 1489, 1745] and by
our own investigations. The half-life of nuclear histones in brain
tissue, according to one study [1489], is three times greater than
in liver. Substantial tissue differences are found in the relative
rates of renewal of histone and nonhistone nuclear proteins: In
the nuclei of brain tissue the rate of renewal of nonhistone nuclear proteins is three to four times greater than the rate of renewal of
histones (Fig. 12), and in the nuclei of liver cells, three to six times
greater [578], whereas in nuclei from Ehrlich's ascites carcinoma cells
it is only 30-35% greater [578]. The half-life of the nuclear histones, calculated in experiments in which labeled lysine remained
for four days in the experimental animals, is 10 or 11 times longer
than the half-life of the other nuclear proteins in brain tissue, but
only four times longer in liver tissue [1489].

Our observations on the low rate of renewal of proteins of the
histone fraction of brain nuclei are in agreement with the literature [90, 264, 578, 1489] indicating comparatively slow in vivo
renewal of nuclear histones in nerve and other tissues. Experiments in vitro also showed [1745] that histones of calf thymus nuclei incorporate labeled leucine 15-17 times less intensively than
proteins of nuclear ribosomes.

The results of Burdman's investigations [874] also revealed
differences in the rate of renewal of proteins of the nuclear subfractions (proteins soluble in buffered salt solution, acid-soluble
ribonucleoproteins, acidic chromatin proteins, and residual proteins). Protein fractions isolated from neuronal nuclei, moreover, have a higher rate of renewal than the corresponding fractions of the glial nuclei; the acidic chromatin proteins of neurons
and glial cells, which incorporate labeled amino acid at virtually
identical rates, are exceptions.

Whereas 10-12 years ago the only data available were those
relating to the intensity of renewal of the total proteins of the
brain and its principal parts [113, 461, 481, 483, 1036, 1301, 1474,
1521, 1524, 1732, 1736], the individual microstructures of nerve

tissue [346, 914, 1002, 1524], and certain protein fractions
isolated from whole tissue of the CNS and PNS [287, 288, 473, 587,
1223, 1224], we now have abundant evidence of the metabolic activity of protein fractions and protein complexes performing their
manifold functions in the various intracellular structures.

Our investigations [2, 33-36, 457, 458, 478, 621] and others
[844, 874, 918, 919, 1729] described in this chapter and previously
reflect the considerable heterogeneity of composition and metabolic activity of proteins of the cell organelles of brain tissue and
their component microstructures. Depending on the functions they
perform, these proteins and their complexes are renewed at widely
different rates — from very high to very low. The comparatively
high level of radioactive label found in the structural proteins of
all membranous formations, including membranes of the most
stable components, must be regarded as evidence that all the proteins of nerve tissue of adult animals, both soluble and structural,
are in a state of continuous and dynamic change, and undergo
more or less active intravital renewal. The view held formerly
that nerve tissue of adult animals contained metabolically inert
protein structures must be rejected.

The neurochemist is faced with the very laborious but theo-
retically and practically important task of isolating the many proteins of nerve tissue and investigating their composition, physicochemical properties, structural organization, and metabolic activity. Information of this kind will help elucidate the role of proteins in the functions of the various structures of the nervous system, including its specific functions. Some progress in this direction has already been made with the study of the acidic protein,
S-100 [997, 1064, 1065, 1153, 1396-1398, 1400, 1485, 1667], basic
proteins with encephalitogenic activity [991, 1110, 1278, 1291,
1325, 1357], and other specific brain proteins [256, 1750-1752].
The first encouraging results have also been obtained in investigating the role of some acid brain proteins in the formation of
long-term memory [674].

Chapter 6

Protein Metabolism of Nerve Tissue in Various Functional States

One of the most important tasks in modern functional neurochemistry is the elucidation of the biochemical and physicochemical bases of the physiological functions of the nervous system, including its specific functions. Research into the possible connections between intracellular metabolism and function of nerve tissue structures is essential not only for the study of the specific functions of the brain, but for understanding many of the forms of pathogenesis of the nervous system, and for discovering ways of restoring the disturbed functions to their normal state. Though research aimed at establishing the connections between the intensity of metabolism in nerve tissue and function began long ago and has progressed considerably, the elucidation of the biochemical and physicochemical basis of the specific functions of nerve cells is, to all intents and purposes, just beginning.

It is impossible in this chapter to describe fully the experimental material relating the processes of intracellular metabolism to the functional activity of the brain and its microstructures. Only the most important and relevant facts will therefore be presented and discussed, with a strong emphasis on protein metabolism. Greater attention will be paid to the data concerning the protein metabolism of the brain during the excitation and inhibition of nervous activity, during the application of certain physiological stimuli to the structures of nerve tissue, and also during natural (hibernation) and artificial hypothermia. Changes in brain protein metabolism in other functional states and in certain pathological processes will be described in less detail.

169

Excitation and Inhibition

   Since the activity of nerve tissue is based on the processes of
excitation and inhibition, metabolism in these fundamental functional states has been the subject of many investigations [103, 128,
443, 444, 449, 459, 514, 562, 1013, 1251, 1472, 1738]. Until recently, however, attention was directed mainly to carbohydrate-
phosphorus metabolism and some aspects of the nitrogen metabolism of nerve tissue. The investigation of the metabolism of
proteins and their complexes with nucleic acids, lipids, and other
compounds in various functional states, including excitation and
inhibition, began only 15 to 20 years ago. The development of
these investigations took place as a result of the large-scale introduction of radioactive labeling into practical biochemistry.
   Technical difficulties have beset the study of functionally determined changes in the metabolism of proteins and other biologically important compounds in the nervous system in states of
excitation and inhibition. The chief of these difficulties are the
extreme structural and functional complexity and heterogeneity
of the tissue, the absence of adequate or near-physiological models of states of excitation and inhibition, and the inadequate selectivity
and sensitivity, or complete absence, of methods for detection and
quantitative determination of conversions of protein and other
structures that form the biochemical basis of the specific functions of nerve tissue.
   Surveys of the literature [449, 562, 1251, 1738] provide evidence of the great variety of experimental approaches and of the
very large number of investigations of the way in which states of
excitation and inhibition of nervous activity are reflected in the
metabolism of the brain. Some workers [410, 454, 482, 560, 710,
908, 1388, 1474, 1605] have investigated the effect of excitation
and inhibition on the intensity of renewal of the proteins of nerve
tissue with radioactively labeled materials whereas others [115,
125, 193, 239, 240, 542, 1699, 1703] have studied the effects on
the structure and physicochemical properties of the tissue proteins. In most investigations methionine-^{35}S was used as the labeled
amino acid [99, 226, 408, 482, 517, 561, 1605]; less frequently
glycine-^{14}C [105, 113, 410, 665] and tyrosine-^{14}C [105, 515] were
used; and in some cases leucine-^{14}C [908, 1228], proline-^{14}C [950],
cysteine-^{35}S [335], and a digest of labeled chlorella proteins [1315]
have been used.

Proteins and protein complexes from various parts of the CNS and peripheral nerve proteins have been selected for study. In the latter case the effect of electrical stimulation on the incorporation of labeled amino acids into proteins was studied in experiments in vivo [542, 1699] and in vitro [1161, 1315, 1619]. In studying the intensity of renewal of nerve tissue proteins the state of excitation was usually evoked by electrical stimulation [408, 410, 514, 516, 560, 950, 1044, 1052] or by administration of pharmacological agents [99, 225, 239, 514, 1767]. Physiological (conditioned-reflex) stimulation was used in only a few investigations [105, 115]. Nervous activity as a rule was inhibited by means of drugs [113, 226, 240, 335, 547, 665, 731]. In some experiments sleep, arising after the animals were excited to exhaustion, was used [710, 1605]. Widely different changes in the intensity of protein metabolism were found in states of excitation and inhibition of nervous activity. During excitation of the nervous system by electrical stimulation the intensity of brain protein renewal was found to be increased [408, 410, 516], reduced [516, 561, 1044], or not significantly changed [514, 561, 950]. Some particularly interesting results in connection with these findings were obtained by Pevzner

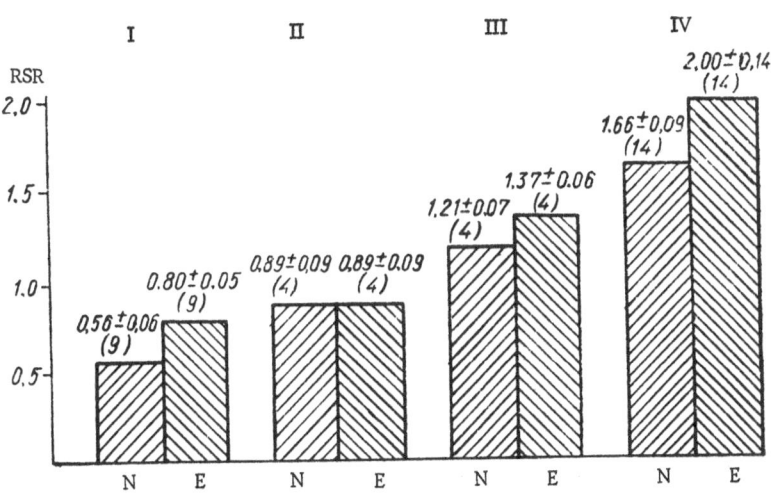

Fig. 13. Effect of amphetamine on intensity of incorporation of methionine-^{35}S into brain proteins of albino rats in experiments of different durations: I) 2.5, II) 4, III) 12, IV) 24 h; N) normal, E) excitation.

and co-workers [500] who, in the course of their investigations, found a substantial increase (40-50%) in the protein content in the cytoplasm of cells of the sympathetic ganglion which had been exposed to electrical stimulation.

More contradictory results were obtained when the effect of pharmacologically induced states of excitation on the metabolism of brain proteins was investigated. In this case also an increase [225, 514], a decrease [560], or no change [561] in the intensity of protein renewal was found in the structures of nerve tissue.

In our own investigations [454, 459] excitation of the CNS was induced in albino rats by subcutaneous injection of amphetamine. The daily dose of amphetamine (0.6 mg/100 g body weight) was divided into three injections (40, 35, and 25% of the dose, respectively), given at intervals of 8 h. When the stimulant was given in this way the animals remained in a state of relatively uniform excitation throughout the 24 h period. Methionine-^{35}S was injected subcutaneously in a dose of 0.09 μCi/g body weight 20-30 min after the first injection of amphetamine. The animals were killed at various times after the injection of labeled amino acid. The results were expressed in units of relative specific radioactivity (RSR) of the proteins, which was calculated as the ratio of the specific radioactivity of protein sulfur to the specific radioactivity of the sulfur of the acid-soluble brain fraction.

Only in experiments lasting 2.5 h was excitation of the CNS accompanied by a statistically significant (P < 0.01) increase in the intensity of incorporation into total brain protein (Fig. 13). In the experiments lasting 24 h a tendency toward an increase in the value of RSR (P $\approx$ 0.05) was observed in the experimental group compared with the control animals. In the experiments lasting 4 and 12 h no effect of excitation of the CNS on the rate of incorporation of the labeled amino acid into the brain proteins was observed.

In most investigations into the effect of pharmacological inhibition of nervous activity on protein metabolism a decrease in the intensity of incorporation of labeled amino acids into brain proteins was found [113, 410, 482, 517, 1044]. A decrease in the intensity of incorporation of leucine-^{14}C into proteins of ribosomal preparations obtained from the brain of rats treated with morphine also was found in experiments in vitro [908]. However, in some investigations no significant changes in the intensity of renewal of the brain proteins were observed during inhibition of the CNS by drugs. Fridman-Pogosova [665], who studied the effect of narcosis induced by a mixture of urethane and veronal, found no change in the intensity of incorporation of methionine-^{35}S into the brain proteins of rabbits though an increased rate was found in the spinal cord. In our own investigations [459, 1472] of the effect of narcotic sleep for 24 h, induced by injection of a mixture of medinal and urethane into albino rats, no change in the rate of renewal of the brain proteins was observed.

The processes of excitation and inhibition of nervous activity are accompanied by considerable changes in the structure and physicochemical properties of the tissue proteins. Functionally determined reorganization of the protein structures of nerve tissue is indicated by the following facts:

1. A change in the isoelectric point, the adsorption properties, and solubility of nerve tissue proteins during pharmacological excitation and inhibition of nervous activity [239, 240] and also during seizures induced by electrical stimulation [514-516].

2. Changes in the degree of amidation of brain proteins under the influence of excitation and inhibition of nervous activity [115, 125].

3. The substantial increase in the content of ionized groups in the brain proteins after electrical stimulation of the optic nerve [1703].

4. The increase in the content of SH groups in soluble proteins and in nonprotein thiol compounds of the motor cortex after electrical stimulation of the sciatic nerve [1500] and in neurons of the superior cervical sympathetic ganglion during electrical stimulation of its central branch [542].

5. Conformational changes, probably due to the rupture of hydrogen bonds, in proteins of peripheral nerves (electrical stimulation in vitro) and the cerebral cortex (electrical stimulation in vivo) which can be detected through a change in the absorption spectra of protein solutions [1699].

6. Alterations in the content of the individual electrophoretic fractions of water-soluble proteins in different parts of the brain during prolonged (10 days) pharmacological inhibition of nervous activity [193], during seizures evoked by electrical stimulation [515], and in chronic alcoholic poisoning [614].

7. Structural-metabolic reorganization of nerve cells observed during inhibition; this reorganization is expressed as a redistribution of enzymes, ribonucleoproteins, and

other components within a particular cell in a direction opposite to that characteristic of a state of cellular excitation; microstructural reorganization of some elementary subcellular structures, especially mitochondria [271], is also involved.

The fact that all these structural changes in the proteins of nerve tissue are reversible (the original structure and properties of the proteins are restored soon after the cessation of excitation or inhibition) must be regarded as conclusive evidence supporting the functional basis for the reconstruction of protein molecules. However, there are no grounds as yet for asserting that it is this reconstruction of the protein molecules and this alone that constitutes the molecular mechanisms of excitation and inhibition. All that can be said with confidence at present is that states of excitation and inhibition are accompanied by definite changes in the structure of proteins along with changes in other biologically important substances.

Characteristically changes in ultrastructure during excitation and inhibition have been recorded at the level of a single neuron [119, 200, 1611]. During amphetamine excitation the number of ribosomes in the neurosecretory cells of the supraoptic nucleus of the rat hypothalamus is increased and more of them are fixed to the endoplasmic reticulum [119], whereas during chlorpromazine inhibition they are reduced in number and pass into the free form. These morphological changes are in full agreement with the view of activation (during excitation) and weakening (during inhibition) of the intensity of protein synthesis. The synthetic capacity of cells is known to correlate with the total number of ribosomes fixed to the membranes, where they incorporate labeled amino acids more actively than free ribosomes [1120, 1440]. Chlorpromazine also induces destructive changes in the mitochondria [200] and convulsants give rise to similar changes in the Nissl's granules, the nuclei, and the nucleoli [1611]. It is also most interesting that the ratio of polysomes to free ribosomes in brain falls under the influence of electroconvulsive seizures, and the intensity of protein synthesis decreases [1324, 1716].

During excitation or inhibition of the activity of the nervous system substantial changes also take place in the intensity of renewal of the brain phosphoproteins (PP). A special feature of these complex proteins is that they contain a phosphoric acid residue attached to the serine of the polypeptide chain rather than through a nucleotide or lipid component, as in nucleoproteins or lipoproteins [330]. A second distinguishing feature of the PP is their ability to exchange their phosphorus quickly with inorganic phosphate of the tissue: the rate of renewal of PP phosphorus is one or two orders of magnitude higher than that of the phosphorus of RNA, DNA, and phospholipids [108, 297, 330].

Data on the effect of excitation of nervous activity on the intensity of metabolism of the brain PP are contradictory. In response to unconditioned-reflex and conditioned-reflex excitation [103] and also to nikethamide convulsions [225], a higher intensity of PP renewal was found in the rat brain. In slices of guinea pig cerebral cortex an increased intensity of PP renewal was observed [1113, 1114] after electrical stimulation. This effect, it subsequently turned out [1115], is exhibited only by PP of the nuclear fraction. By contrast with these data, during excitation of nervous activity by leptazol and phosphacol no significant changes were observed in the intensity of renewal of the brain PP [18].

Less contradictory findings were obtained on the effect of inhibition of nervous activity on the PP metabolism of brain tissue. The rate of renewal of PP in the brain is not significantly changed either during inhibition of nervous activity by the diethylaminoethyl ester of benzylic acid [18] or during prolonged (24 h and 9 days) narcotic sleep [18, 595].

In experiments on slices of rat cerebral cortex attempts were made [326, 329] to find a specific effect of stimulants (amphetamine, leptazol, strychnine), depressants (medinal, phenylurethane), and tranquilizers (chlorpromazine, reserpine) on the intensity of PP renewal. The effect of these substances on this process was not found to be directly connected with their pharmacological effects in vivo, nor was it due to the direct action of the drugs on PP metabolism. The changes taking place were produced by the effects of the drugs on tissue respiration, glycolysis, and oxidative phosphorylation, i.e., on processes leading to ATP synthesis.

During excitation induced by convulsants other changes take place in the nervous system: hydrogen bonds are disturbed in the structures of ribosomal RNA [935], the total RNA content is reduced [580, 899], and the content of free nucleotides is increased [1611].

This very heterogeneous picture of the changes in protein metabolism detected experimentally in nerve tissue during excitation and inhibition by pharmacological agents is not unexpected and can easily be understood if the following circumstances are borne in mind.

A complex physiological situation (determined by the underlying biochemical situation) is observed in the structures of nerve tissue in states of excitation and inhibition. In the functioning brain at any given moment some cells or groups of cells are in a state of excitation whereas others are inhibited; the degree of excitation and the depth of inhibition vary in different cells (what Pavlov called a "functional mosaic"). In order to create a more "homogeneous" functional state in brain tissue, investigators have been compelled to make use of diffuse excitation and seizures, sometimes resulting in complete exhaustion, or very deep inhibition, not at all corresponding to natural excitation and inhibition of nervous activity.

Besides their unphysiological nature, these experimentally induced states are also characterized by certain metabolic features which hinder the interpretation of the results. Under such conditions the probability of a nonspecific effect of excitatory and inhibitory substances or physical agents on tissue metabolism is increased. The preparations used often differ in their action on various enzyme systems, on the permeability of the cell membranes, the body temperature, the intensity of respiration, and the circulation of the blood, as well as on other physiological parameters of the organism and the metabolic processes in its organs and tissues.

In analyzing the experimental data with excitation or inhibition experiments, factors which have to be taken into account include the level of excitation and the depth of inhibition, the duration of the state studied, interaction between excitatory and inhibitory processes, and possible fatigue and exhaustion of the nervous system. Depending on these factors the intensity of the processes of protein synthesis and breakdown, as well as the rate of other metabolic conversions, may change significantly. Unfortunately it is not always possible to allow for these states precisely in experiments on animals.

For instance, in most cases [225, 408, 410, 514, 516] giving moderate excitation of nervous activity with the aid of electrical stimulation or drugs the intensity of renewal of the brain proteins was increased. However, in animals with features of overexcitation induced by electrical stimulation, the rate of renewal of the brain proteins was indistinguishable from that of control rats in a state of relative rest [408]. During the development of exhaustive excitation, leading to fatigue, the rate of incorporation of labeled amino acid into the brain proteins fell sharply [710, 1605]. Experiments showed [408, 410] that the longer the period of stimulation, the greater the degree of increase in the intensity of renewal of the brain proteins during moderate excitation by electrical stimulation. The intensity of protein metabolism rises considerably both during single and during repetitive electroconvulsive seizures, but it remains within normal limits if the seizures are repeated at intervals of a few hours, and falls considerably after the seizures [514]. During sleep for many hours, induced by sodium amytal, the intensity of incorporation of labeled amino acids into the brain proteins falls considerably [106, 482, 517], whereas during sleep for 30 min arising naturally after exhausting excitation, it rises sharply [710, 1605].

It must also be remembered that the intensity of protein renewal in the brain tissues differs significantly in animals of the same species but of different breeds [1043, 1044]. Even in animals of the same line, the rates of incorporation of labeled precursors into the compounds of brain tissue show considerable individual variation. However, the functional dependence of protein metabolism has been investigated by different workers in animals of different species, and sometimes of differing age.

Another fact to be taken into account when characterizing the changes in protein metabolism in nerve tissue and other organs under the influence of various pharmacological agents, toxins, poisons, or physical factors is that the metabolic disturbances observed are by no means always due to an increase or decrease in the intensity of the specific functions of the particular cell. Changes in protein metabolism are often evoked by the action of the test substances on intracellular protein synthesis or proteolysis either directly or indirectly, i.e., through a change in the intensity of synthesis and breakdown of RNA, the permeability of the cell membranes to metabolites, the activity of the biochemical transport mechanisms of the cell, the energy metabolism, respiration, and the circulation.

The dependence of tissue protein metabolism on the general state of the organism, more specifically on respiration, the circulation, temperature regulation, etc., has been discussed in the literature and deserves special attention [312, 360, 361, 562, 1044, 1520]. Protein biosynthesis requires the expenditure of energy and the participation of the numerous enzyme systems supplying the cell with energy. Naturally, therefore, disturbances of the physiological functions mentioned above, manifested clearly in states of excitation and inhibition induced by pharmacological agents, are bound to affect the levels of protein metabolism. Nerve tissue, with no significant reserves of high-energy compounds or glucose as a source of energy, is particularly sensitive to the disturbance of these functions and for that reason a decrease in the supply of these and other substances, especially oxygen, is reflected in nerve tissue first of all.

The dependence of tissue metabolism, including protein metabolism, on the physiological functions of the organism has been confirmed experimentally by several investigations [312, 361, 562, 1044]. Richter [1044], for example, showed that in rats deeply anesthetized with ether or pentobarbital the body temperature falls and may reach 28°C. Besides the pharmacological inhibition of nervous activity, this hypothermia itself may well be a cause of the reduced intensity of brain protein renewal found in these animals. Richter postulated that the changes in the intensity of brain protein renewal observed in these experiments may be associated with a decrease in the uptake of oxygen by the nerve tissue. However, in experiments in vitro the decrease in intensity of leucine-^{14}C incorporation into ribosomal proteins in the brain of rats previously receiving morphine was not accompanied by a decrease in the body temperature [908].

Rozengart and Maslova [562] found a clear correlation between the intensity of protein renewal and blood pressure during convulsions evoked by injection of picrotoxin or diisopropylfluorophosphate in the cerebral cortex and diencephalon of cats. A sharp decrease in protein renewal was observed only in animals whose blood pressure fell considerably. It has also been shown [1106] that the ability of free amino acids to pass through the blood vessels of the pia mater is increased if the blood pressure in the cerebral vessels is increased and during diffuse inhibition induced by electrical stimulation of the cortex. The permeability to free amino acids is reduced by asphyxia. Under normal conditions there is no direct correlation between the blood flow and the intensity of protein renewal in the different parts of the brain [1375].

Depending on the type of functional activity of the tissue structures of the CNS and PNS (excitation, inhibition, anoxia, different
stages of individual development) substantial changes take place in
the content of free radicals or, more precisely, of unpaired electrons [7]. The rate of formation of unpaired electrons in biological systems, according to most investigators [583, 584], is determined by the character and the intensity of intracellular biochemical processes. Functionally determined changes in paramagnetic
properties found in parts of the nervous system [7, 583] (cerebral cortex, cerebellum, brain stem, cervical and lumbo-sacral
divisions of the spinal cord) in general adequately reflect the
changes in intensity of metabolic processes and physiological functions in the corresponding functional states. Since free radicals
are formed in living cells during biochemical conversions of various substrates, including proteins, data on the nature and content
of radicals and the establishment of the relationship between the
rate of their formation and the functional state of the tissues may
help shed light on the physical and chemical mechanisms of many
biological processes, including the mechanisms of nervous excitation and inhibition.

Physiological Stimulation

Functionally determined changes in the size of nerve cells,
the content and intensity of metabolism of proteins, nucleic acids,
and free nucleotides, and the activity of certain enzymes directly
or indirectly connected with protein metabolism have been found
in the intracellular structures of anatomical regions of nerve
tissue by cytospectrophotometric, electron-microscopic, histochemical, isotopic, and other methods. It is significant that these
changes can be found in response to physiologically adequate factors such as photic [65, 67, 101, 856, 1333, 1337, 1618], acoustic [67,
101, 604], and vestibular [286, 496, 1147, 1148] stimulation, and
during the ordinary diurnal rhythm of physiological activity [175].
The content of protein and RNA, for example, increases significantly in the ganglionic neurons of the frog retina after photic
stimulation [65, 67]. More prolonged stimulation, giving rise to
fatigue, leads to a sharp decrease in the content of cytoplasmic
RNA in these neurons. The RNA content in the pyramidal cells
of the mouse forebrain rises during the afternoon and evening, i.e.,
during a period of increase in their functional activity, and it declines considerably at night [175]. Under the influence of high-

frequency flashes the intensity of incorporation of labeled lysine into proteins of the monkey visual cortex is increased [1618]. Prolonged deprivation of photic stimulation by suturing the eyelids together for 4-7 months, on the other hand, leads to a decrease in the rate of incorporation of leucine-^{14}C into proteins of visual cortical neurons and to a decrease in the size of the nerve cells by 20% in 14-day rats [1337]. In rabbits kept in darkness for a long time, protein and ribonucleoprotein synthesis in the ganglionic cells of the retina also falls sharply [856, 1333].

After exposure for 3 h to acoustic stimulation the renewal of RNA and phospholipids in the auditory cortex rises, whereas in the motor and visual cortex no change is observed [604]. The activity of enzymes of the succinic oxidase system in the neuronal mitochondria in auditory and visual cortex rises under the influence of the corresponding acoustic or photic stimulation and returns to normal immediately after the end of this stimulation [101]. Mitochondria of the Purkinje cells and neurons of the dentate nucleus are exceptional in this respect.

The content of protein and RNA and the activity of cytochrome oxidase and succinate oxidase are considerably increased in the neurons of Deiters' nucleus in rats exposed to vestibular stimulation by rotating on a special turntable (for 25 min daily for 7 days) [1147, 1148, 1154].

The following data apparently provide the most convincing evidence of a cause-and-effect relationship between the specific functions of neurons and their biochemical properties. Experiments have shown [643] that the content of cytoplasmic RNA is more than doubled in the bipolar cells of the outer layer of the frog retina during exposure of the eyes for 1 h to fast and slow photic stimuli. These cells respond by a change in spike generation to fast and slow flashes. On the other hand, in the bipolar cells of the inner layer of the retina the spikes and RNA content change only during exposure to slow flashes; during exposure to fast flashes no change occurs in either the biopotentials of these cells or their RNA content.

Changes in protein metabolism similar to those described above cannot be detected if the investigations are carried out on the tissues of the whole brain and not on single cells or structures responsible for a given function. For example, in monkeys exposed to photic or electrical stimulation no appreciable changes were found in the content and intensity of renewal of the proteins of the subcellular fractions of cerebral hemispheres [1378], although under analogous conditions considerable activation of protein synthesis was observed in the occipital cortex (areas 17, 18, and 19) [1618]. Characteristically, the degree of activation depends on the intensity, frequency, and duration of the photic stimulation. The effect is more marked in response to blue and green light and less marked in response to yellow and red light.

The connection between the protein metabolism of nerve tissue and the functional activity of the CNS and PNS is confirmed by studies of increased neuromuscular activity. Making animals swim or run for a long time are the methods most frequently used to increase the functional activity of the cells. Under these conditions substantial changes in the intensity of ammonia formation and in the degree of amination of the proteins [1161, 1723, 1724], the protein and RNA content [141, 498, 1161], and also the rate of incorporation of labeled amino acids into proteins [763, 1222] are found in the nerve cells.

The content of protein amino nitrogen in rat brain tissue falls significantly after prolonged (4.5 h) swimming, whereas the free ammonia concentration rises [1723, 1724]. Under these conditions, just as during prolonged electrical stimulation, analogous changes take place in the rat sciatic nerve [1161]. It is interesting that in neither of these situations did short periods of stimulation (15 min) cause any change in these indices of protein metabolism in a peripheral nerve. In connection with the dependence of protein metabolism on the level of functional activity in the organ observed, special attention must be paid to recent findings [153] showing that the content of functionally active protein groups (COOH, SH, NH_2) is highest in neurons with more complex functions (receptor, motor) and that their concentration is lower in neurons performing associative, relaying, or analytical functions.

During increased functional activity substantial changes also take place in the rate of incorporation of labeled amino acids into brain proteins [763, 1222]. These changes vary in different parts of the brain (cerebral hemispheres, cerebellum, brain stem) and they depend on the duration of loading [1222]. Whereas the intensity of protein renewal in the brain stem of rat was higher than in controls during a 4-h period of swimming, after which it fell, in the cerebellum during the same period two alternate decreases and increases in the intensity of protein renewal were observed. The author cited [1222] suggested that these complex changes in different parts of the brain could be explained by different competitive relations between structural and functional metabolism in the tissues studied. However, in this particular case specific features of activation or inhibition of the activity of functionally different neurons in different parts of the brain and at different stages of increased motor loading may also play an important role. This hypothesis is supported by experiments showing [763] that intensity of incorporation of labeled leucine into motoneurons of the lumbar and cervical segments of the rat spinal cord rises considerably during training (running for 2 h on a wheel). If, however, the labeled amino acid is injected into the animals after the end of training, the intensity of incorporation of the label into the motoneurons falls. During the investigation of 15 functionally different parts of the brain an increase in the rate of incorporation was found in both sensory and motor cells, but in the latter the increase was much greater.

It is also noteworthy that the RNA content in the motoneurons of the anterior horns of the spinal cord rises considerably in animals swimming for a long time [71, 498]. Restoration of the normal RNA level in these cells takes place 4 h after the end of swimming. In hypodynamia induced by keeping mice for 3 weeks under conditions of restricted motor activity, no substantial changes in the RNA content were found in motoneurons of the anterior horns of the spinal cord or the spinal ganglia, whereas the RNA content rose appreciably in the glial satellite cells of the spinal ganglion toward the end of the second week, but then fell sharply toward the end of the third week of hypodynamia [72].

The changes in intracellular metabolism during physiological stimulation take place only within certain limits and in accordance with the levels and functional rhythm of the loads placed on the tissue structure. These observations are in harmony with the concept of restoration of the properties of a functional cell within the limits of its working activity [677].

The Formation of Behavioral Responses

(Learning)

Functionally determined changes in protein and nucleic acid metabolism in brain tissue are found not only during the more or less physiologically adequate forms of stimulation specified above, but also during the formation of behavioral responses and skills in animals, i.e., during learning — a specific function of the nervous system.

Investigations [674, 1151] have shown, for instance, that in rats with experimentally produced behavioral responses, such as learning to obtain food from a special device with the paw not usually preferred, two fractions of acid proteins exhibit a higher intensity of intravital renewal than the same protein fractions of the corresponding tissue structures of untrained control animals. These protein fractions were isolated from the pyramidal neurons of the hippocampus by means of a microelectrophoretic method and they migrated in an electric field immediately after the S-100 protein specific for nerve tissue. During learning, increased synthesis of protein S-100 [674] and intensification of the incorporation of labeled uridine into polysomes of brain tissue were observed, in the absence of any such activation of incorporation into polysomes of the liver and kidneys [741, 742].

Changes in the nucleotide composition (an increase in the adenine content and a decrease in the cytosine content) of nuclear RNA of nerve cells and PNA of the glia [1150] had previously been found in the course of training rats to balance on a stretched wire. In the control rats, stimulated by rotating them in a special wheel, the ratio between the bases in RNA was unchanged although the actual RNA content was increased. However, rotation in a wheel can hardly be accepted as a suitable control for experiments to study training in balancing. It is therefore difficult to say at present whether the changes observed in the composition of RNA were in fact due to the learning process or whether they are the result of nonspecific activation of the functions of nerve cells during balancing. These misgivings are strengthened by the fact that during certain forms of nonspecific stimulation, such as during leptazol convulsions and electrical stimulation of the brachial plexus, an increase in the adenine content is also found in the nucleic acids of the cerebral cortex [1053]. The fact that structural changes can be found in the proteins and nucleic acids of the brain during exposure to physiological factors such as the formation of behavioral responses is of great interest and importance to the elucidation of the specific functions of these compounds in nerve tissue.

It is impossible in this section to give a complete account of the divergent views and experimental data [122, 674, 751, 752, 793, 795, 1007, 1008, 1314, 1513, 1701], which at times are incomplete and contradictory, or of the mechanisms of formation, storage, and reproduction of memory and the role of protein and RNA in these processes. We shall therefore examine only some of the experimental data that confirm the existence of structural changes that arise in proteins and nucleic acids of the brain during learning, as well as changes in the metabolic activity of these particular components. A more detailed account and an analysis of the experimental material gathered up to the present on the role of proteins and nucleic acids in the processes of learning and remembering, and a discussion of the hypotheses regarding the mechanisms of learning, can be found in several accessible publications [6, 56, 180, 275, 429, 568].

Artificial Hypothermia and Hibernation

Since in man and higher animals the CNS, in conjunction with the endocrine system, plays a leading role among the mechanisms controlling metabolic processes in the body, and since proteins constitute the basis of the biochemical structure and behavior of nerve tissue, the investigation of brain protein metabolism in hypothermic states is an extremely important stage in elucidating the pattern of intracellular biochemical processes taking place at lowered temperatures.

Hypothermic states are characterized by a rapid slow down of the vital processes and, in particular, by deep inhibition of the functions of the CNS. To study tissue metabolism during hypothermia two states that differ in the extent to which they can be described as physiological have been used: hibernating animals, in which natural, physiological hypothermia can be produced, i.e., hypothermia requiring no drugs or other agents affecting the bodily functions; or animals that do not hibernate, in which artifical hypothermia is induced by lowering the external environmental temperature or by a combination of drugs and physical cooling.

From the biological point of view hibernation is a unique adaptation of the animal organism, formed in the process of evolution, to survival under unfavorable conditions by means of a low body temperature. This remarkable state of prolonged, infrequently interrupted, deep sleep is regarded [248, 650, 711, 712, 726] as a special functional state characterized by extremely depressed physiological functions (respiration, circulation, excretion, etc.), by a sharp fall of the body temperature, a decrease in the intensity of metabolism, marked hypofunction of the endocrine system, and deep inhibition of nervous activity. On natural awakening of the animals all these functions are usually restored to normal very quickly. Hibernating animals can easily be made to awaken artificially, by raising or lowering the ambient temperature. Artificial awakening is accompanied by a very rapid increase in the intensity of all physiological functions and biochemical processes and by sharp reexcitation of activity of the CNS.

The physiological decline in the intensity of all vital processes in mammals during hibernation, the rapid increase in the intensity of these processes on awakening, and the gradual transition following awakening to the characteristic level of wakefulness are particularly interesting in connection with the functional biochemistry of nerve tissue. This interest is justified by the fact that under ordinary experimental conditions, without the use of gross, unphysiological procedures, it is virtually impossible to induce deep inhibition or sharp excitation of the activity of an animal's nervous system. Nor can one induce a sharp decline or elevation of the functions of other organs and study the course of functionally determined changes in the chemical composition and intensity of the biochemical processes of the brain. For these reasons hibernating animals are a useful, indeed unique, test object.

The study of metabolic processes, including protein metabolism, in the tissue structures of the nervous system, as in the tissues of other organs, during natural hibernation and artificial hypothermia induced by a combination of physical and pharmacological agents is of tremendous theoretical and practical interest for medicine [341, 503, 570]. The limited information in the literature on the character of protein metabolism in nerve tissue in the two types of hypothermia mentioned will next be described briefly.

In hibernation and in artificially induced hypothermia, as well as in animals during the transition from the hypothermic to the normothermic state, considerable changes take place in the content of free amino acids in the metabolic reserves of the brain tissue. These changes not only reflect the character and intensity of the metabolic conversions of amino acids themselves, but they are a consequence of changes in the rates of synthesis and breakdown of the tissue proteins. The changes taking place apply chiefly to amino acids such as glutamic and aspartic acids, GABA, and glutamine, which are functionally important for nerve tissue. It is these amino acids whose dynamics have been studied in greatest

detail at the various stages of hibernation and artificial hypothermia [123, 152, 726-728, 1232, 1381, 1385]. The content of most
other amino acids in the reserves of nerve tissue is upset much
less during these hypothermic states. Appreciable changes are
found only for alanine [128, 129], tyrosine, and aspartic acid [1328].
More detailed information on these matters will be found in the
surveys of Émirbekov [727, 728].

It can be concluded from the data in the works cited above
that during hibernation, as during artificial hypothermia, the degree of amination of the brain proteins and the content of glutamic
and aspartic acids, glutamine, and GABA in the free amino acid
pools of the brain tissue are considerably reduced. The content
of ammonia, however, in brain tissue is increased severalfold
in these states. The degree of the changes observed depends both
on the depth and on the duration of the natural or artificially induced hypothermia. Investigations have shown [152, 726, 727, 1232,
1385] that the character and disturbance of the free amino acid
content in hypothermia differ in morphologically and functionally
different parts of the CNS and in animals of different ages. It is
particularly interesting [1385] that in hibernating animals the
GABA level changes most in those brain structures (the brain
stem, diencephalon) that are believed to be responsible for the interchange between normothermic and hypothermic states.

During artificial awakening of the suslik, a large short-tailed
ground squirrel, from hibernation, the free GABA concentration
in the brain falls, while the glutamic acid and glutamine levels are
not significantly changed. During rewarming of artificially cooled
susliks, however, the concentration of glutamic acid and, in particular, of glutamine in the brain rises [1381]. Natural awakening
from hibernation and artificial rewarming after hypothermia are
thus accompanied by different changes in the levels of amino acids
in the nerve tissue. This is quite understandable, for the functional states of animals during natural hibernation and in artificially
induced hypothermia, and also in the transition from each of these
two states to normothermia, can only be compared conventionally
as regards both the degree of hypothermia and the character and
mechanisms of their onset [341]. An increase in the content of
histidine and glutamic acid and a particularly sharp increase (more
than twofold) in the content of aspartic acid and tyrosine are found
in the brain of the garden dormouse after natural awakening; the
GABA and glutamine content, on the other hand, are lowered at

this time [1328]. Restoration of the normal ammonia level in brain tissue with the return of the animals to a normothermic state develops gradually and continues for several days [726, 728].

The changes in the concentration of nitrogenous metabolites in the brain with a change in body temperature in hibernating and nonhibernating animals are accompanied by a significant decrease in the degree of amination of the proteins [123, 152, 726]. In susliks, after a long period of hibernation, the concentration of amino groups in the total protein falls by 22% in the cerebral hemispheres and by 48% in the cerebellum [726]. Artificial lowering of the body temperature in rats, on the other hand, is accompanied by a much greater decrease in the concentration of amino groups in the brain proteins than is artificial cooling of waking susliks [123, 152, 726]. In the latter case, moreover, the degree of amination of the proteins falls less than in susliks during natural hibernation or in rats during artificial hypothermia. It is interesting to note that when susliks awaken from a long period of hibernation, there is a further decrease in the content of protein amino groups in the cerebral hemispheres, whereas their content in the proteins of the cerebellum rises substantially [726]. It is also important that changes in the degree of amination of the proteins during hypothermia differ in character in morphologically and functionally different parts of the brain.

The varied character of the changes in nitrogen metabolism and in the content of amino groups in the brain proteins during physiological (hibernation) and artificially induced hypothermia, as well as on recovery from these states, suggests that the mechanisms of their onset are not identical [726, 1381].

The intensity of renewal of the tissue proteins, including the proteins of nerve tissue, during natural hibernation and artificial hypothermia changes more sharply than does the metabolism of low-molecular-weight nitrogenous substances and the degree of amination of the proteins. For instance, if the body temperature of rabbits is lowered by 8°C by means of physical cooling or a combination of chlorpromazine and physical cooling, the intensity of incorporation of methionine-^{35}S into proteins of the gray matter of the cerebral hemispheres falls by 18-19%, but into proteins of the white matter of the cerebral hemispheres and the cerebellum it falls by only 4-9% [382]. When the body temperature of rats anesthetized with ether and pentobarbital falls to 28°C the rate of intravital renewal of total brain proteins falls by nearly 70% [1044].

In rabbits whose body temperature was lowered to 25 or 24°C by cooling or by the combined action of sedative, ganglion-blocking drugs, and physical cooling, the intensity of protein renewal in the whole brain and in various parts of the CNS was one-half or one-quarter that in normal animals [421, 731]. Since the level of the label was not determined in the metabolic reserves of the tissues studied in these two investigations and since this may be increased in hypothermic animals [105], there was in this case actually a greater reduction in the intensity of intravital renewal of the brain proteins.

It is pertinent [16, 421] that at the same depth of hypothermia (24-25°C) the intensity of renewal of the tissue proteins falls much less in the liver, kidneys, heart, adrenals, intestine, and other organs than in the brain. During deeper artificial hypothermia (with the body temperature down to 20°C) the rate of incorporation of methionine-^{35}S, glycine-^{14}C, and tyrosine-^{14}C into the proteins of the brain and blood of rats falls even more — by 15-20 times [105, 106].

One should note that the free amino acids of brain tissue in hypothermic rats contain twice as much labeled tyrosine as in animals with a normal body temperature. These differences are evidently explained, at least in part, by slower metabolic conversions of the amino acid and decreased incorporation into tissue proteins during hypothermia [106].

Like the protein of nerve tissue, the intensity of renewal of the brain lipoproteins falls during artificial hypothermia, and the lower the body temperature of the animals the greater the fall. For instance, the intensity of intravital renewal of lipoproteins is reduced by about half when the body temperature of albino rats falls to 33-26°C, and with a further deepening of the hypothermia to 20-13°C falls to almost one-third [104, 109].

For a long time there was no information in the literature on the intensity of protein renewal in nerve tissue in hibernating animals. Skvirskaya and Silich, working in Palladin's laboratory in 1955, obtained the first data on the rate of incorporation of radioactive phosphorus into tissue phosphoproteins of the brain and spinal cord of the suslik (Citellus suslicus). Radioactive phosphorus in the form of $Na_2H^{32}PO_4$ was injected subcutaneously into waking, sleeping, and artificially awakened animals in a dose of 0.1 μCi/g body weight 4 h before sacrifice and the rate of incorporation of the ^{32}P into the phosphoproteins (PP) was determined. The results showed [594] that during hibernation the intensity of renewal of phosphorus of the brain PP was so low that in some experiments it could not be determined. The rate of renewal of PP in the brain and spinal cord of the susliks rises sharply during artificial awakening. Moreover, the level of radioactive label in PP in the brain tissue of artificially awakened animals is about the same as in waking animals, whereas in the spinal cord tissue the level is much lower.

A little later Belik and Krachko [25, 30] investigated the intensity of renewal of total proteins and proteins of subcellular fractions of suslik brain tissue in the same three functional states. Methionine-^{35}S was injected subcutaneously into the experimental animals in a dose of 0.09 μCi/g body weight. The experiments continued for 18 h. During investigation of the metabolic activity of the total brain proteins of the waking susliks the experiments were conducted in June, i.e., after their emergence from a state of hibernation, and the intensity of renewal of protein in the subcellular fractions was studied in September, i.e., before the animals fell into hibernation again. It is clear from the results given in Fig. 14 and Table 7 that the intensity of incorporation of labeled amino acids into total proteins and into proteins of all the sub-

TABLE 7. Specific and Relative Specific Radioactivity
of Proteins of Subcellular Fractions of Brain
Tissue in Hibernating, Artificially Awakened,
and Waking Susliks (M ± m)

Functional state of animals	No. of exp.	Subcellular fractions					
		nuclear	heavy mito-chondria	light mito-chondria	micro-somal	soluble	washings
Specific radioactivity (SR), pulses/min/mg nitrogen							
Hibernation	3	6±2	3±1	5±1	8±1	7±1	3±1
Artificial awakening	8	522±43	302±21	457±33	579±52	539±39	256±20
Awake	15	376±17	241±12	385±16	486±19	456±20	204±10
Relative specific radioactivity (RSR)							
Hibernation	3	1.0±0.4	0.5±0.1	0.8±0.2	1.2±0.2	1.1±0.1	0.6±0.3
Artificial awakening	8	372±44	214±22	328±42	404±38	378±35	183±20
Awake	15	235±16	151±10	242±17	302±22	284±16	127± b

$$RSR = \frac{\text{pulses/min/mg protein nitrogen}}{\text{pulses/min/mg nitrogen of acid-soluble fraction}} \times 100.$$

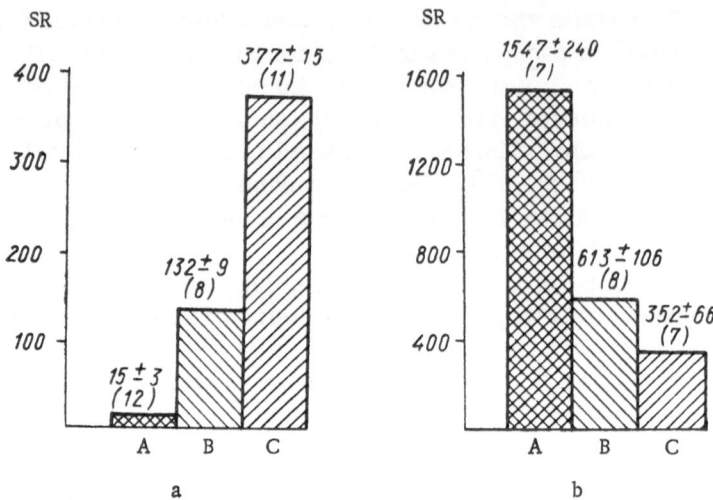

Fig. 14. SR (pulses/min/mg nitrogen) of protein (a) and acid-soluble fraction (b) of brain tissue of hibernating (A), waking (B), and artificially awakened (C) susliks.

cellular fractions of the brain is extremely low in hibernating susliks. In individual experiments in which the animals were in a state of deep stupor, incorporation of the radioactive label into the brain proteins virtually did not take place. Labeled methionine was incorporated into protein many times more intensively in the waking susliks than in the sleeping animals. The highest rate of incorporation of amino acids into brain proteins was found in artificially awakened animals.

The question arose whether these sharp differences in the SR of the brain proteins in the functional states studied are attributable to differences in the permeability of the BBB for methionine-^{35}S. The answer to this question was negative, for the highest level of radioactive label in the brain was found in the hibernating susliks (Fig. 14). In the waking and, in particular, the artificially awakened animals, in which the brain proteins were labeled most intensively, a very low content of labeled free amino acid was found in the brain. On the basis of the RSR values for proteins of the subcellular fractions, i.e., indices allowing for differences in the content of the label in the tissues of the animals studied (Table 7), it can be concluded that the very low level of labeled amino acids in proteins in the hibernating susliks and the high levels in aritificially awakened animals are due, not to functional differences in per-

meability of the BBB to methionine-^{35}S, but to the levels of activity
of the intracellular systems of protein synthesis in these groups of
animals.

No significant functionally dependent differences in the intensity of protein renewal in the subcellular fractions were found in
these investigations. In each group of animals studied the proteins
of the various subcellular fractions were characterized by an unequal intensity of renewal. The highest level of radioactive label
in the animals of all groups studied was found in the proteins of the
microsomal, nuclear, and soluble fractions, and the lowest level
in proteins of the "washings" fraction, containing intact cells, cell
membranes, myelin structures, and nerve fibers.

The results thus indicate considerable differences in the intensity of renewal of the proteins of nerve tissue in susliks in
different functional states. The extremely low level of renewal
of brain proteins observed in the hibernating susliks is due to a
sharp decrease in the functional activity of the tissue structures
and a very low temperature in them. For example, with an external environmental temperature of 1-5°C the temperature
within the brain tissue of hibernating susliks varies between 3 and
5°C [186]. The more intensive incorporation of radioacitive label
into brain proteins in artificially awakened susliks compared with
waking animals can evidently be attributed to the sudden excitation
of physiological functions, including the activity of the CNS, at the
moment of awakening and the associated activation of metabolism,
particularly of protein synthesis.

The sharp decline in the intensity of protein metabolism in
the brain tissue during the period of hibernation, together with the
sharp decrease in the intensity of other biochemical processes in
the organs of hibernating animals [339, 411, 650, 1477], plays a
decisive role in the economic utilization of vitally essential materials during the animal's long stay under unfavorable external
environmental conditions.

The rate of renewal of nucleic acids in nerve tissue also decreases under conditions of hypothermia, but much less so than
the rate of renewal of proteins, lipoproteins, and phosphoproteins.
These differences are particularly marked during artificial hypothermia in rabbits [731] and much less so in susliks in a state of
hibernation [594]. After artificial awakening of susliks, the intensity of renewal of RNA phosphorus in the brain and spinal cord
rises sharply, although it does not at once reach the waking levels
[594].

In connection with the sharp changes in intensity of protein, lipoprotein, PP, and nucleic acid renewal in the brain tissues of the suslik during hibernation and artificial awakening, the observations of Tyulenev and Belik [633] are interesting. They found that the potential ability of inorganic pyrophosphatase (3.6.1.1) to split pyrophosphate, a substance formed intensively during the synthesis of the compounds mentioned above, is virtually unchanged in the brain tissue of hibernating and artificially awakened susliks, but not in the waking animals. During a change in the functional state, only certain changes in the ratio between the activities of the free and bound forms of this enzyme are observed. Investigations showed [633, 634] that with reference to some kinetic characteristics the mitochondrial inorganic pyrophosphatase of the suslik brain differs significantly from that of the rabbit brain.

Another fact to be noted is that the sedatives and ganglion-blocking drugs that are usually used in conjunction with physical cooling to lower the body temperature of warm-blooded animals either have no effect on the metabolism of proteins, nucleic acids, and other substances in nerve tissue or give rise to only very slight disturbances [382, 383, 731]. Narcotic sleep lasting 17–19 h, induced in waking susliks by administration of a medinal—urethane mixture, not only does not lower the intensity of renewal of the tissue phosphoproteins of the CNS, but actually raises it a little [594].

Comparison of the data given above and of other findings pertaining to the effect of hypothermic states on various aspects of metabolism in the brain tissue leads to the conclusion that the rates of renewal of proteins and their complexes with lipids, nucleic acids, and other substances [25, 30, 105, 109, 421, 594, 731] are lowered during hypothermia by a much greater degree than the rates of reactions related to energy and other types of metabolism [92, 104, 128, 384, 427, 682, 722].

Protein and Other Forms of Starvation

For a long time there was no general agreement in the literature on the effect of protein and other types of starvation on the content and metabolism of protein in brain tissue. The data obtained experimentally were often contradictory and allowed no definite conclusions to be drawn regarding the character of the changes in intensity of function of the biochemical system for synthesis and breakdown of the tissue proteins. It was later found

that the nature of the changes taking place in protein metabolism
during starvation depends on many factors that are not always
taken into consideration.

For example, a deficiency of proteins and other nutrients in
the diet affects protein metabolism differently in the brain tissues
of adult and young, developing animals. Investigations in Palladin's
laboratory [470], by the radioactive labeling method, showed that
in adult rabbits starved for 14 days or more, during which time
they lost about 30-35% of their body weight, the intensity of renewal of water-soluble and water-insoluble proteins of the brain is
indistinguishable from that in normally fed adult animals. In this
connection note that the activity of inorganic pyrophosphatase
(3.6.1.1) — an enzyme splitting the pyrophosphate formed in the
cell during the synthesis of protein, nucleic acids, lipids, several
coenzymes, and other biologically important compounds — likewise
shows no change in the brain tissue of adult animals during starvation for 2 weeks [632]. Under conditions of general starvation the
intensity of renewal of the proteins of these fractions in young
rabbits (weighing 1-1.2 kg) fell, in different experiments, by 22 to
58% [470].

Similar experiments showed [599] a marked decrease in the
intensity of protein renewal in the subcellular fractions of brain
tissue of young starved rabbits compared with this parameter in
control, normally fed animals. The intensity of protein renewal
in the various subcellular fractions has also been found to change
unequally. The SR [599] and RSR (Fig. 15) levels show a somewhat smaller decrease (by only 18-20%) in the intensity of protein
renewal in the nuclear fraction than in the other subcellular fractions of the brain (by 30-38%).

The character of the disturbance of protein metabolism in the
brain tissue of adult animals depends on the degree and duration
of protein and calorific deprivation. Whereas during prolonged
(7-14 days or more), exhaustive starvation of adult animals no
significant changes take place in the content and intensity of metabolism of brain proteins [461, 470, 1375], in the earlier stages of
starvation (before the 3rd day) some increase in the incorporation
of labeled amino acids into brain proteins is actually observed
[281]. The intensity of protein renewal in the brain and its various
parts also falls if the diet is lacking or deficient in vitamins A, B,
C, and E [201, 461, 655].

Particularly important disturbances of protein metabolism
arise in brain tissue if the diet is deficient in protein or other

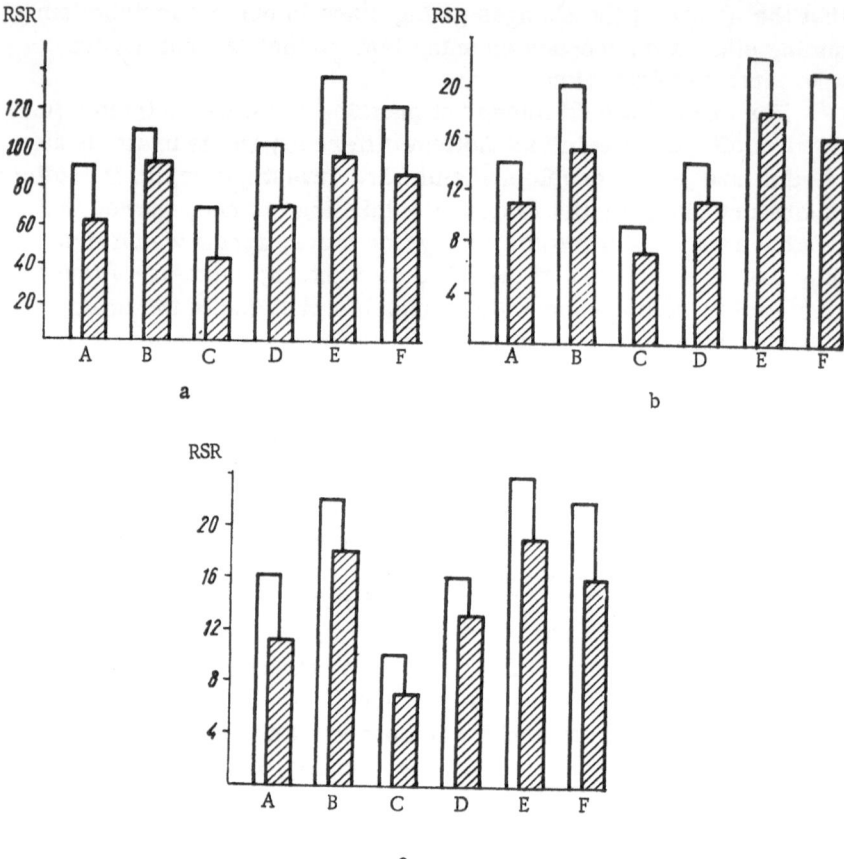

Fig. 15. RSR of proteins of homogenate and subcellular fractions of rabbit brain tissue under normal (unshaded columns) and various functional conditions (shaded): a) starvation, duration of experiments 5 h [599]; b) action of serotonin, duration of experiments 2.5 h [600]; c) action of iproniazid, duration of experiments 2.5 h [602]; A) homogenate; subcellular fractions: B) nuclear, C) heavy mitochondria, D) light mitochondria, E) microsomal, F) soluble.

$$RSR = \frac{SR \text{ of proteins of homogenate or subcellular fraction}}{SR \text{ of acid-soluble brain fraction}} \times 100.$$

nutrient factors during the period of embryonic development [1521, 1522, 1802]. A considerable decrease in the number of cells and in their protein content is observed in the brain tissue of rats born after the mothers were kept on a low-protein diet throughout pregnancy and for 1 month beforehand [1802]. No appreciable changes take place in protein metabolism in the brain of young rats

born from normally fed mothers but kept on a low-protein or protein-free diet during the first 3 days of life, whereas the intensity
of protein and RNA biosynthesis is substantially reduced in the
liver, kidneys, heart, and skeletal muscles [760].

The results obtained by workers cited in this section demonstrate that the protein metabolism of the brain — this most vitally
important organ — of adult animals is disturbed by a much lesser
degree in starvation than it is in the tissues of other organs. One
possible mechanism determining this resistance of nerve tissue
to harmful effects may be activation of the tricarboxylic acid cycle
during starvation and the utilization of α-ketoglutaric acid in transamination reactions [1281]. If amino acids are deficient in the
tissue, this compensatory mechanism may play an important role
in maintaining the protein metabolism of the brain at the normal
level.

Characteristically, despite the high sensitivity of the brain
to hypoxia, the intensity of protein renewal in nerve tissue falls
less during oxygen deprivation than in the liver or the blood plasma [308]. Another important fact is that protein metabolism in
the cerebral cortex is depressed much less than in the white
matter of the hemispheres, in the spinal cord, or in other parts
of the CNS. It can be concluded from these facts that the compensatory and adaptive mechanisms are more highly developed in
nerve tissue than in other organs, and in the cerebral cortex than
in the phylogenetically older and functionally less complex parts
of the nervous system.

Hormonal Effects

Although the experimental data on the regulatory role of
hormones in tissue protein metabolism are few in number and
contradictory in nature, there is no doubt that protein metabolism
in the tissues of the nervous system, as in other organs, is controlled by hormonal systems [278, 641, 733, 787, 1251, 1514, 1515,
1522].

It should first be pointed out that during embryonic development and in the earliest stages of postnatal life, when cell differentiation and growth takes place, protein and other types of metabolism in the brain tissues are particularly sensitive to hormonal
pathology [786, 787, 1339, 1514, 1521, 1522]. Moreover, at these
stages of ontogeny, protein synthesis is sensitive to the action
not only of growth hormone, but also of other hormones.

The data so far obtained on the effects of hormones on brain protein metabolism is concerned chiefly with the effects of thyroid hormones and insulin. The intensity of protein renewal in the brain tissues of adult animals falls in experimental hyperthyroidism [99, 355] (although some workers [1380, 1629] found no changes), but rises significantly in hypothyroidism [732, 733]. However, in rats thyroidectomized immediately after birth or at an early age the intensity of protein synthesis and breakdown in the cerebral cortex is considerably reduced [787, 1048]. Hormone therapy of thyroidectomized young animals stimulates protein synthesis in the brain tissue [1048]. In experiments in vitro on adult animals receiving thyroid hormones the intensity of incorporation of labeled amino acids into the brain proteins also was reduced [1054]. However, thyroxine stimulates protein renewal in vitro in the brain tissues of growing animals [1054, 1200, 1629]. In other investigations [1380, 1630-1632, 1671] thyroid hormones had a stimulant rather than inhibitory effect on the intensity of renewal of liver proteins of adult animals; conversely, the renewal rate of proteins in the liver of thyroidectomized animals was lowered. However, the rate of incorporation of phenylalanine-^{14}C into proteins of microsomal preparations of the brain of hypothyroid and normal rats was practically identical [1441].

In rats with experimental insulin hypoglycemia the intensity of incorporation of amino acids into brain proteins is reduced, whereas in the liver under these same conditions no inhibition of protein biosynthesis was observed [1044]. Numerous investigations (see [282]) in vivo and in vitro have shown that insulin does not inhibit, but stimulates biosynthesis of the tissue proteins in the liver, kidneys, diaphragm, skeletal muscles, and gastric mucosa. In diabetes the intensity of incorporation of labeled amino acids into the tissue proteins is reduced, but it can be easily restored to normal by injection of insulin.

To conclude, a deficiency or excess of certain hormones in the body can give rise to changes in various physicochemical properties of the brain proteins [636, 637]; in particular, the degree of their amination, their electrophoretic mobility, and other characteristics are altered.

Though these observations confirm the effect of some hormones on protein metabolism in the tissues of the brain and other organs, the molecular mechanisms of this effect remain unknown. At present we have only preliminary indications that the stimulant action of thyroid hormones on protein synthesis in the liver is unconnected with activation of amino acids [1671] and that their site of action may be DNA-dependent RNA polymerase (2.7.7.6) and the stage of transfer of aminoacyl tRNA to ribosomes [1632]. It has been postulated [787] that the system for protein biosynthesis in brain tissue in growing thyroidectomized animals is disturbed at the translation stage.

Effects of Biologically Active Substances and Psychotropic Drugs

The many and diverse activities of the nervous system under normal and pathological conditions and its regulatory functions in the body are largely determined by low-molecular-weight biologically active substances such as GABA, serotonin, catecholamines, and acetylcholine [19, 173, 191, 563, 608, 629, 640, 642]. In order, therefore, to investigate the connection between the biochemical properties of nerve tissue and its functions, it is extremely important to have information about the effect of these biologically active substances on the protein metabolism of the brain and on the neurotropic conditions [453] controlling their metabolism.

There is information in the literature on the effect of some of these substances on the protein and nitrogen metabolism of the brain. It has been shown that under the influence of serotonin, iproniazid (see Fig. 15), and lysergic and 2-bromolysergic acids [1221] the intensity of incorporation of labeled amino acids into total proteins and into the various protein fractions of the tissues of the whole brain and its parts is reduced. Iproniazid considerably reduces the glutamine level in the brain, does not affect the concentration of ammonia and of protein amino groups, but appreciably increases the metabolism of the labile amino groups [619, 620, 622]. Acetylcholine lowers the rate of renewal of PP, lipoproteins, and RNA in brain slices [683, 684]. GABA is a specific component of nerve tissue with a role in blocking synaptic transmission [563, 615], and in regulating bioelectrical activity and permeability of cell membranes to metabolites, including amino acids [18, 84, 85, 735]. This substance also plays an important role in amino acid metabolism and in metabolic formation and elimination of ammonia and the amination and deamination of tissue proteins [86, 150]. It is interesting to note that GABA specifically stimulates the incorporation of labeled amino acids into proteins of brain ribosomal preparations although it has no such effect on the analogous system from liver tissue [1674].

Serotonin, acetylcholine, iproniazid, transamine, and other biologically active and psychotropic substances, as well as GABA, cause changes in some physicochemical properties of the proteins in the tissues of the brain and its various parts [191], disturb the permeability of cell membranes to certain free amino acids and other metabolites [85, 600, 601, 630, 735], and alter the content of free amino acids [620, 695, 1414, 1560, 1617] and nucleotides [236, 307, 1221] in the tissue. It is noteworthy that serotonin and iproniazid, for example, differ in their action on permeability of the BBB to various amino acids. They do not affect the permeability to methionine-^{35}S but increase the permeability to lysine-^{14} 1.5-2 times [601]. These metabolic and structural disturbances unquestionably play an essential role in the changing nitrogen [612, 620, 622] and protein [287, 601, 683, 684] metabolism in nerve tissue under the influence of these substances.

Protein Metabolism in Some Pathological States

Protein and amino acid metabolism in brain tissue is disturbed in a number of inherited diseases accompanied by mental retardation and other mental disorders. These diseases include phenylketonuria (phenylpyruvic oligophrenia), Wilson's disease (hepato-lenticular degeneration), Hartnup disease, Lowe's syndrome, citrullinuria, hyperglycinemia, galactosemia, and schizophrenia [252, 344, 1189, 1251, 1521]. Important changes in the physicochemical properties of the soluble brain proteins, expressed as an increase in the content of protein fractions with high electrophoretic mobility, has also been found in epilepsy and cerebral edema [1198, 1521]. In sclerotic leukoencephalitis and experimental allergic encephalomyelitis, on the other hand, the content of the fractions of soluble proteins with low electrophoretic mobility is increased [370, 1183]. However, in guinea pigs with experimental allergic encephalomyelitis no abnormality in the content of protein fractions extracted from the tissue with physiological saline could be detected in any of the five regions of the CNS studied [1800]. The rate of renewal of soluble and total proteins in regions of the CNS investigated in this disease was unchanged.

Considerable changes in the physicochemical properties of proteins soluble in 0.9% sodium chloride solution have also been

found [519, 520] in the cerebral hemispheres of rabbits with
experimental cholesterol atherosclerosis. In this condition the
content of proteins with maximal and minimal electrophoretic mobility is reduced, the content of proteins with average mobility is
increased, and the A/G ratio is reduced by almost half. Characteristically in the brain of rabbits with this condition no marked
changes have been found in the intensity of protein renewal, although in most tissues except skeletal muscles the intensity of
protein renewal is considerably reduced [282].

Viruses, toxins, and other pathogenic factors have a marked
effect on protein biosynthesis in the CNS [546, 1371, 1508, 1780].
In chickens and rats diphtheria toxin inhibits the incorporation of
glycine-^{14}C into proteins of peripheral nerves in vitro. The inhibition is more marked in the tissues of growing animals than in
adults [1371]. The content of protein nitrogen is reduced and that
of nonprotein nitrogen considerably increased in the brain and
spinal cord of rabbits administered diphtheria toxin. Mouse encephalomyelitis virus stimulates the incorporation of ^{14}C from glucose into most amino acids of brain proteins in day-old mice in
vitro but inhibits its incorporation into lysine and histidine [1508].
No significant changes in the rate of incorporation of valine-^{14}C
and phenylalanine-^{14}C into the proteins of various subcellular brain
fractions of mice with scrapie could be found [1387]. The rate of
incorporation of glycine-^{14}C and tyrosine-^{14}C into proteins of the
whole brain and its various parts is slowed during the development
of transplanted carcinoma in vivo [174, 1780].

The rate of incorporation of leucine-^{14}C into proteins is also
considerably reduced in brain slices of rats poisoned with mercury
preparations [1798]. Under these conditions protein biosynthesis
in the brain tissue is disturbed before the intensity of oxygen consumption, of aerobic carbohydrate breakdown, and of succinate
dehydrogenase activity falls. The activity of Mg^{++}-activated
ATPase and aldolase in brain tissue, on the other hand, is unchanged even in the final stage of poisoning when the neurological
symptoms are clearly defined. Important disturbances of the
physicochemical properties, composition, and metabolism of brain
proteins also take place in oxygen poisoning in vivo [147, 148, 336].

Effect of Ionizing Radiation

Ionizing radiation induces considerable morphological, functional, and biochemical disturbances in nerve tissue [313, 377,

687]. Depending on the dose and duration of action of radiation and on the initial state of the animal, changes in the nervous-system protein metabolism may vary considerably and are often contradictory in character. The extensive data on this problem are sufficiently described in the surveys cited above. Suffice it to say here that the responses of nerve cells to ionizing radiation are based on disturbances of various links in the chain of metabolism. It is important to emphasize that changes in the composition and metabolism (synthesis and breakdown) of nerve tissue proteins are on the whole in harmony with the morphological and functional disturbances taking place in the tissue structures at each stage of exposure to ionizing radiation. These changes concern (1) the composition of the protein fractions and macrostructure [662, 686], (2) the content of functionally active groups [24, 123, 190], (3) the amino acid composition of the proteins [1767], (4) the permeability of the BBB to free amino acids [162], (5) the content of these amino acids in the brain [22, 24], (6) the enzymic and metabolic activity of the proteins [378, 513, 518, 552, 686, 687], and (7) the physicochemical properties and structural organization of protein—lipid membranous structures of cell organelles [701]. As a result of the action of ionizing radiation, changes in the macrostructure and content of amino, SH-, and other functionally active groups take place in the tissue proteins of the nervous system [24, 123, 190, 688] which not only appreciably reduce the electrophoretic mobility of the proteins and produce changes in their quantitative content and amino acid composition, but also lead to the appearance of new (as reflected in their electrophoretic mobility) protein fractions [24, 400, 552, 613, 688].

Numerous investigations [24, 313, 377, 687] have shown that earlier views regarding the insensitivity of the tissues of the nervous system to ionizing radiation are incorrect. Some stages of metabolism in nerve tissue (oxidative phosphorylation, tissue respiration, the tricarboxylic acid cycle, etc.) were found to be more radiosensitive, whereas others (glycolysis, protein and nucleic acid metabolism) are comparatively radioresistant. The distinguishing feature of nervous tissue is thus its simultaneous sensitivity and resistance to ionizing radiation.

Protein metabolism, like other metabolic processes and physiological functions, is more radiosensitive in the early stages of individual development, i.e., during the period of differentiation and development of the tissue structures [22, 313, 613, 686, 1767].

Functionally Determined Changes

in Free Amino Acid Content

Not all functionally dependent changes in the intensity of protein metabolism in the tissues of the nervous system are caused by disturbances of intracellular biosynthetic mechanisms. The rate of renewal of the tissue proteins also depends on the levels of free amino acids in the metabolic reserves. Despite their relative stability in the tissue of the nervous system [1539], they may vary substantially in certain functional and pathological states of the organism [669, 1172, 1339, 1560, 1570-1572]. In the other chapters of this book data are given on age and other functional variations in the free amino acids content of the brain and active amino acid transport through cell membranes. We shall therefore simply list investigations confirming the concentration variations of free amino acids in nervous tissues as a result of certain physiological and pathological states and the influence of various drugs.

The content of several amino acids in the metabolic reserves of nerve tissue is substantially changed by the action of chlorpromazine [227, 695, 1457, 1617], reserpine [1262, 1414, 1616, 1659], serotonin [600, 601], iproniazid [601, 620, 1414], and other psychotropic drugs [1414, 1560, 1617, 1659], methionine − sulfoximine [1015, 1016, 1677], and other convulsants [1181, 1457, 1675], during spreading cortical depression [1228], in drug-induced excitation and inhibition of nervous activity [422, 423, 669, 695, 907, 1063, 1262, 1560], in anoxia [1675], hypoxia [957], and hyperoxia [521], in hypothermia [726, 727], hypoglycemia [906, 920, 1204, 1369, 1457, 1559], hypothyroidism [1339], under the influence of ethyl alcohol [1097, 1098] and of electrical stimulation [1099, 1181, 1457], and also as a result of the administration of a nonprotein diet [1280, 1332] or a diet lacking in or with an excess of individual amino acids [194, 1046, 1522, 1542] or vitamins [1181, 1369, 1626, 1676]. These changes vary in degree in different parts of the brain and sometimes are even opposite in direction [194, 1616, 1617]. Under the influence of chlorpromazine, for example, the content of glycine, glutamine, and glutamic acid rises in the mesencephalon but falls in the hypothalamus. Conversely, under the same conditions the aspartic acid content rises in the hypothalamus but falls in the frontal lobes while the combined content of cystine and cystathionine falls in all six parts of the brain studied [1617].

The most significant functionally determined changes, incidentally, take place in the content of glutamic and aspartic acids,
GABA, and glutamine — amino acids which are closely linked metabolically with each other and with the tricarboxylic acid cycle.

Chapter 7

Brain Protein Metabolism in Ontogeny

The individual development of organisms is a problem of great theoretical and practical importance. Progress in its study is vital to the solution of many other problems in the medical and biological sciences, including such matters as the prolongation of human life and the ability to control the development of livestock and plants. Despite its great importance, this problem is still far from being solved. Progress is needed in aspects of the morphological evolution of organs and tissues and the evolution of their physiological functions and in the ontogenetic development of biochemical systems responsible for the specific direction and intensity of metabolic reactions characteristic of a given organ or tissue.

In this chapter we shall examine data obtained from a study of the literature and from our own investigations into age-related changes in the chemical composition of proteins, their regional distribution, and the intensity of their metabolism in the brain in the course of development of the individual organism, especially in the postembryonic phase. Considering the close connection between protein metabolism and nucleic acids and also the exceptionally high content of lipids in nerve tissue and their important role (as complexes with proteins) in the activity of the CNS, we shall begin by briefly summarizing the information on their content in the brain at the various stages of postnatal development.

Age-Related Changes in the Chemical Composition of the Brain

Quantitative Changes in the Brain Proteins in Ontogeny. Proteins and their complexes with other biologically important compounds play the leading role in the processes of embryonic and postembryonic development of all tissues of the body, including those of the nervous system. That is why the attention of investigators has been drawn to proteins from the very beginning of the study of the biochemical basis of individual development of animals.

Before the widespread introduction of isotopes into biochemical research, age changes in protein metabolism were chiefly measured by determinations of the content of nitrogen and total proteins in organs, by changes in the qualitative composition and physicochemical properties of proteins and their complexes with other biologically important compounds, and by the quantitative and qualitative composition of the end products of dissimilation. The results of these early investigations have been adequately described by Nagornyi and collaborators [399]. These investigations marked an important stage in the elucidation of qualitative and quantitative changes in metabolism taking place in ontogeny and yielded valuable experimental material essential for the creation of a general theory of ontogeny [395-399, 415].

Naturally, before starting to investigate age changes in brain protein metabolism it was necessary to make sure that such changes in fact exist. The first and fundamental indicator of these changes was the protein content in the brain at various stages of ontogeny. There is abundant evidence in the literature [37, 75, 83, 161, 373, 566, 815, 1022, 1206] to show that the protein content, calculated per wet weight of brain tissue, increases with age, although in old age its concentration falls [199, 1127, 1128].

In her investigation of the protein content in the brain of the albino rat during the postnatal period of development, Golubitskaya [161] found a steady increase in concentration with age, especially during the first 30 days of life. The greatest increase in the protein content in human [1127] and rabbit [37] brain was also found in the early stages of individual development. Later, as the number of mitoses falls, as the neurons cease to divide, as the rate of growth of the nerve cells and their processes decreases and the intensity of myelination of the nerve tissue diminishes, the protein concentration calculated per wet weight of brain tissue rises more slowly.

In the early stages of postnatal development there is a very rapid increase in the total content of solid matter in the brain tissues [373, 566, 1206, 1406, 1451]. This increase can be attributed only partly to an increase in protein content. It is due to a much greater degree to the rapid increase with age in the content of lipid components. As a result of the higher rate of increase in the content of solids compared to protein content with age, the protein content expressed per dry weight of brain tissue falls appreciably with age [37, 373, 566, 1206]. The most marked changes in the content of protein and dry matter in the brain tissues of rabbits takes place during the first month of postnatal life (Table 8). For instance, the mean daily increase in the protein content calculated per wet weight of tissue during the first 11 days of life is 1.1%, falling to 0.8% during the next 20 days, and to only 0.08% after the age of 1 month [37].

Age changes in the protein content in the various parts of the brain mainly follow a course parallel to that of those in the whole brain: an increase in the protein content calculated per wet weitht and a decrease calculated per dry weight of tissue. In addition, the changes in different parts of the brain do not coincide in time; they are determined by the morphological, functional, and biochemical features reflecting the maturation of these various parts. Phylogenetically older parts of the brain develop earlier in ontogeny than phylogenetically younger parts [443, 447, 904, 1185]. Evidence in support of this statement is given by age changes observed in the content of total nitrogen [1127, 1185, 1406], protein [447, 904, 905, 1094, 1127], water [91, 1077, 1127], and other components in the various parts of the brain compared. Furthermore,. the protein content expressed per wet weight of tissue is lower in the gray matter of the brain in the various stages of embryonic

TABLE 8. Content of Protein and Dry Matter in Brain
Tissue of Rabbits at Different Ages (M ± m)

| Age groups of animals | Protein | | | Dry weight, as percent of wet weight of tissue |
| | percent of wet wt. of tissue | | percent of dry wt. of substance | |
	by Gachev's method [116, 395]	by Lowry's method [1072]		
Newborn	5.8±0.19 (9)	6.7±0.06 (3)	51.2±1.07 (8)	11.6±0.13 (11)
Capable of sight	6.7±0.14 (8)	7.5±0.00 (3)	46.9±0.81 (8)	14.3±0.20 (13)
One month	7.8±0.21 (9)	8.7±0.15 (3)	40.5±1.35 (8)	19.5±0.25 (14)
Adult	7.7±0.19 (9)	9.9±0.47 (3)	35.3±0.76 (8)	21.8±0.27 (18)

The number of experiments is shown in parentheses.

and postembryonic development than in the white matter. More
detailed information on age changes in the chemical composition
of the various parts of the brain is given elsewhere [219, 815].

In most species of experimental animals the sharp changes
in the protein and nitrogen content in the tissue of the whole brain
and its parts terminate by the end of the first month of life [75,
905, 1022, 1094, 1206, 1451, 1715]. The rapid decrease in the
water content and the increase in the content of dry matter also
largely come to an end at the same time [754, 905, 1022, 1077,
1127, 1206, 1451].

More marked age changes have been found [1023, 1503] in the
content of "proteolipid protein," a component of myelin, a structure specific to the nervous system. Differences in these changes
in the gray and white matter of the brain are of particular interest.
For instance, the concentration of proteolipid protein (calculated
per wet weight of tissue) in the white matter of the brain is 13
times higher in infants aged 18-23 months than in those aged 1
month, and 1.3 times higher in adults than in infants aged 18-23
months. In the human gray matter, on the other hand, throughout
postnatal development the proteolipid content rises by little more
than twofold. The content of proteolipid protein in the gray and
white matter of the brain differs only slightly (by 1.3 times) in
infants under the age of 1 month; at the age of 18 months this difference has increased to eightfold, and in the adult to as much as
ten- or elevenfold. Approximately the same difference in the
content of proteolipids was found previously in the gray and white
matter of the adult human brain [764, 765].

Two different groups of proteins may be present in the proteolipids in the white matter of the brain. One group is formed during
intrauterine development and is common to both the gray and white
matter, whereas the second is formed intensively, chiefly in the
white matter, during postnatal ontogeny, and it is incorporated into
the myelin structures. This hypothesis is based on observations
[1502, 1503] of different age changes in the content and amino acid
composition of proteolipid protein in the gray and white matter of
the brain. Substantial age changes in the content of individual
protein fractions of white matter myelin from human brain have
also been found [989]. With age, the ratio between basic and proteolipid proteins in myelin increases because of an increase in
the content of the former and a decrease in the content of the latter.

Definite age changes in the composition of the protein fractions in nerve tissue have been found by using various modifications of electrophoretic protein separation [375, 454, 473, 854, 979, 989, 1057, 1358, 1458]. In the course of postnatal development the number of electrophoretic protein fractions in the brain increases. Some redistribution of proteins takes place among these fractions. With paper electrophoresis a new protein fraction migrating toward the cathode appears, and the separation of the protein fractions improves. All these changes reflect the character of the ontogenetic formation of the composition, structure, and physicochemical properties of the nerve tissue proteins. Unquestionably this differentiation of the proteins corresponds to the functions they perform at each stage of individual development. This is confirmed by changes in the other physicochemical properties of the tissue proteins: an increase in resistance to the action of proteolytic enzymes with age, a decrease in their reactivity, an increase in the orderliness of their structure, and changes in other indices of the functional activity of protein molecules [76, 80, 399, 415, 577].

In the course of embryonic and postembryonic development an increase in the concentration of protein S-100, specific to nerve tissue, has been found [1806] in the tissues of the human spinal cord and various parts of the brain. This protein appears sooner in the course of embryonic development in phylogenetically old parts of the brain than in the cerebral cortex and cerebellum. In general its appearance and accumulation parallel the morphological and electrophysiological maturation of the brain structures. In adult animals the concentration of this protein in different parts of the brain is three or four times higher than in newborn animals, and in the cerebellum the increase may be as much as six to eight times.

Much experimental evidence has been obtained to confirm the existence of quantitative and qualitative age changes in the supramolecular complexes formed by brain proteins with nucleic acids, lipids, and other compounds [75, 79, 80, 102, 419, 585]. These changes can be briefly summarized as follows. The relative content of protein and lipids in lipoprotein and nucleoprotein complexes increases with age whereas the content of nucleic acids falls. This agrees closely with the changes in the total content of these compounds in the nervous system. An increase in the content of phosphoinositides is observed with age in the phosphoinositide—

protein complexes, and the intensity of the intracellular metabolic processes reduced.

Substantial age and other functionally determined changes have recently been discovered in the chemical composition [165, 295, 425, 487, 505, 748, 927, 1757], the structural organization [505, 508, 699, 701, 737, 1409, 1757], and the functional activity [737, 748, 927, 1176, 1200, 1757] of the cell organelles. These differences are concerned chiefly with the number, size, and shape of the cell organelles [425, 508, 699, 701, 1409] and also the distribution of protein [505, 748, 927], nucleic acids and free nucleotides [425, 487, 748, 927, 1757], and lipids [295, 487, 1541]. There are also differences in the activity of ATPase and other enzymes [737, 1176] between the fractions of the subcellular structures and in the intensity of respiration and oxidative phosphorylation (505, 508, 737, 927, 1757].

Ontogenetic Changes in the Amino Acid Composition of the Brain Proteins. Age changes in the amino acid composition of the brain proteins were first studied about 30 years ago [47, 73, 198, 631, 838]. Since then more refined and reliable methods of determining amino acids in protein digests have become available in biochemistry. These methods have been used to study the amino acid composition of the proteins of the whole brain and its parts [1765-1767]. More recently data have been obtained for the amino acid content in highly purified nerve tissue proteins of adult animals [300, 304, 479, 892, 992, 1191, 1291, 1307]. However, until recently our knowledge of the processes determining the protein composition of the nervous system during ontogeny remained incomplete, fragmentary, and at times contradictory. The available data reveal only some features of the age changes in the amino acid composition of the tissue proteins of the brain and its parts.

Earlier investigations [73] revealed no significant age changes in the arginine content in proteins of the whole human brain. Some decrease in the arginine content (after 9 years of age) was found [198] in the proteins of the human cerebral hemispheres, with an increase in the cystine content (from birth to the age of 9 years), and a gradual decrease in the tryptophan content with age. The cystine content in proteins of the cerebellum rises quickly until the age of 9 years, whereas its increase in proteins of the cerebral hemispheres is less marked. The arginine and, in particular, the histidine content in rat brain proteins rises considerably with age but the ratio between the arginine concentration and the histidine concentration falls steadily [904, 905]. No such pattern is observed in the guinea pig brain; the histidine content in the proteins of the white matter of the brain and spinal cord falls considerably in ontogeny [1765, 1767].

Numerous investigations [574, 1541, 1765-1767] have shown that brain proteins at all stages of individual development have a high content of glutamine and glutamic and aspartic acids. This distinguishes the proteins of the nervous system essentially from the proteins of other organs. Because of the important role of these amino acids in brain activity a change in their content in the tissue proteins and the free pool during the development of the CNS functions in the course of ontogeny has attracted particular attention as a subject for research. This interest grew considerably after the metabolic link between the dicarboxylic amino acids and GABA had been found and the important role of the latter in brain activity demonstrated. Investigations showed [574] that at the age of 1 month, i.e., when the animal had acquired vision, the content of glutamic and aspartic acids in the proteins of the cerebral hemispheres of rats is rather higher than at the age of 12 days (before the acquisition of vision). The content of these amino acids is lower in the brain proteins of adult rats. However, there is evidence that the content of dicarboxylic and of some other amino acids in proteins of the cortex of the adult mouse is not significantly different from their content in newborn mice [1541].

The most systematic investigations of the amino acid composition of the brain proteins were undertaken by Wender and Waligora [1765-1767]. They determined the content of 13 amino acids by chromatography in the proteins of the gray and white matter of the cerebral hemispheres and also of the spinal cord of guinea pigs at various stages of individual development. Animals of seven age groups were investigated: embryos at the age of 8-9 weeks of development, young animals aged 3, 5, 9, and 12 days and 2 months, and adult animals.

Substantial differences were found in the amino acid composition of the proteins of the various parts of the nervous system and at different stages of individual development, and a parallel was found between the age changes in the content of some amino acids in proteins of the white matter of the brain, on the one hand, and the intensity of myelination on the other hand. The amino acid composition of proteins of the gray and white matter of the brain

display the greatest differences in the stages of embryonic development. Compared with proteins of the gray matter of the brain, the
proteins of the white matter have a higher content of histidine,
serine, and valine but a lower content of aspartic and glutamic
acids, threonine, arginine, tyrosine, phenylalanine, and leucine.
During the first days of postnatal development the changing con-
tents of these amino acids follow different patterns in proteins of
the gray and white matter of the brain. The amino acid composition of proteins of the gray matter does not change significantly
at this time, but in proteins of the white matter there is a marked
increase in the content of glutamic acid, arginine, tyrosine, phenylalanine, and leucine, but a decrease in the content of histidine,
serine, and valine. These changes in the amino acid composition
of white matter proteins coincide with the period of intense myelination of the tissue. Later in postnatal life changes in the amino
acid composition of the proteins are very small, whereas in the
gray matter of the brain the content of cysteine, arginine, and
serine rises significantly.

The most marked changes in the amino acid composition of
the proteins in the white matter of the brain thus take place during
the embryonic and early postembryonic period of development,
whereas in the proteins of the white matter the largest changes
occur later. These differences are probably due to differences in
morphological and functional maturation of the parts of the brain
studied. The white matter of brain and spinal cord differ only
slightly in the amino acid content of their protein structures at
all stages of individual development, which indirectly confirms the
view that age changes in amino acid composition are connected
with morphological and functional maturation.

Evidence of a connection between age changes in the qualitative composition of the proteins of the brain as a whole and its
various parts and the formation of the functions of the CNS in
ontogeny is given by the observations of Dobrynina [199]. The
work of Sadikova and Kudryashova [574], who demonstrated the
appearance of new N-terminal amino acids (tryptophan, lysine,
and leucine) in proteins of the cerebral hemispheres of adult rats,
evidently points to the same conclusion.

To conclude this section the ontogenic changes in the amino
acid composition of the proteolipid protein, regarded as a specific
protein for nerve tissues, are briefly reviewed. The first data
on age changes in the amino acid composition of proteolipids

from human white matter were obtained recently by Prensky and
Moser [1502]. It was later shown [1503] that the amino acid composition of the proteolipids from the gray matter does not change
significantly with age. In amino acid composition the proteolipids
of the gray matter resemble those of brain white matter in embryos
and in infants in the first month of life. The proteolipids of brain
gray matter in children aged 18 months and over differ from those
of the white matter of embryos and infants in the first month of
life by their higher content of phenylalaine and tyrosine and their
much lower content of aspartic acid, proline, and leucine.

The data on age changes in the amino acid composition of
proteins of the brain and its various parts are far from complete
and are sometimes contradictory. Nevertheless, it can be concluded that the differentiation of nerve tissue and rapid growth of
neurons, glial cells, and other brain cells, together with the less
marked changes in the structural formations of brain tissue in
the later stages of postnatal development, are accompanied by a
considerable reorganization of the complement of tissue proteins
and by changes in their amino acid composition and physicochemical and biological properties. The greatest changes in the amino
acid composition of the total brain proteins take place in embryo-
genesis and in the early stages of postnatal development and they
apply primarily to the most important amino acids of nerve tissue.

Free Amino Acids of the Brain in Ontogeny.
Amino acids participate in many functions of the nervous system.
Processes such as the formation of new proteins at the stage of
differentiation and growth of the tissue, protein breakdown and
resynthesis in the course of renewal of the tissue structures, the
formation and elimination of ammonia, the regulation of homeostatic reactions of the brain, and the formation and metabolic
conversions of many biologically active compounds, including neu-
rohormones, are closely bound with the amino acids and their
derivatives. A particularly important role in the activity of the
CNS is played by amino acids and their derivatives such as glutamic, aspartic, and N-acetylaspartic acids, GABA, and glutamine.
Compared with other organs, these compounds occupy a special
place in the brain tissues by virtue both of their high content and
of their important metabolic role [84, 87, 89, 112, 615, 755, 1254, 1340,
1639, 1686, 1730].

The multiple functional role of amino acids has been examined
in several easily available publications [274, 340, 344, 488, 766,

1659, 1730]. Therefore in this section we briefly summarize the information on the free amino acids of the brain and changes in their content during ontogeny, i.e., only the data necessary to assess research into the metabolism of brain proteins at different stages of individual development.

It follows from the investigations cited above that considerable changes in the intensity of biochemical conversions of amino aicds take place in the brain tissue during individual development. It was therefore natural to expect that there would be substantial changes in the concentrations of the brain free amino acid pool with age. Such changes have in fact been found for many amino acids [11, 50, 385, 491, 753, 755, 801, 956, 1246, 1286, 1339, 1384, 1535, 1715]. The data on ontogenetic changes in the free amino acid content in the various regions of the brain have been obtained by methods that differ in their accuracy and sensitivity and therefore are often difficult to reconcile. Nevertheless, some definite conclusions can be drawn from the available material.

It is now reliably established that in animals of different species practically all the amino acids are present in the free form in the tissues of the brain. The concentration of amino acids such as glutamic and aspartic, moreover, is much higher than the concentrations of the other amino acids and is an order of magnitude higher than in the blood plasma [344, 886]. Brain tissue also contains comparatively high concentrations of glutathione and taurine [47, 344, 1659]. Amino acid derivatives such as GABA and acetylaspartic acid are even considered specific for nerve tissue, for they are found in other organs in negligible amounts [1659, 1662]. Glutamine, glutamic acid, and GABA together account for about 45% of the total amino nitrogen in the brain [768]. Each of the amino acids tryptophan, leucine, isoleucine, methionine, phenylalanine, valine, arginine, histidine, tyrosine, and proline account for 1-6% of the free amino nitrogen of the brain tissue [1597]. Free amino acids are present not only in the cell cytoplasm, but also in subcellular structures (nuclei, mitochondria, myelin, synaptosomes) of nerve tissue [1334].

The most substantial changes in the level of free amino acid reserves of the brain with age are those in the content of glutamic and aspartic acids, GABA, and taurine [112, 491, 651, 801, 829, 1286, 1339, 1659, 1715]. The content of these same amino acids changes most in different functional states [112, 130, 344, 359, 362,

766, 957, 1659, 1686]. Most of these free amino acids are formed very intensively in the tissue of the brain, just as in other organs, by intracellular metabolism, and they play a more active part than the other amino acids in metabolic reactions not directly connected with intravital protein renewal [84, 488, 615, 1254].

Most of the increase in the concentrations of glutamic and aspartic acids, GABA, and glutathione in brain tissue is complete by the end of the first month of postnatal life [491, 753, 801, 829, 1339, 1715]. However, in animals of different species these amino acids reach maximal levels at different times [755, 824]. Differences observed between species are evidently due to differences in the time of development of the cells and the formation of the brain functions. The maximal increase in the concentration of these compounds takes place during the most rapid development of the bodies, dendrites, and axons of the neurons. Significantly, the ammonia concentration in brain tissue also undergoes the greatest changes during the first month of postnatal development [149, 670]. During this period the ammonia concentration falls steadily, and according to the authors cited this is evidence of the formation of ammonia-fixing systems of the brain. However, other investigators [492] found that the content of free ammonia in the brain tissue increased with age.

A high concentration of glutamine is also found in the brain tissue and it changes insignificantly in the postnatal period of development [149, 727, 801, 824, 829, 956, 1127, 1286, 1659]. The age stability of the glutamine concentration is difficult to explain, especially in the early stages of postnatal development of the brain, when the concentrations of glutamic acid, GABA, and glutathione, linked metabolically with glutamine, change substantially. It has been claimed [818, 824, 825, 827, 830, 1254, 1735] that brain tissue contains an "active" pool of glutamic acid located mainly in the bodies of the neurons and that the level of the amino acid in this compartment changes only slightly with age. The amino acid of this pool is also exchanged very slowly with the amino acid of the general tissue glutamic acid pool. During ontogenetic development a less active compartment or pool of glutamic acid is formed in the brain and increases in size. Glutamine is metabolically linked chiefly with the active pool of glutamic acid.

It can be concluded from recent experimental investigations [1039, 1040, 1281, 1459, 1460, 1506, 1612] that several other amino acids, as well as certain metabolites of the Krebs cycle, also have spatially separate compartments in the brain. Other mechanisms responsible for the stable level of glutamine in the brain tissue during individual development may also exist. A key role in all these mechanisms must be played by glutamine, with its participation in the fixation of ammonia and its involvement in tissue metabolism. This role of glutamine, investigated in various tissues [586, 651-654, 656, 1225, 1226], is particularly important for the normal functioning of cells of the nervous system [128, 149, 260, 277].

Amino acids such as alanine, glycine, lysine, serine, and threonine, which are present in brain tissues in much smaller concentrations than glutamic and aspartic acids, exhibit less marked ontogenetic changes of concentration [491, 1659, 1715]. Many of the amino acids listed are metabolically less active than glutamic and aspartic acids and their concentrations thus vary only slightly in different functional states of the organism.

Analysis of the results cited above permits a general conclusion that the content of glutamic and aspartic acids and their derivatives, and also of tyrosine in the brain of experimental animals, rises gradually after birth, then remains at the maximal level (with slight fluctuations), and falls slightly in old age. The concentrations of these amino acids reach maximum levels mainly in the early stages of postnatal development, i.e., at the end of the period of intensive morphological, functional, and biochemical maturation of the brain. The concentrations of the other amino acids are less subject to age changes and do not follow a general pattern. Age changes in the concentrations of most amino acids in the tissues of the brain and other organs have been described in more detail by Parina [488].

Age Changes in Nucleic Acids in the Brain Tissue. There is abundant factual evidence in the literature [69, 75, 264, 321, 350, 425, 566, 596, 747, 857, 858, 1094, 1341, 1406, 1451, 1640] to show that the content and metabolism of nucleic acids in the brain tissue undergo substantial changes during the embryonic development of the animal, in the course of its postnatal development, and during tissue regeneration in the adult. In the embryonic period and in the early stages of postnatal development the highest concentrations of RNA and DNA in brain tissue are found. Later the concentrations of RNA and DNA in the brain, calculated per wet weight of tissue or per weight of dry substance, per nucleus, or per cell nucleus, fall steadily, the decrease being more marked for DNA [75, 350, 596, 1341, 1451, 1466]. As a result of this, the DNA/RNA ratio falls sharply with age and the RNA/protein ratio falls less than the DNA/protein ratio. The most marked changes in these indices take place during the first month of life [350, 596, 1451, 1466]. The level of nucleic acids characteristic of the adult organism is established in animals of different species after different periods of postnatal development; these periods coincide with the times of morphological and functional maturation of the brain [321, 350].

Different parts of the brain differ significantly in their content of nucleic acids [350, 1341]. Their RNA and DNA content corresponds to the morphological and functional features and also to the intensity of the biochemical processes, and it changes at different rates with age. As DNA content falls rapidly, the RNA/DNA ratio rises with age and the degree of this rise differs sharply in different parts of the nervous system.

The nucleotide composition of the RNA and DNA of the brain and its parts does not change significantly in the period of postembryonic development of animals [592, 596]. The content of free nucleotides, on the other hand, undergoes substantial age variations [967, 1330, 1336, 1755, 1757], and these correspond generally with changing metabolic activity. In the chicken brain, for example, the total content of free nucleotides falls throughout the postembryonic period of development. The sharpest decrease, nearly 4% daily, is observed between the 6th and 10th days of life [1336]. Sharp changes in the content of free nucleotides in rat brain tissue are observed at two stages of postnatal growth. In the first days after birth there is a sharp rise in the content of nucleoside triphosphates, and in the 4th week of development a critical decrease in the content of UTP and even more so of CTP. A decrease in the CTP content is also found during the embryonic development of the chick. CTP is regarded as a factor which might limit protein synthesis in the brain tissue [1755].

A more detailed analysis of the extensive literature confirming the close connection between the intensity of metabolic conversions of nucleic acids and the functional activity of the neurons is given by Pevzner [496].

Nerve Tissue Lipids in Ontogeny. In the comparatively early stages of research a structural role was usually ascribed to the lipids of nerve tissue. Their participation in vital metabolic functions and in the transport of metabolites through cell membranes has now become apparent [244, 289, 291, 294, 590, 660, 673, 1119]. The lipids in nerve tissue play a particularly important role in the adult brain, since they constitute almost half, and in some parts more than half, of the dry weight of the tissue.

The total content of lipids in nerve tissue rises steadily during ontogeny. The rise is particularly rapid in late embryonic and early postnatal development because of the rapid growth of the tissues, the rapid development of the dendrites and axons, and the intensive myelination [293, 295, 495, 761, 798, 1022, 1140, 1451, 1709, 1763]. The content of individual lipids varies unequally with age [290, 295, 1277]. In the adult rat brain, for instance, the content of inositides, calculated per wet weight of tissue, is about twice the content in 3-day-old animals while the content of phospholipids is 3.5 times, that of acetylphosphatides 6 times, and that of sphingomyelin 10 times higher than in the young [1027, 1028]. In mouse brain tissue in the postnatal period of development the content of total lipids increases almost fourfold, and that of proteolipids, sixteenfold [1022]. Some lipid components (diphosphoinositide, sphingomyelin, cerebrosides, etc.) are absent in the early stages of individual development and appear only in the last stage of embryonic development or in the first weeks of postnatal life, in connection with their role in myelination [494, 761, 798, 1541, 1709].

A special pigment, lipofuscin, containing 30% more lipids than the rest of the cytoplasm, appears in the nerve cell cytoplasm in the postnatal period of development and increases in amount [673, 1510]. This pigment includes in its composition insoluble proteins which are metabolically relatively inactive [1609].

The degree of the age increase in brain lipid content is considerably higher than that for proteins [373]. As a result of this, the content of total lipids and their separate components per unit weight of tissue protein rises sharply with age. The intensity of this age increase in the lipid concentration differs in different parts of the nervous system. The increase is more rapid in the brain stem, the white matter of the cerebral hemispheres, and the spinal cord [761, 1022, 1127] than in other regions.

The changes in the lipid content in animals of different species occur at different times and parallel the morphological and functional maturation of the brain structures. The most intense changes occur in the early stages of postnatal development [1027, 1028, 1451, 1709].

Changes in Brain Protein Metabolism
in Postnatal Ontogeny

Intensity of Protein Renewal. Before the method
of radioactive isotopes was used on a wide scale in biochemical
research, the investigations of Nagornyi, Nikitin, Bulankin, and
their collaborators [395-397, 413-417] revealed the general direction of the ontogenetic changes in the protein synthesis and
breakdown which are the basis for renewal of cell structures.
The results of these and other investigations are summarized in
recent monographs [399, 488] and there is therefore no need to
describe them here. It will suffice to state that the general rule governing age changes in the ratio of synthesis to breakdown of tissue
components was established by these investigations. This rule
states essentially that in the early stages of individual development
synthesis predominates over breakdown. Hence there is a steady
increase in the mass of the protein structures in the developing
organism.

This predominance diminishes with age due to a continuous
decrease in the rate of biosynthesis and a much less marked increase in the intensity of protein breakdown. In the later stages
of ontogeny the intensity of these mutually interdependent phases
of protein metabolism gradually becomes equalized and in old
age the level of synthesis falls below that of breakdown. This age
pattern is typical of all the components of the cell, including
protein structures. The increase or decrease in the content of
tissue proteins in the various stages of ontogeny is therefore
determined not so much by their absolute rates of synthesis and
breakdown as by the relative intensities of these processes.

These general rules governing age changes in protein metabolism established in the course of earlier research were subsequently confirmed and extended in investigations in vivo and in
vitro with the use of amino acids labeled with radioactive and
stable isotopes.

Orekhovich et al. [435] obtained conclusive evidence that the
intensity of renewal of the tissue proteins is higher in newborn
rats than in the fully grown animals. The original experimental
technique used, feeding the animals on heavy water for several
days, excluded for all practical purposes any effect of the BBB on
the admission of the label to the organs tested and thus enabled
the true rates of renewal of the tissue proteins to be determined
in animals of different ages. The rate of renewal of proteins in the

brain, muscles, skin, and internal organs in normal adult rats
and in the mothers of newborn rats was found to be much lower
than in the offspring. In similar experiments Davydova et al. [182]
also found a higher intensity of renewal of total brain proteins in
young animals.

Somewhat later Gaitonde and Richter [1044] showed, by intracisternal injection of methionine-^{35}S into 15-day-old albino
rats 3 h before sacrifice, that the SR of the brain proteins is considerably higher than in adult animals. The value of RSR for the
brain proteins was also higher in the brain of the young animals,
although the sulfur content in the acid-soluble fraction was almost
one-third higher than in the adult animals [1043].

Similar results were obtained by Panchenko et al. [308, 481,
483] when they studied the intensity of protein renewal in different
parts of the brain of kittens aged 1 month and of adult cats into
which methionine-^{35}S was injected subcutaneously. The SR of the
proteins in the parts of the brain studied, except the cerebral cortex, were higher in the young animals.

In these investigations data on age changes in the intensity of
renewal of the total brain proteins were obtained on animals of
two age groups. The character of the changes in the intensity of
renewal of the brain proteins in the intervening stages of postnatal
development remained unstudied. Work was accordingly carried
out in Palladin's laboratory to determine [446, 454, 459] the intensity of renewal of the total (unfractionated) brain proteins of
rabbits at four stages of postnatal development: newborn, just
acquiring vision (11-12 days), 1 month, and adult (6-7 months).
These age groups were chosen to obtain comparable data on the
intensity of renewal of brain proteins at stages of postnatal development of the animals characterized by substantial quantitative and
qualitative changes in the morphological substrates, the chemical
composition, the physiological functions, and the biochemical processes of the tissues [197, 264, 366, 399].

The results of these experiments (Fig. 16), expressed as RSR*
values for the brain proteins, confirmed the general rule that the
intensity of renewal of cell structures decreases with age [459].
The marked age decrease in the intensity of renewal of the total
brain proteins terminates toward the end of the first month of life,

---

*$$\text{RSR} = \frac{\text{pulses}/\text{min}/\text{mg protein sulfur}}{\text{pulses}/\text{min}/\text{mg brain ASF}} \times 100.$$

i.e., at a time when the postembryonic formation of the morphological structures and growth of the cellular elements of rabbit nerve
tissue are largely complete [197]. After the age of 1 month the
changes in the intensity of renewal of the brain proteins are minimal. For instance, whereas the mean daily changes in RSR of the
brain proteins calculated from data in Fig. 16 for the first and
second consecutive 10–day periods of postnatal brain development
are 2-4%, after the age of 1 month they fall to 0.1-0.2%. The
changes in the intensity of renewal of the total brain proteins in
postnatal ontogeny are in good agreement with age changes in the
protein content in nerve tissue (Table 8).

The age decrease in the intensity of renewal of total brain
proteins is more rapid in experiments in which the label remains
for a longer time in the body of the experimental animals (Fig. 16).
For instance, whereas in experiments lasting 2.5 h the mean
diurnal decrease in RSR of the brain proteins calculated for a
period of 7 months is 0.12%, in experiments lasting 14 and 24 h
it is 0.30 and 0.46%, respectively. The provisional conclusion can
be drawn from these findings that the age decrease in the intensity
of renewal of the total brain proteins is determined chiefly by a
decrease in the rate of synthesis and also, possibly, by an increase
in the content of metabolically more stable protein structures.
This conclusion was subsequently confirmed by the investigation

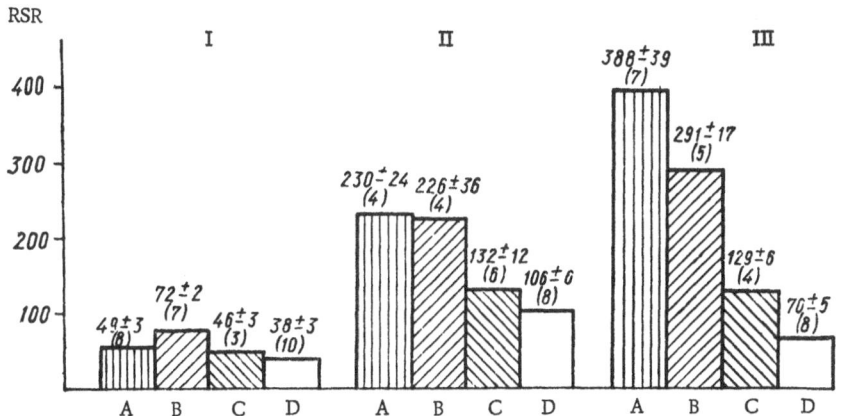

Fig. 16. RSR of total brain proteins of rabbits of different ages: I-III) duration of experiments 2.5, 14, and 24 h, respectively; A) newborn, B) just acquiring vision, C)
aged 1 month, D) adult animals.

of age changes in the metabolic activity and content of proteins in subcellular brain fractions.

In experiments lasting 2.5 h (Fig. 16) on rabbits just acquiring vision some increase in RSR of the brain proteins was observed compared with that in newborn animals. This was probably due to activation of protein metabolism in the period of emergence of the new, visual function. In experiments lasting 14 and 24 h, in which not only metabolically active, but also slowly renewed protein structures are labeled, no increase is found in RSR of the brain proteins in rabbits recently acquiring vision. Schreier [1596] also observed activation of the biosynthesis of protein and nucleic acids in the rat brain on the 10th day of postnatal development. In connection with these results it is pertinent to mention the observations of Sadikova and Kudryashova [574], who found the highest concentration of dicarboxylic amino acids in rat brain proteins in the period after acquisition of vision.

Some interesting results from this point of view were obtained by Brattgard [856], who studied changes in the ganglionic cells of the retina during the development of rabbits over a 10-week period under normal conditions and in darkness. Under normal conditions, when the retina was exposed to photic stimulation, the mass of ganglionic cells increased during this period by 100%, whereas in darkness these cells did not develop. Cells not receiving adequate excitation contained only half as much protein and negligibly small amounts of ribonucleoproteins. The earliest investigations demonstrating a connection between nerve cell functions developing in ontogeny and biochemical processes, notably protein metabolism, were carried out by Gorodiskaya [167]. She observed an increase in the intensity of proteolysis in the visual centers and tracts of the brain during exposure of the retina to photic stimulation and a decrease in the intensity of proteolysis in the absence of such stimulation. Interesting results were also obtained in Palladin's laboratory by Fedorov [646]. He found (Fig. 17) that the activity of inorganic pyrophosphatase in the brain tissue of young rabbits just acquiring vision is 30-50% higher than in embryos and newborn rabbits and 120-130% higher than in 1-month-old and adult rabbits. All these facts demonstrate the close connection between the biochemical properties of nerve cells, including their protein metabolism, and their functional state.

The most significant age changes in the intensity of renewal of the total brain proteins thus take place during the first month

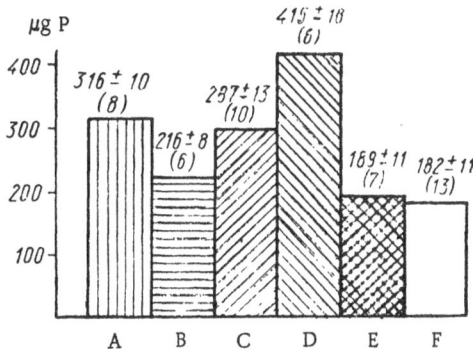

Fig. 17. Activity of inorganic pyrophosphatase (µg phosphorus removed during incubation at 37°C for 30 min at pH 7.8, per mg protein) of rabbit brain tissue at different stages of individual development: a) 20- to 23-day embryos, B) 26- to 29-day embryos, C) newborn, D) just acquiring vision, E) aged 1 month, F) adult.

of postnatal life, i.e., at a period of development that is characterized by the greatest changes in chemical composition, intensive development of the cell structures, and the formation of enzyme systems and physiological and biochemical functions [197, 292, 366, 399, 1451].

The data described above are compatible with the results of other investigations in which methionine-^{35}S and ^{14}C-amino acids were used to study protein renewal. Richter [1520, 1521] found a constant decrease with age in the rate of renewal of total rat brain proteins. A particularly rapid decrease was observed during the first 1.5-2 months of postnatal development. Since methionine-^{35}S was injected intracisternally in these experiments, the observed age decrease in the intensity of protein renewal could not be attributed to a decrease in the permeability of the BBB for the amino acid during ontogeny. Recent experiments have shown [1371, 1645, 1646] that the intensity of incorporation of glycine-^{14}C and lysine-^{14}C into proteins of the CNS and PNS of different species of animals also decreases with age. A particularly sharp decrease was found in the peripheral nerves. In the nerve tissue of growing chickens and rats, for example, the rate of protein synthesis · is 4 and 10 times higher, respectively, than in adult animals [1371].

The decrease in the renewal rate of brain proteins with age was also confirmed by studies of the half-life of the proteins. For example, the rate of renewal of total brain proteins in mice aged 10 days was found to be 2.0-2.5 times higher than in animals aged 4 months [1254, 1255]. In rats aged 1 and 24 months the half-life of the total brain protein is 1.21 and 3.66 days, respectively [79]. Substantial age differences in the incorporation of amino acids into the structures of the nervous system have also been discovered [1024, 1162] by autoradiography.

The highest intensity of metabolism of phosphoinositide − protein complexes in the rat brain is found at the age of 2 to 3 weeks, after which the intensity decreases.

Age changes in the activity of the biochemical systems of the brain for synthesizing amino acids from glucose [1009, 1508, 1541, 1620, 1621, 1781], although not entirely consistent, nevertheless indicate a decrease in the synthetic capacity of nerve tissue during ontogeny. It has even been postulated, on the basis of these findings, that some enzyme systems connected with the protein metabolism of nerve tissue function in the early stages of individual development during the period of rapid differentiation and growth and then disappear as the intensity of these processes declines [1738]. However, the total activity of the aminoacyl RNA-synthetases of the brain tissue of albino rats does not change with age [490].

The higher intensity of protein synthesis in the brain tissue of young animals has been confirmed in vitro. Greenberg et al. [1080] first found higher incorporation of labeled glycine by brain homogenates from embryonic and young rats than by brain homogenates from adult animals. Schepartz [1588] investigated the incorporation of 13 labeled amino acids by brain homogenates from mice aged 1 day and 7 weeks and found that most amino acids are incorporated rapidly into the brain proteins of the younger animals. The intensity of incorporation of glycine-^{14}C into proteins of the subcellular brain fractions from young (15-day) rats was much higher than in adult animals. However, the age differences almost disappeared if a system generating adequate amounts of ATP was added to the incubation mixture instead of preparations of the mitochondria [1200].

A higher rate of incorporation of labeled amino acids (leucine-^{14}C and valine-^{14}C) was also found in nonmitochondrial preparations (microsomes, ribosomes, soluble fraction) of the whole brain of newborn rats and rabbits compared to similar preparations from adult animals [845, 1412]. The intensity of incorporation of phenylalanine-^{14}C and ^{3}H-amino acids by nonmitochondrial mouse brain preparations shows a threefold decrease during the first 17 days after birth [1169]. This decrease is not due to the permeability of the cell membranes and it is evidently under the control of intracellular mechanisms. Cell-free preparations of the brain from young rats incorporate leucine-^{14}C more rapidly than the same preparations from the brain of adult animals [1054]. Kinetic studies showed [748] a higher synthetic ability of microsomal preparations from the brain from 4-day-old (weight about 7 g) and weanling rats (weight 50 g), reflected in the incorporation of valine-^{14}C, than for preparations from adult animals.

Other investigations [775, 875, 1171, 1370, 1394, 1409, 1464, 1625] on homogenates, tissue slices, cell suspensions, and subcellular fractions from tissues of the CNS and PNS demonstrated decreases in the rate of incorporation of labeled amino acids into proteins with age.

Suzuki et al. [1642] observed substantial age differences in the incorporation of a labeled amino acid into proteins of the white matter of the brain but found no such differences in the gray matter. Lim and Adams [1295] observed a higher intensity of incorporation of labeled amino acid in vitro into protein of ribonucleoprotein particles of the brain of adult animals than of those 4 days old.

Recent experiments in vitro showed [1764] that the activation of cysteine, observed in homogenates of the gray and white matter of the guinea pig brain during the last days of embryonic and the first days of postnatal development, disappears completely after the 9th day of life. The rate of activation of glutamic acid by these preparations, on the other hand, increases considerably with age.

Experiments in vitro [1792] showed that the age decrease in the intensity of incorporation of amino acids into brain proteins may be due, at least in part, to a decrease in the activity of enzymes activating amino acids, and also to functional disturbances of the ribosomal systems and of nuclear RNA. In this connection, as experiments in vivo [15, 747] and in vitro [1086, 1168] have shown, the biosynthesis of nucleic acids in brain tissue also decreases with age, particularly rapidly in the early stages of postnatal development. These observations correlate with the marked age decrease in the content of soluble RNA in the rat cerebral cortex [747]. The ontogenetic changes observed in nucleic acid metabolism are probably one cause of the age decrease in the intensity of renewal of the total proteins of nerve tissue.

One should not overlook the possibility that the activity of enzymes activating amino acids is not the limiting factor in the renewal of brain proteins in the adult organism, for there is no complete parallel between the intensity of incorporation of labeled amino acids into proteins of different parts of the CNS and the concentration of activating enzymes in them [1653].

The results of these investigations and those obtained by other workers using labeled amino acids in experiments in vivo and in vitro confirmed the earlier view that the intensity of renewal of the total proteins of brain tissue diminishes with age.

A different and rather unexpected dependence of brain protein renewal on age was discovered by the present writers [31, 457, 1471] in the course of an investigation of the incorporation of methionine-^{35}S into proteins of subcellular brain fractions from rabbits of different ages: newborn, just acquiring vision, aged 1 month, and adult. The principal criteria of the rates of renewal of proteins of the subcellular structures in vivo were the values of SR and RIR. This latter parameter expresses the ratio between SR for the proteins of each subcellular fraction and SR of the proteins of the original brain homogenate, as a percentage. By expressing the results in this way it was possible to eliminate the differences between the figures for the intensity of renewal of the tissue proteins in vivo caused by age differences in the content of the labeled amino acid in the brain free amino acid pools.

The intensities of protein renewal in the various subcellular fractions follow different age patterns (Table 9). For instance,

TABLE 9. Specific Radioactivity (SR) and Relative Intensity of Renewal (RIR) of Proteins of Subcellular Fractions from Brain Tissue of Rabbits of Different Ages 2.5 h after Subcutaneous Injection of Methionine-^{35}S (M ± m)

Age groups of animals	Parameters	No. of exp.	Nuclear	Mito-chondrial	Micro-somal	Soluble	Washings
Newborn	SR	11	303±25	218±20	301±22	408±37	284±25
	RIR	11	98±3	70±2	98±3	131±4	98±5
Just acquiring vision	SR	12	323±42	202±28	317±40	363±40	245±32
	RIR	12	123±4	76±2	121±4	142±5	94±3
1 month	SR	8	408±30	289±20	521±32	445±32	205±19
	RIR	9	132±2	94±4	173±4	145±2	66±2
Adult	SR	12	487±40	363±26	724±54	529±36	202±1
	RIR	12	139±3	106±2	213±7	156±4	60±24

the renewal rates of proteins in the principal subcellular fractions (nuclear, mitochondrial, microsomal, and soluble cytoplasmic) increase significantly with age. The degree of this increase varies in the different fractions. The smallest increase in the intensity of protein renewal is observed in the soluble fraction, and the largest in the microsomal fraction. The intensity of renewal of proteins of the "washings" fraction, on the other hand, decreases with age. It will be recalled that the "washings" fraction consists of supernatants obtained during the repeated washing of the original nuclear fraction in 1 M sucrose solution (Fig. 10). This fraction is very heterogeneous morphologically; it mainly contains the slowly renewed tissue structures, such as cell membranes, myelin fragments, nerve fibers, unbroken cells, and mitochondria.

The most marked age changes in the intensity of protein renewal in the rabbit brain fractions studied takes place during the first month of life (mean daily measurements of RIR over that period). Later these changes become less marked. The highest degree of increase in the rate of protein renewal in the nuclear and soluble fractions is observed during the first 11-12 days of life, whereas in the mitochondrial and microsomal fractions it continues during the first month of life. The greatest decrease in the rate of protein renewal in the "washings" fraction takes place during the first 10-30 days of postnatal development.

These investigations yielded the somewhat unexpected result that the intensity of protein renewal in some subcellular brain fractions rises significantly with age. Accordingly values of RSR were calculated for the proteins of each subcellular fraction. In order to calculate this index, which is the most objective measure of the intensity of protein renewal [602, 1044], values of the SR of sulfur in the brain free amino acids were obtained for each age group. In these experiments a study of age changes in the rate of renewal of total brain proteins [459] was used though a certain degree of arbitrariness was involved. The intensity of protein renewal of the principal subcellular fractions, reflected in the RSR values, was found to increase during the first month of postnatal life. In adult animals the RSR values of proteins of the microsomal, nuclear, mitochondrial, and soluble fractions were also considerably higher than the corresponding values for newborn rabbits, but about the same as (microsomal fraction) or even lower than (nuclear, mitochondrial, and soluble fractions) those for animals aged 1 month.

Do the results of investigations of age changes in the rate of protein renewal in the subcellular fractions contradict the earlier data for total brain proteins? To answer this question one must remember that a considerable age decrease in the intensity of protein metabolism was discovered in the "washings" fraction. It is also necessary to consider age changes in the protein content in the subcellular fractions studied (Table 10). With age, especially during the first months of life, the protein content was found to decrease in the chief subcellular fractions (nuclear, mitochondrial, microsomal, and soluble) and to rise sharply in the "washings." If the total protein content in the principal subcellular brain fractions at each age group is taken as unity, the protein content in the washings is: in newborn animals 0.4, in those just acquiring vision 0.5, at the age of 1 month 1.0, and in adult animals 1.1.

The results of the earlier investigations of Bulankin et al. [83, 585] also show that the content of so-called metaplasmatic proteins, i.e., metabolically stable supporting elements of cell structures, increases in rat brain with age. The ratio between the structural and metabolically active soluble proteins in the tissues changes with age in favor of the less active proteins. Gaitonde et al. [1042] recently also found a sharp age decrease in the content of metabolically active water-soluble proteins in rat brain tissue and an increase in the content of proteolipid protein and insoluble proteins, which are metabolically more stable.

The intensity of renewal of the metabolically more active proteins participating in the numerous reactions of intracellular metabolism thus increases with age whereas their level in the

TABLE 10. Distribution of Protein (percent of total protein content in all fractions for each age group) among Subcellular Fractions of Brain Tissue of Rabbits of Different Ages (M ± m)*

Age groups of animals	Nuclear	Mito-chondrial	Micro-somal	Soluble	Washings
Newborn	14.4±0.44	25.6±0.54	4.0±0.30	28.0±0.32	28.0±0.83
Starting to see	14.5±0.39	22.6±0.26	3.6±0.19	26.0±1.32	33.2±0.93
1 month	6.6±0.10	19.6±0.07	3.3±0.00	20.1±0.00	50.4±0.16
Adult	5.9±1.07	18.9±0.81	2.9±0.15	19.5±0.35	52.9±1.59

* Mean of 3 experiments.

brain tissue decreases. The rate of renewal of the metabolically less active proteins, responsible chiefly for structural functions in the cell, on the other hand, falls with age whereas their content in the cell rises sharply. These observations show that the general rule of a decrease in the intensity of renewal of total brain proteins with age established previously is explained by the appearance of slowly metabolized protein structures in the course of ontogeny and a sharp increase in their content. Considering that the brain, a continuously and actively functioning organ, loses many of its nerve cells irreversibly in postnatal ontogeny [53, 66, 215, 331, 762, 864, 974, 1082], the activation of protein renewal of the nuclear, mitochondrial, microsomal, and soluble fractions observed with age can be regarded as a unique mechanism of compensation for the functions of the dying cells. In this connection allowance must also be made for the possible conversion of a proportion of the diploid neurons into tetraploid, and sometimes octaploid, neurons [66, 1121, 1269]. Polyploidization is accompanied not only by an increase in the DNA content in the nuclei, but also by a substantial increase in size of the cells and an intensification of their functions. Characteristically, the number of polyploid cells in the tissue of the brain and other organs rises with age [66]. The fact that most polyploid nerve cells are found in the cerebral cortex, moreover, correlates with the level of morphological and functional differentiation of nerve tissue [581]. Compensatory hypertrophy of the bodies of the nerve cells (and, in particular, of their dendrites) takes place in response to the death of other neurons [215] and also during their active function [286].

An important conclusion which can be drawn from the results shown in Tables 9 and 10 is that the activation of protein metabolism found by the study of the rate of renewal of total brain proteins in animals just acquiring vision is due chiefly to the increased rate of protein renewal in the soluble, mitochondrial, microsomal and, in particular, the nuclear fractions. The latter accounts for the greater part of the increase in the intensity of renewal of the total brain proteins during the first 11-12 days of postnatal life. It is in this fraction that the greatest age increase in the intensity of renewal of proteins, without any accompanying change in their percentage content, is observed, whereas in the mitochondrial and soluble fractions the protein content falls and the degree of age increase in the intensity of incorporation of labeled amino acid rises less than in the nuclear fraction.

The intensity of incorporation of serine-^{14}C into proteins of the subcellular brain fractions of rats aged 3 and 12 days, measured 19 h after injection of the label, was investigated by Abdel-Latif et al. [738]. They found that the specific radioactivity of the proteins of all subcellular fractions studied (nuclear, mitochondrial, microsomal, soluble, and the fraction of nerve endings) is higher in younger animals. The myelin fraction, in which the specific radioactivity of the proteins is much higher at the age of 12 days, is an exception.

A superficial examination of these data gives the impression that they contradict the results of our investigations (Table 9). However, if the results of Abdel-Latif's investigations are to be compared with ours, some essential differences in the methods must be considered. These include the unequal duration of the experiments (19 and 2.5 h), the different amino acids used (serine-^{14}C and methionine-^{35}S), the method of administration (intracerebral and subcutaneous injection), the species (rats and rabbits), and the ages compared (the 3rd and 1st day after birth). The 3rd day of postnatal development of these animals is "critical" for several metabolic indices [1451]. Furthermore, in our experiments, in three of the subcellular brain fractions (mitochondrial, soluble, washings), which together contain more than 80% of the tissue protein (Table 10), the value of SR of the proteins was lower in the rabbits aged 12 days than in the newborn rabbits. Hence, if these technical differences are taken into account when the results are analyzed, there is no contradiction between the findings.

No data corresponding to our own for age changes in the protein content in all subcellular brain fractions at different stages of postnatal development could be found in the literature. However, the observations published by various authors for certain fractions and for particular age groups are in general agreement with the results described above. For example, a steady decrease with age has been found in the number of polysomes in the cells [944, 1409], in the protein content in the microsomal and pH 5-soluble fractions [1412], in the total nitrogen content in the soluble fraction of whole rat brain tissue [1176], and also in the content of protein and RNA in the microsomal fraction of the rat cerebral cortex [748].

Information given in the literature on the mitochondrial fraction of brain tissue is somewhat contradictory. Dahl et al. [927], for instance, found that the concentration of mitochondrial protein in the rat brain does not change during the first 5 days after birth, then rises until the 21st day, and remains at this high level until the 50th day of postnatal life. Jordan et al. [1176] found a higher total nitrogen content in the mitochondrial fraction of the adult rat brain than in rats at the age of 4 days. However, these workers are inclined to attribute these differences to the higher concentration of free nitrogen in the brain mitochondria of adult animals.

According to Pigareva [505], the lowest protein concentration in the mitochondria of the rabbit cortex and brain stem is found in adult animals, and the highest at 15-30 days of age. Gregson et al. [1082] found that the number of mitochondria in rat brain tissue, calculated per cell, almost doubles between birth and the adult stage.

The increase in the quantity of slowly metabolized protein structures in the brain tissue observed with age agrees well with the results of investigations by Thompson and Ballou [1678], Buchanan [871], and Lajtha and Toth [1263], who found a high content of metabolically stable proteins in the tissues of the nervous system of adult animals. Other observations [1023, 1042, 1503] show an increase with age in the brain content of proteolipids, which are known to have an exceptionally low rate of renewal [941, 1022]. It must be emphasized that the observed age changes between the amounts of rapidly and slowly metabolized proteins in the nerve tissue correspond to morphological transformations taking place in the brain tissue during postnatal ontogeny. The age changes in the biochemical properties of the brain enumerated in this section occur in animals of different species on different scales. That is, the changes take place at different times corresponding to the patterns of individual development of the particular species [292, 344, 399, 1206, 1451].

In conclusion, the divergent course of the age changes in the intensity of incorporation of labeled amino acids into proteins of subcellular fractions in experiments in vivo and in vitro is probably due to the considerable differences in the metabolic situations existing in the experiments concerned. It must be especially emphasized that even in analogous experiments, the greatest care must be exercised when results obtained in vitro are assessed with a view to their extrapolation to the intact organism.

The Synthetic Capacity of Nerve Tissue. One
of the surprising features of neurons is that, unlike other somatic
cells, they very quickly lose their ability to divide by mitosis and
many of them thus function in the body virtually throughout life.
A second and no less important distinguishing feature of the neuron is that it is the largest cell in the body. Furthermore, the
numerous dendrites and the axon, with a length attaining 1 m or
more in man and in large animals, make the neuron a functional
unit of such complexity that its long life span can be maintained
only by virtue of its well-developed synthetic systems, providing
for the continuous synthesis and renewal of all its structural
components, including proteins.

There is much factual evidence in the literature of the existence of this synthetic system in the nerve cell. It includes: (1)
the presence of a large nucleus with well developed nucleoli, whose
important role in synthetic processes, particularly in the synthesis of protein and nucleic acids, is well known, [67, 257, 286, 331,
347, 366, 671, 673]; (2) the high nucleo-cytoplasmic ratio in nerve
cells and the negligible change in this ratio with age compared
with the cells of other organs [47, 67]; (3) the presence of a well-
developed endoplasmic reticulum, rich in ribosomes [46, 64, 67,
257, 347, 1566]; (4) the relatively high concentration of nucleic
acids in nerve tissue and its maintenance at a high level throughout
the postembryonic development of the brain [69, 170, 350, 596,
1094]; (5) the extremely high intensity of respiration of the brain,
where more than 20% of the oxygen utilized by the organism is
taken up by the brain of adult animals, and even more in the early
stages of individual development [344, 399]; (6) and finally, the long
functional life of neurons whose protein structures are in a permanent dynamic state [127, 214-216, 399].

However, all these facts are presumptive rather than absolute
evidence of the high metabolic activity of the proteins and other
components of the nerve cell and of the existence of a powerful
synthetic system in it. The first experimental data contradicting
the earlier notion of the metabolic inertia of the brain proteins
[1081, 1669] were obtained mainly in experiments on adult animals,
into which labeled amino acids were injected intracisternally, i.e.,
bypassing the BBB [1030, 1044, 1520]. In experiments performed
in this way the level of radioactivity of the nerve tissue proteins
was found to be much higher than the level of radioactivity of other
tissue proteins. However, this is an unnatural way for a labeled

amino acid to enter the brain, and under these conditions the other organs used for comparison are separated from the labeled compound by the same barriers acting in the opposite direction. In this case, therefore, it would be wrong on the basis of the SR value to conclude that the metabolic activity of the brain proteins is higher than that of proteins in other organs. The low level of label in the proteins of those organs could be due to the slower entry of labeled precursor into the free amino acid pools of that organ rather than to a low synthetic capacity of the intracellular systems of the tissue studied.

The results of these experiments provide firm evidence of the existence of highly active mechanisms of protein synthesis in the brain tissue. Although cisternal puncture evidently has no significant effect on the rate of protein synthesis in the brain tissue [1387], it does appreciably increase the intensity of protein breakdown [21].

Convincing evidence of the comparatively high rate of renewal of the brain proteins was obtained in Palladin's laboratory [31] in which labeled amino acids entered the tissues by the natural route, i.e., from the blood stream. The intensity of renewal of total proteins of the homogenate and proteins of the subcellular fractions of the brain was compared with that of the blood serum proteins. The results were expressed as RIR of the proteins, the ratio between the specific radioactivity of proteins of the subcellular brain

TABLE 11. SR and RIR of Total Brain and Blood Serum Proteins of Rabbits of Different Age 2.5 h after Subcutaneous Injection of Methionine-^{35}S (M ± m)

Age groups of animals	SR of proteins, pulses/min/mg nitrogen		RIR of proteins
	brain	blood serum	
Newborn	312±29.6 (11)	178±14.4 (11)	56±3.5 (10)
Just acquiring vision	263±33.4 (12)	183±14.4 (11)	75±5.3 (10)
1 month	304±23.1 (8)	464±28.8 (8)	149±9.7 (7)
Adult	339±21.6 (12)	457±23.4 (10)	143±6.5 (8)

$$RIR = \frac{\text{SR of blood serum proteins}}{\text{SR of total brain proteins}} \times 100.$$ Number of experiments shown in parentheses.

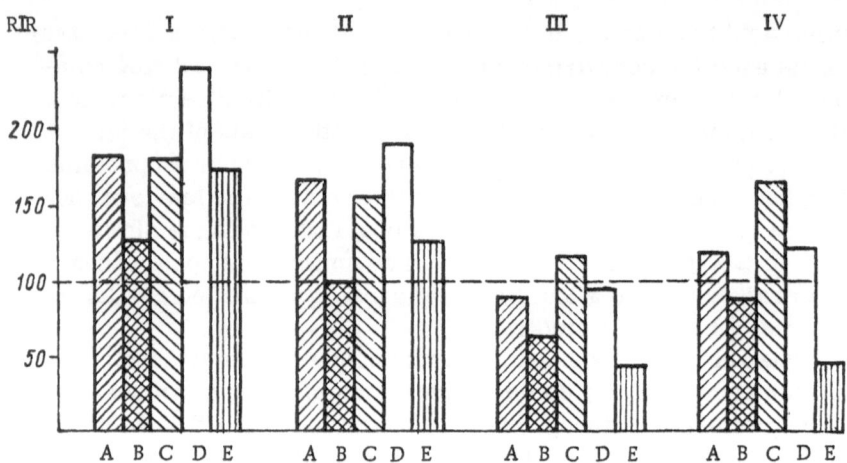

Fig. 18. RIR of proteins of subcellular fractions of brain tissue from rabbits of different ages 2.5 h after subcutaneous injection of methionine-^{35}S [31]: I) newborn, II) just acquiring vision, III) aged 1 month, IV) adult animals; subcellular fractions: A) nuclear, B) mitochondrial, C) microsomal, D) soluble, E) "washings." [In each age group of the animals SR of the blood serum proteins is taken as 100 conventional units (broken line).]

fractions and the specific radioactivity of the blood serum proteins, in percent. They disproved the previously held view of the metabolic inertness of brain tissue proteins. In newborn rabbits and those just acquiring vision, in which the BBB is forming [254, 913, 952, 1245], the rates of renewal of the total protein (Table 11) and proteins of the subcellular fractions (Fig. 18) of the brain were found to be much higher than those of the blood serum proteins. In the newborn rabbits and young rabbits just acquiring vision the renewal rate of the blood serum proteins is 44 and 25%, respectively, below that of the total brain protein. The intensity of protein renewal in the metabolically active subcellular fractions (nuclear, microsomal, soluble) of newborn rabbits is 80–140% higher than the rate of renewal of the serum proteins.

With age, the relative difference between the higher renewal rate of brain proteins compared to blood serum proteins diminishes gradually. This is due largely to the marked age increase in the rate of renewal of blood serum proteins (Table 11), confirmed by Salganik's observations [576]. The rate of renewal of serum proteins is 40–50% higher than that of the total brain proteins in one-month-old and adult rabbits, but the rate of renewal of proteins of the nuclear, soluble and, in particular, microsomal brain fractions is about equal or even higher than that of blood serum proteins.

When the rates of renewal of the serum proteins and total brain proteins of adult animals are compared, it must be borne in mind that the blood serum contains none of those metabolically inactive structural proteins that are so abundant in nerve tissue. Blood serum chiefly contains proteins performing active metabolic, enzymic, protective, and transport functions. Another fact to be considered is that the increase with age in the level of serum protein labeling is due mainly to delayed utilization and not to intensified synthesis [81, 82, 486, 488].

The high potential capacity of adult nerve tissue to synthesize protein is evident from Smerchinskaya's findings [601, 602]. She showed that if the dose of labeled amino acid is increased three-fold, the content of the label is increased by about the same degree not only in the free amino acid pool but also in the proteins of the brain.

In other experiments on young animals [182, 1247] a higher rate of renewal of total brain proteins than that of the blood plasma, liver, and kidney protein was revealed. In connection with these findings and the data described above, it is interesting that the protein content in the microsomal and pH 5-soluble fractions of brain tissue of newborn rats is several times higher than in the liver, an organ in which many serum proteins are synthesized [1412]. The protein content in these fractions of brain tissue falls with age, but in the liver, on the other hand, rises. As a result, in adult animals the content of these proteins in brain tissue is several times less than in liver tissue. These figures agree closely with the results of our own investigations which revealed an age decrease in the protein content in the principal subcellular fractions (Table 10) and in the intensity of renewal of the total brain proteins (Fig. 16), a marked increase with age in the intensity of renewal of the serum proteins (Table 11), and a higher rate of metabolism of the brain proteins compared with that of blood serum proteins in newborn animals and in animals just acquiring vision (Fig. 18).

The high potential capacity of neurons for active biosynthesis of proteins and nucleic acids is manifested particularly clearly during the regeneration of injured neurons [67, 673, 857, 858, 1002, 1091, 1146]. For instance, the intensity of incorporation of lysine-^{14}C into proteins of a regenerating neuron in the nucleus of the

hypoglossal nerve at various stages of recovery from mechanical
crushing is two or three times higher than in the normal cell [858].
Regenerating nerve cells in the nuclei of the facial nerve also
incorporate methionine-^{35}S much more intensively than intact
control cells [1002]. As might be expected, the potential ability of
neurons to synthesize protein is greater in young animals. The
intensity of incorporation of methionine-^{35}S into proteins of the
regenerating spinal motoneuron in rats aged 1-3 months is considerably higher than during repair of the corresponding cells in
animals aged 24-28 months [1091].

These experimental results indicate that brain tissue possesses extremely active mechanisms of protein biosynthesis. The
high synthetic powers of brain tissue, moreover, are found not
only in the early stages of postnatal development, when active differentiation and cell growth are taking place, but also in the later
stages of ontogeny, and even in adult animals, where the dominant
process is of self-renewal of previously formed tissue structures.
At each stage of individual development of the organism the activity
of the biochemical systems responsible for the intravital renewal
of tissue proteins is exhibited in accordance with the functional
needs of the organ though dependent on the presence of the appropriate precursors in the metabolic reserves of the brain. The
supply of precursors to the metabolic reserves of brain tissue at
each stage of ontogenetic development is controlled by the BBB
and the mechanisms transporting metabolites through the intracellular membranes.

Metabolic Heterogeneity of the Brain Proteins.
The general principles of individual development of the organism
[61, 100, 292, 366, 397, 399, 507, 1451] result from differentiation
and growth of tissues, the formation of structural and functional
units of organs, and the establishment of physiological functions
and biochemical systems, including the processes of renewal of
protein structures. These principles suggest that every tissue,
especially one so morphologically and functionally heterogeneous
as nerve tissue, contains very many proteins and protein complexes, with varied rates of metabolic activity, and that the differences in the rates of renewal of proteins subserving different functions increase with age.

Convincing evidence of the existence of metabolic heterogeneity
of the tissue proteins and of its increase with age was obtained
after the large-scale introduction of the labeled precursors into

practical biochemical research. The first relevant data were described by Orekhovich et al. [435]. They found that in newborn animals the proteins of all the organs they studied are renewed at practically the same rate, whereas in adult animals considerable differences are found in the rates of renewal both of the total proteins of the various tissues and of individual protein fractions in the same tissue. The results of investigations by Panchenko et al. [308, 483] are also evidence of an increase in the heterogeneity of the tissue proteins, as regards the rates of their renewal, with age. They found that the rates of protein renewal in different parts of the brain are more uniform in kittens aged 1 month than in fully grown cats.

The relative uniformity of the rates of renewal of the protein structures in the brain tissues of young animals has also been demonstrated autoradiographically [914]. Diffuse incorporation of methionine-^{35}S into the protein structures of the brain was found in day-old rats, whereas in old animals a higher rate of renewal of the proteins was found in areas of the brain containing chiefly neurons (the cellular layers of the cortex) or neurosecretory cells (the supraoptic nuclei of the hypothalamus). Clear differences in the intensity of protein synthesis in different parts of the nervous system, in different types of cells, and even in different parts of the same nerve cell have been demonstrated autoradiographically by Gracheva [170].

The increase in the metabolic heterogeneity of the brain proteins with age is clearly demonstrated by our observations [459] of the intensity of renewal of total brain proteins at four stages of the postnatal development of the rabbit. The conclusion that the rates of renewal of the individual brain proteins become more heterogeneous with age follows from the fact that the longer the labeled amino acid remains in the body of the experimental animal, the greater the degree of age decrease in the rate of renewal of the total brain proteins (Fig. 16). Several factors may be responsible for this relationship, and the most important of them are as follows.

The intensity of renewal of the total brain proteins is a mean index reflecting the rates of renewal of many individual proteins, each of which is renewed at its characteristic speed. Depending on the functions performed by the individual proteins, some are metabolized very rapidly, others more slowly, while some are almost metabolically inert. An important aspect in interpreting the results shown in Fig. 16 is that the total content of each of the individual proteins, each with differing metabolic activity, varies considerably in animals of different ages, and that the ratio between the rapidly and slowly metabolized proteins shifts with age in favor of the latter. Another important fact is that the brain tissues of adult animals do contain very rapidly metabolized proteins (Table 9), but their relative content decreases with age (Table 10).

Consequently age differences in the rate of renewal of the total brain proteins ought to appear in short-term and longer-term experiments depending on whether rapidly or slowly metabolized proteins predominate in the brain tissue at the particular stage of its development. In short-term experiments the radioactivity of the brain proteins, for obvious reasons, is determined chiefly by the incorporation of the labeled amino acid into the rapidly metabolized proteins, whereas in experiments of longer duration it is due to incorporation of the label into both rapidly and slowly metabolized proteins. Since rapidly metabolized proteins are found in the brain tissue in the early and later stages of development, it is not by accident that the age differences in the rate of renewal of the total brain proteins in short-term experiments are of slight degree. During the first 7 months of postnatal life the RSR of the brain proteins in experiments lasting 2.5 h falls by 22% (Fig. 16). With an increase in the duration of the experiments the degree of the age decrease in the intensity of renewal of total brain proteins increases steadily, so that in experiments lasting 14 and 24 h the decrease during the period from birth to the adult state is 54% and 82% respectively. This increase in the decline of protein synthesis rates with age is attributable chiefly to the increase with age in the content of slowly metabolized proteins in the brain and the gradual decrease in the rate of their renewal.

There are, of course, other metabolic situations in the brain tissue which also might affect the character of the age decrease in renewal rates of total brain proteins in long-term experiments. These include changes in the permeability of the BBB for labeled amino acids at different stages of development [346, 898, 1126, 1246, 1259], changes in the ability of the labeled amino acid to be reused for the renewal of proteins in experiments of different duration [1263], and certain other conditions.

Our conclusion that there is an increase in the content of slowly metabolized proteins in the brain tissue with age and that the intensity of their renewal decreases considerably with age is supported by the findings of Lajtha et al. [1253, 1255]. From the spectrum of renewal rates of total brain proteins obtained in experiments ranging from 2 to 60 min on mice aged 10 and 100 days, they concluded that slowly metabolized protein fractions are virtually absent from the brain tissues of young animals and that metabolically active brain proteins of young and adult animals do not differ significantly in their renewal rates. The longer half-life of brain proteins in short-term experiments on adult animals compared with the half-life in young animals and a sharp increase in the difference between the two groups in this parameter in longer-term labeling experiments are regarded by these workers as evidence of considerable metabolic heterogeneity of the adult brain proteins and of an increase in this heterogeneity with age.

The increase with age in the differences in metabolic activity of brain proteins is more clearly apparent when the intensity of renewal of proteins of the subcellular fractions is studied. The results of our investigations, Table 9, show that in newborn rabbits, i.e., in the earliest stages of postembryonic development, when the predominant process in the brain tissue is synthesis to provide for growth of the organ, the metabolic heterogeneity of proteins in the subcellular fractions studied is very slight. At this stage of development the formation de novo and renewal of all the intracellular protein structures proceed relatively uniformly. Later, with the decline of "growth synthesis" [78, 399, 418] and a gradual increase in the intensity of anabolic and catabolic processes to provide for the renewal of protein structures formed earlier ("self-renewal synthesis") the metabolic differences between proteins of the various subcellular fractions gradually increase (in rabbits just acquiring vision and at the age 1 month), to reach a maximum in adult animals.

The increasing degree of metabolic activity of the brain proteins with age is accounted for, first, by a decrease in the rate of renewal of the more inert protein structures, the content of which rises with age, and second, by an increase in the rates of renewal of proteins belonging to metabolically active intracellular structures, the content of which decreases with age in the brain tissue. These age changes, in opposite directions, in the renewal rates of the proteins and their content in the subcellular fractions of the brain is clearly visible in Fig. 19 [31]. The RIR values of proteins from the subcellular brain fractions are expressed in this figure

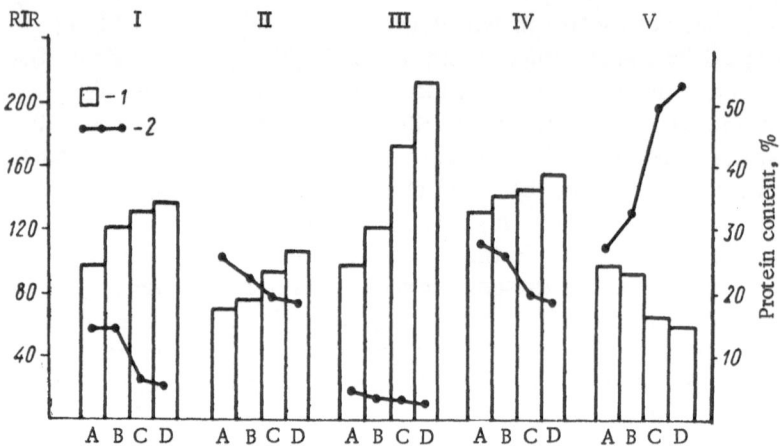

Fig. 19. Age changes in the intensity of renewal and content of protein in subcellular fractions of rabbit brain tissue: 1) RIR values of proteins 2.5 h after injection of methionine-^{35}S; 2) change in protein content (percent of total content
in homogenate); I) nuclear, II) mitochondrial, III) microsomal, IV) soluble, V)
washings; A, newborn, B) just acquiring vision, C) aged 1 month, D) adult animals.

as the ratio between SR of the proteins of each subcellular fraction and SR of the proteins of the original homogenate, in percent.

Similar opposite age changes in the intensity of renewal are
also observed for the serum protein fractions [1164]. For instance,
the intensity of renewal of γ-globulin in adult rats is two or three
times higher than in rats aged three to four weeks, but the rate of renewal of the serum albumins, on the other hand, decreases by about half.

Our results showing an age increase in the content of slowly
metabolized proteins in brain tissue agree with the observations
of Folch-Pi [1022], who found an almost sixteenfold increase in
the concentration of the nearly stable proteolipids in mouse brain
tissue during the first 6 months of postnatal development. The
total content of these and other very slowly metabolized protein
structures in brain tissue is evidently very small, as is the content of proteins with a very high rate of renewal. More than 90%
of the proteins of nerve tissue have a mean half-life of about 10-20
days [1263].

The increase in the metabolic heterogeneity of proteins with
age, observed in experiments on the whole brain and on its subcellular fractions, can evidently be explained as follows. In the
earliest stages of postnatal development, when proliferation and

intensive growth of cells are taking place [197], the formation
of protein de novo takes place uniformly and intensively in all
intracellular structures. Processes of breakdown and resynthesis
of structures formed previously are negligible at this stage of
development; hence the high intensity and comparative unformity
of the rates of incorporation of the label into the proteins of all
subcellular fractions.

Later, in connection with the slowing of the rate of division
of the cells and the decrease in the growth rate of the tissue, the
period of structural differentiation of the protoplasm gradually
develops, accompanied by a marked differentiation of the biochemical functions of the cell and its organelles. The intracellular
structures thus formed do not, however, remain unchanged. They
stay in a dynamic state and their proteins and other components
are renewed depending on the function they perform in the cell;
i.e., there is a change from "metabolism of growth" to "metabolism
of function." With the end of cell division and growth, differences
in the metabolic activity of the proteins are seen more clearly.
In addition to the actively metabolized proteins, others of low metabolic activity appear. These are slowly renewed proteins and protein components occurring in the various intracellular structures.

As a final word, in order to understand the biochemical
mechanisms of the age changes in synthesis rates of the various
protein structures, some of which are changing in different directions, and to elucidate the biochemical basis of nervous activity,
it is equally important to determine the functional state determining the metabolism and intracellular localization of both slowly
and comparatively rapidly metabolized proteins and protein components.

Changes in Activity of Intracellular

Proteinases and Other Enzymes in

Postnatal Ontogeny

In the various stages of ontogenetic development the continuous process of protein reconstruction differs in its character. The
relative contributions of the mutually dependent processes of
synthesis and breakdown in this reconstruction are unequal. For
example, in the embryonic and early postembryonic periods of
development the intensity of protein synthesis characteristically
is higher than the intensity of protein breakdown. This relation-

ship between the rates of these two processes results in a rapid
increase in the content of the tissue proteins. During later development of the organism the intensities of protein synthesis and protein breakdown gradually become more equal and the conditions
are created for maintenance of nitrogen equilibrium while preserv-
ing the dynamic state of the proteins. Later, the general ability
of tissues to synthesize proteins gradually diminishes whereas
the intensity of destructive processes increases. Finally, a period
arrives in individual development when the processes of protein
breakdown become predominant.

What changes take place in the activity of the enzyme systems
responsible for the catabolic phase of protein metabolism in the
various stages of individual development?

Ansell and Richter [769] first drew attention to the lower activity of cathepsin in the cerebral cortex of the newborn infant
than in the adult human. The hypothesis also was put forward on
the basis of indirect evidence [1529, 1541] that in the period of
growth of nerve tissue the processes of protein breakdown are
depressed, and that this leads to an increase in the content of tissue
proteins in the early stages of development.

Experiments in Palladin's laboratory [27, 37, 1473] showed
that the activity of acid and neutral proteinases in rabbit brain
tissue at four consecutive stages of postnatal development — at
birth, on the acquisition of vision, at 1 month, and in the adult —
is dependent on age. Acid proteinase activity was determined by
Anson's method [770], and neutral proteinase activity by the hydrolysis of protamine at pH 7.2 [28].

The activity of these enzyme systems, calculated per wet
weight of tissue and per weight of protein, increases in postnatal
ontogeny (Fig. 20). The most rapid increase in proteolytic activity
is observed during the first 11-12 days after birth. For instance,
if the enzyme activity is calculated per wet weight of tissue the
mean daily increase in the first 11-12 days of postnatal life is 3.5
and 2.5% for acid and neutral proteinases, respectively. During
the next 20 days of development of the brain the mean daily rate
of increase in activity of the intracellular proteinases falls by
about half, and after the age of 1 month becomes extremely
small. Calculated per wet weight of tissue, the activity of acid
and neutral proteinases in the brain tissue of adult rabbits is 184
and 182%, respectively, of the level in newborn animals. The increase is much smaller if the enzyme activity is expressed per
weight of tissue protein.

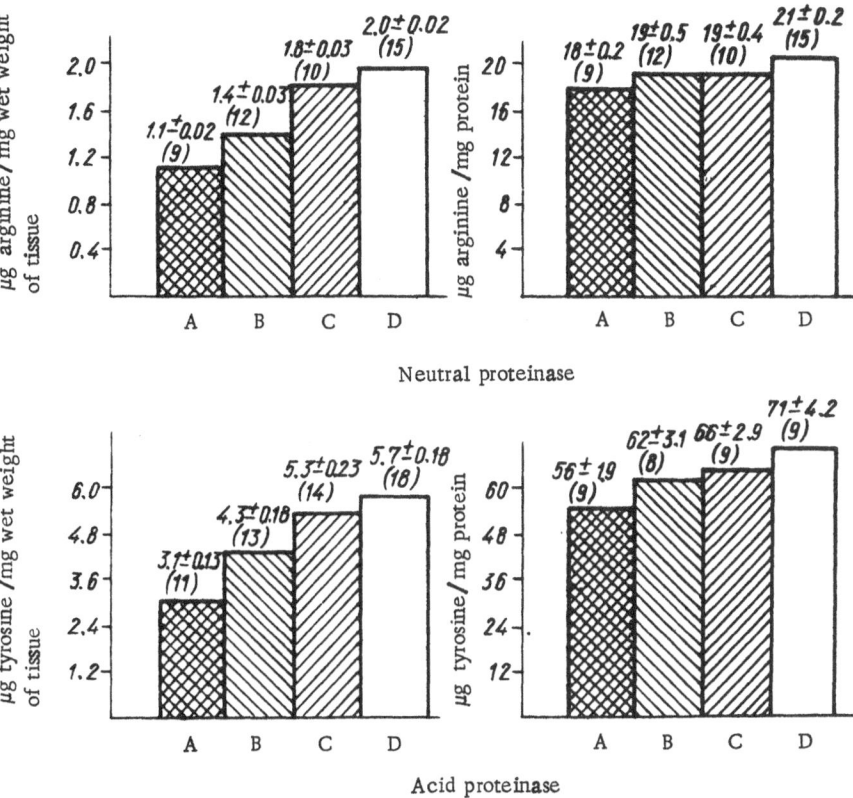

Neutral proteinase

Acid proteinase

Fig. 20. Age changes in activity of neutral and acid proteinases of rabbit brain tissue:
A) newborn, B) just acquiring vision, C) aged 1 month, D) adult rabbits.

If calculated per dry weight of brain tissue substance, the activity of acid and neutral proteinases is almost unchanged with age, for the increase in enzyme activity almost parallels the increase in the weight of solids; only an extremely small increase in the activity of these two enzymes is found in rabbits just acquiring vision, when expressed per weight of dry substance. Mention has already been made of the paper by Gorodiskaya [167], in which it was stated that proteolysis is activated in the visual centers and tracts of the brain during exposure of the retina to photic stimulation.

Similar investigations of the age dependence of intracellular proteinase activity were carried out recently by S.S. and H. Oja [1453], who determined the proteolytic activity of brain homogenates from rats aged 1, 5, 14, 60, and 300 days at pH 3.0, 6.5,

and 8.5. They used proteins denatured with urea as exogenous
substrate — hemoglobin, casein, and endogenous brain proteins,
as well as a number of synthetic substrates of "trypsin" and
"chymotrypsin" type. The rate of protein hydrolysis by the enzyme systems of the brain is approximately trebled at pH 3.0 and
about doubled at pH 6.5 and 8.5 in the course of postnatal development. These findings agree closely with our own results on rabbits.

The investigations of Lajtha and Marks [1257] revealed particularly sharp and fluctuating changes in the activity of acid and
neutral proteinases in rat brain tissue during the most active
phases of development — in the late embryonic period and in the
first 25 days after birth. Less marked changes were found in
the later stages of postembryonic development.

The data on age changes in proteolytic enzyme activity in
other animal tissues so far available [284, 285, 488, 489] are few in
number and contradictory in nature and no final conclusions regarding
the role of intracellular proteinases in the renewal of tissue proteins
at different stages of postnatal development of the brain can be drawn
from them. Analysis of the data given in the surveys by Parina [488,
489] shows that the activity of the same enzymes in different
tissues and even in different intracellular structures and the activity of different enzymes in the same tissue change in different
ways in ontogeny. The activity of the intracellular proteinases
may increase, decrease, or remain unchanged with age.

The activity of the other enzyme systems of the brain, directly
or indirectly linked with cell differentiation and growth and with
renewal of the tissue proteins, also changes substantially in postnatal ontogeny. For instance, adenosinetriphosphatase (3.6.1.3)
activity in the tissue of the whole brain [192, 721, 852] and its
various parts [292, 721] and in the subcellular structures [737,
1176] of the brain rises appreciably with age.

Age-related increases in enzyme activity have also been found
for cytochrome oxidase (1.9.3.1), succinate dehydrogenase
(1.3.99.11) [74, 292, 504], acid (3.1.3.2) and alkaline (3.1.3.1)
phosphatases [915], and also for glutamate decarboxylase (4.1.1.15),
glutaminase (3.5.1.2), and aspartate-aminotrasferase (2.6.1.1).
The latter enzymes are concerned with the metabolism of glutamic
and aspartic acids, which are of great functional importance in
nerve tissue [801]. The activity of inorganic pyrophosphatase
(3.6.1.1), an enzyme hydrolyzing inorganic pyrophosphate formed

in the tissues during the synthesis of practically all biologically important compounds, diminishes with age (Fig. 17), although a high level of activity is found in rabbits just acquiring vision [646]. A decrease in glutamine synthetase (6.3.1.2) activity has also been found in rats aged 360 and 720 days compared to rats aged 30 and 90 days [166]. The activity of most of the enzymes listed above changes most rapidly during the first month of postnatal development. More detailed information and an analysis of the data on ontogenetic changes in tissue enzyme systems are given in the surveys of Parina [488, 489].

How can the data given in this section for the age decrease in the intensity of renewal of total brain proteins be reconciled with the increase in proteolytic activity observed in brain tissue during postnatal development? If the system of tissue proteolytic enzymes in fact is responsible for the catabolic phase of protein metabolism in vivo, and there is no doubt that it is, the age decrease in the intensity of protein renewal, starting in the early stages of individual development, should presumably be accompanied by a corresponding decrease in the activity of these enzymes, but no such decrease is observed.

However, this disagreement is only apparent. When the results are compared, allowance must be made for differences in protein metabolism and the role of proteolytic enzyme systems in it at different stages of individual development. For instance, the whole embryonic period and the early stages of postembryonic development are characterized by intensive protein synthesis, providing for tissue differentiation and growth. At these stages of ontogeny the formation of all the components of the cell structures, including proteins, takes place intensively de novo, and their breakdown is negligible. Synthesis in the early stages of ontogeny has been called "growth synthesis" [418]; it diminishes sharply with age, especially at the beginning of the postembryonic period, and becomes negligible at the stage of maturity.

The period of rapid decline in the intensity of growth synthesis coincides with an increase in the intensity of "repair synthesis," providing for self-renewal of structures formed earlier. This type of synthesis, also called "self-renewal synthesis," decreases only very slightly with age. By contrast with growth synthesis, it requires the participation not only of biochemical systems of protein synthesis, but also of enzyme systems responsible for protein degradation to free amino acids. If these special features of the age changes in the two principal forms of synthesis are taken into account, the apparent incongruity between the direction of the age changes in enzyme activity of the protein catabolism systems and the rate of incorporation of labeled amino acids into brain proteins disappears.

Comparison of the available information on the development of the morphological structures of brain tissue [215, 399, 544] with the data on ontogenetic changes in protein concentration, the intensity of incorporation of labeled amino acids into the protein, and the activity of intracellular proteinases in brain tissue leads to the conclusion that in the various stages of individual development of the brain the level of activity of the proteolytic enzyme systems depends on the rates of cell differentiation and growth, on the intensity of physiological renewal of protein structures, and on the irreversible breakdown of some proteins in pathological processes.

CHAPTER 8

Catabolism of Nerve Tissue Proteins

The dynamic state of the tissue proteins is maintained by synthesis of protein molecules de novo to replace those broken down. The life span of protein molecules varies and is determined by a complex group of factors, including the functional characteristics of the organ and the type of its cells and their organelles. At present very little is known about the biochemical mechanisms of protein breakdown and virtually nothing is known about the causes of such breakdown and the regulation of its intensity. Our knowledge on the catabolic phase of protein metabolism is very limited when compared with the facts that have accumulated in the field of protein biosynthesis in the last decade.

In this chapter the basic information on enzyme systems of protein catabolism, their distribution, their tissue and intracellular localization, and their possible biological role will be briefly described.

Classification and Distribution of Intracellular Proteolytic Enzymes

Information on existing classifications of the many intracellular proteolytic enzymes and their physicochemical properties can be found in readily accessible surveys [135, 171, 196, 692]. In their surveys, Lajtha, Marks, and Waelsch [1248, 1251, 1345, 1738] gave a classification and described the distribution and properties of intracellular endo- and exopeptidases of nerve tissue. Accordingly, these particular problems will be discussed only to the extent required for a logical exposition of the material and for an analysis of the data in the literature on the possible biological role of this important class of enzymes of protein metabolism.

As regards the classification of the proteolytic enzymes, all that need be said is as follows. The Commission on Enzymes of the International Biochemical Union has included about 40 proteolytic enzymes in the modern nomenclature [426] (in the class of hydrolases, subclass peptide hydrolases). In the literature, the classification suggested by Bergmann [821] is used, according to which all proteolytic enzymes can be divided into endopeptidases and exopeptidases. The former, by acting on peptide bonds at a distance from the terminal amino acids, split proteins and polypeptides into shorter peptides, and thus increase the number of terminal amino acids exposed to the action of exopeptidases. The latter remove terminal amino acids either from the carboxyl end (carboxypeptidases) or from the amino end (aminopeptidases) of the polypeptide chain.

Many endopeptidases and exopeptidases exist, each hydrolyzing only particular peptide bonds depending on the enzyme's characteristic specificity. The accessibility of a peptide bond to the action of a given enzyme is determined by the nature of the chemical groups and side chains lying close to that bond and not, as was hitherto considered, by the length of the polypeptide chain. The subdivision of the proteolytic enzymes into proteinases, splitting high-molecular-weight proteins, and peptidases, hydrolyzing low-molecular-weight peptides, is thus regarded as unsound [196]. Nevertheless, this classification is still used in the modern literature [171, 210, 1443, 1708, 1710].

Systems of intracellular endo- and exopeptidases, capable of splitting proteins by their combined action into free amino acids, are found both in the gastrointestinal tract (pepsin, trypsin, chymotrypsin, carboxypeptidases, amino peptidases) and in various other tissues (intracellular proteolytic enzymes or cathepsins). The term cathepsins is nowadays given to intracellular proteolytic enzymes hydrolyzing proteins in acid, neutral, and alkaline media [692, 1648], although originally the name was reserved for tissue enzymes with optimum action in an acid medium. Besides the digestive and intracellular peptide hydrolases there is a special system of enzymes with proteolytic action which are responsible for blood clotting and for subsequent lysis of the blood clot thrombin, plasmin, etc.).

The distinguishing feature of all proteolytic enzymes is evidently their biosynthesis in the form of inactive precursors, which are later converted, under certain physiological or pathological conditions, into active enzymes. Inactive precursors are known, for example, for the digestive proteolytic enzymes and also for enzymes of the blood-clotting system. Although no inactive precursors of the intracellular proteolytic enzymes have yet been isolated, this must not be regarded as evidence of their absence in the cells. Since all intact tissues contain assortments of endo- and exopeptidases, yet at the same time the tissue protein structures do not undergo uncontrolled hydrolysis to free amino acids, there is reason to suppose that definite mechanisms for regulating the activity of these enzymes exist in the tissues. These mechanisms may differ depending on the physicochemical properties of the enzymes themselves and also on the tissues, cells, and intracellular structures in which these enzymes are found and on the physiological or pathological conditions of the organs and tissues under which they function.

Let us consider some of the possible mechanisms of regulation of the action of intracellular peptide hydrolases. Some intracellular proteolytic enzymes contained in the tissues in the form of inactive precursors may be activated as the result of a rise of body temperature, changes in the ionic strength of the medium, and other factors [995, 1361, 1363, 1364, 1366, 1700]. Other enzymes are evidently present in the tissues in an active form but can exhibit their action only under definite conditions (for example, enzymes with pH optimum in a strongly acid or strongly alkaline medium). The activity of some proteolytic enzymes in the tissues is inhibited by specific inhibitors and is manifested only after dissociation of the enzyme—inhibitor complex [995, 1363, 1365]. One way in which the activity of intracellular hydrolytic enzymes, including cathepsins, is regulated is by their localization in special cell organelles, or lysosomes, separating the enzymes physically from their substrates [202, 424, 966]. The role of lysosomes in cell function will be discussed in rather more detail below. The complex system of membranous structures in the cell and its organelles, forming metabolic compartments of various types [1733, 1734], suggests that besides lysosomes, other methods of physical separation of the enzymes, including neutral peptide hydrolases, from their substrates, may also exist in the cells.

The maintenance of protein catabolism at levels corresponding to the metabolic situation of particular tissues, cells, and their organelles under particular conditions of activity in the living organism is evidently ensured not only by the mechanisms listed, but also by other, as yet unknown, mechanisms of regulation of proteolytic enzyme systems.

Enzymes catalyzing the hydrolysis of peptide bonds are widely distributed in nature. They are found in microorganisms and in the tissues of plants and animals and they can act over a wide range of pH values — from 2.0 to 10.0. For most intracellular proteolytic enzymes the pH optimum in vitro is either over the pH range of 3.0-5.5 ("acid") or in the range of 6.5 to 7.8 ("neutral"), i.e., at pH similar to that existing in the tissues. Intracellular enzymes hydrolyzing proteins in an alkaline medium are less active than the acid and neutral enzymes. The predominance of the "acid" proteolytic enzymes over the "alkaline" is evidently not accidental, for an acid medium is encountered much more often in the tissues than is an alkaline pH [13, 250, 564].

Historically more attention was paid initially to enzymes hydrolyzing peptide bonds in an acid medium. This was because of the practical importance of autolytic processes taking place in the tissues in an acid medium. Numerous enzyme systems hydrolyzing peptide bonds in an acid medium are found in virtually all the tissues that have been studied: in the spleen [169, 851, 1035], liver [169, 558, 851, 1696], kidneys [220, 558, 692], adrenals [1681], lungs [692, 931], uterus [1783, 1784], heart [851], skeletal muscles [220, 337, 558, 851, 1095, 1627], thymus [692, 1576], thyroid [692, 1693], lymph glands [692], bone marrow [692, 785], and bone [807].

Later, enzyme systems hydrolyzing peptide bonds in a neutral medium, i.e., at pH values close to those existing in the tissues, and in an alkaline medium were detected and studied [749, 832, 931, 1095, 1202, 1218, 1219, 1220, 1242]. Enzymes with optimum action in a neutral medium, as is found in most cells under physiological conditions, play a particularly important role in the renewal of the tissue proteins. It is therefore not by accident that during recent years more attention has been paid to the investigation of enzymes hydrolyzing peptide bonds in neutral medium. Such enzymes have been found in skeletal muscles [1218-1220, 1418], the muscles of the heart [995, 1623] and uterus [1623], in the liver [954, 995, 1389, 1418, 1696, 1712], kidneys [210, 995, 1417, 1418], spleen [987, 995], lungs [931, 995, 1033], intestine

[995, 1443], pituitary [984], pancreas [1418], thymus [1488, 1576], submandibular salivary gland [1525, 1527], the skin [810, 1033, 1361], the crystalline lens [1117, 1202, 1740], the red blood cells [1405], bone [807], and bone marrow [1693]. The relative proteolytic activity in various tissues was based on the hydrolysis of protamine at pH 7.2. The organs studied, in descending order of neutral proteinase activity, were: kidneys, spleen, pancreas, liver, brain, heart, skeletal muscle, and blood.

Proteolytic Enzymes of Nerve Tissue

Enzymes hydrolyzing protein molecules and peptides to free amino acids in the tissues of the nervous system have been known for a comparatively long time [49, 146, 597, 638]. Yet hardly anything was known until recently about the tissue and intracellular localization of these enzymes, their physicochemical properties, the mechanism of their action, their biological role, and the relationship between their activity and the physiological and pathological state of the organism. Our knowledge in this field has grown considerably in the last 10 to 15 years. Investigations have shown that the different parts of the nervous system contain intracellular proteolytic systems with optimal activity in acid [38, 532, 658, 769, 861, 995, 1197, 1248, 1349, 1350, 1478], neutral [27, 28, 262, 745, 769, 995, 1085, 1137, 1349, 1445, 1473, 1492, 1494, 1496], and alkaline [658, 861, 995] media. Each of these proteolytic systems consists of several enzymes, including both endo- and exopeptidases. Because of the presence of these complex groups of enzymes with endo- and exopeptidase action in the cells, tissue homogenates can hydrolyze proteins of high and low molecular weight [12, 27, 658, 743, 1197, 1250, 1350, 1473, 1496] and peptides with various molecular weights [153, 745, 769, 788, 861, 1137, 1316, 1445, 1492, 1494, 1495, 1708, 1710, 1711] to free amino acids. Some of these enzymes also have esterase and amidase activity [745, 936, 984, 1227, 1698, 1699, 1702].

Many endogenous tissue proteins are hydrolyzed by intracellular proteinases less intensively than some exogenous protein substrates [1197, 1349, 1350]. This is clear from the fact that the rate of formation of low-molecular-weight products of protein hydrolysis as a rule rises considerably after the addition of exogenous proteins to incubated tissue homogenates. However, some endogenous proteins are hydrolyzed preferentially by intracellular peptide hydrolases. For example, the neutral proteinase of the lens hydrolyzes the endogenous protein α_2-crystallin more rapidly than hemoglobin and other exogenous proteins [1117, 1202, 1740]. The proteolytic systems of the skeletal and uterine muscles, both with optimum activity at pH 8.5-9.0, also hydrolyze endogenous proteins preferentially [1218, 1219, 1783]. There is therefore reason to suppose that the proteolytic enzymes of some subcellular structures can hydrolyze proteins in other intracellular formations that are not split by their "own" peptide hydrolases [1248]. These results indicate that proteolytic enzymes in different tissues and different cell structures of the same tissue, although possessing identical or very similar pH optima, may differ substantially in their substrate specificity. The absence of total autolysis of the protein structures in the tissues is perhaps attributable in part to this property of the intracellular peptide hydrolases. The fact that certain proteins in the tissues, especially in myelin, are not hydrolyzed is due to their resistance to the action of proteolytic enzymes [744]. Also, among the water-soluble brain proteins, some are more resistant than others to proteolytic hydrolysis [917].

Of the exogenous protein substrates, hemoglobin is hydrolyzed most rapidly by intracellular peptide hydrolases, whereas other proteins (for example, casein, gelatin, egg and serum albumin, fibrinogen, edestin) are hydrolyzed slowly [49, 1034, 1035, 1349, 1350, 1453].

Acid and neutral peptide hydrolases of nerve tissue are more active than the corresponding muscle enzymes but much less active than those of the liver and, in particular, of the kidneys and spleen [1349, 1473, 1564]. Analysis of the experimental data so far obtained reveals some facts of great importance in explaining how enzymes so different in pH optima can function separately in the same cells. The peptide bonds of endogenous proteins are hydrolyzed by brain enzyme systems more intensively in an acid than in a neutral medium [814, 1197, 1349], whereas exogenous proteins are hydrolyzed more intensively by neutral proteinases of nerve tissue [1349]. Neutral brain peptide hydrolases are more sensitive to the inhibitory action of ions of the heavy metals and other inhibitors, and some of them are rather unstable when stored [1350]. Acid proteolytic enzymes of nerve tissue, like those of other organs, do not hydrolyze peptide bonds or do so only on a very small scale at the optimum pH for neutral peptide hydrolases, and the latter, for practical purposes, do not hydrolyze protein substrates in acid medium [1350].

These results suggest that numerous enzymes of these groups of peptide hydrolases (acid and neutral) exhibit their action largely independently in the tissues. The action of the enzymes of each of these groups is exhibited only when optimal conditions are brought about for them to function. Proteolytic enzyme systems with optimum activity in acid medium exhibit their hydrolytic action mainly in pathologically changed and dying cells, in which the pH of the medium is shifted toward the acid side. Neutral peptide hydrolases, on the other hand, are responsible for the hydrolysis of protein molecules undergoing renewal in functioning cells. The facts presented above are also evidence that under conditions similar to those prevailing in normally functioning cells, the potential activity of the neutral peptide hydrolases is utilized only in part and, moreover, under the strict control of the regulatory systems of the cell and in accordance with the demands determined by the functional activity of the intracellular structures. On the other hand, the activity of the acid hydrolases under these conditions is largely in a latent state; they exhibit their maximal activity only in an acid medium.

Most earlier investigations of the catabolism of nerve tissue proteins were concerned with the study of enzymes hydrolyzing protein substrates in an acid medium. Proteolytic systems with a pH optimum in a neutral medium received considerably less study. Information on their activity obtained by different methods does not always agree, and in some cases may even provide a basis for incorrect conclusions. For instance, investigators [769, 810, 968, 1197] who used hemoglobin, gelatin, or casein as exogenous substrates found no enzymes in nerve tissue capable of hydrolyzing these proteins in neutral medium.

It is not always possible to draw consistent conclusions regarding the presence or absence of a particular type of proteolytic activity in a tissue on the basis of results obtained by different methods. Adams and Smith [749] determined proteolytic activity by measuring the increase in optical density at 280 nm and found a proteinase in hog pituitary tissue that hydrolyze hemoglobin at pH 8.3. It was later shown by Tsaryuk [679, 680], and later by Lisowski [1299], that the increase in optical density in protein-free filtrates of brain and pituitary homogenates under these conditions is not due to proteolysis at all, but to ribonuclease activity.

The colorimetric method of determining acid-soluble ninhydrin-positive material as an indicator of tissue proteolytic activity is not sufficiently specific. Besides amino acids and peptides, many other compounds contained in protein-free filtrates of test samples also give a color reaction with ninhydrin, although less strongly [1402, 1589, 1606]. These compounds include certain amines which are functionally important and metabolically active in nerve tissue, such as histamine and tyramine, as well as ammonia, pyruvic, α-ketoglutaric, oxaloacetic, and ascorbic acids and other metabolites. For this reason, results can be very misleading if proteolytic activity is determined by this method with whole brain homogenates which contain enzyme systems which form the compounds mentioned, particularly when the determination is carried out at pH values close to the optimum for the overwhelming majority of these enzymes (pH 6-8).

For work with tissue homogenates, Anson's method [770], based on the determination of the quantity of tyrosine and tryptophan released from proteins, also has disadvantages. As Klein and co-workers [262] showed, tyrosine formed during incubation of brain homogenates under aerobic conditions is not accurately determined by this method, probably because of the rapid metabolic conversions of this amino acid. These workers indicated that only under conditions of strict anaerobiosis can values of the intensity of proteolysis in nerve tissue close to those actually prevailing be obtained by Anson's method.

Analysis of the results of numerous investigations has shown that when determining proteolytic activity in such complex enzyme systems as tissue homogenates and their subcellular fractions it is advisable to use several methods based on different principles, so that the disadvantages of some methods can be balanced against the advantages of others.

In Palladin's laboratory the activity of intracellular peptide hydrolases was determined [27, 28, 1473] in a medium with pH close to the physiological value. The method used was based on determining the arginine released from a protein substrate by the action of tissue proteinases. Protamine sulfate, used previously by Veremienko and Belitser [94, 95] in a micromethod for the determination of the proteolytic activity of trypsin, was used as the substrate. This colorimetric method of determining the activity of tissue peptide hydrolases is simple enough to be of practical use and in addition possesses high sensitivity, specificity, and accuracy.

Some proteolytic enzymes of nerve tissue have been obtained
in a purified form and their physicochemical and biological properties have been studied. Kies and Schwimmer [1197] first suc-
ceeded in partially purifying (by 65-fold) the acid proteinase
(pH 3.5) of calf brain in 1942 by fractionating the tissue with the
aid of acetone, ether, and ammonium sulfate and then dissolving
the residue obtained in 0.2 M acetic acid and dialyzing the solution
against distilled water. Guroff [1085] isolated a neutral proteinase
from the soluble fraction of rat brain and purified it about 30-fold.
This proteinase is activated by calcium ions and sulfhydryl compounds. The purified enzyme preparation has a pH optimum for
the hydrolysis of casein of 7.1-7.3, is inhibited by ions of heavy
metals, and is stable if stored in the absence of calcium ions. This
enzyme is evidently localized chiefly in the soluble brain fraction.
This neutral proteinase is also found in bovine brain and in the
brains of pigs, cats, guinea pigs, and rabbits.

A preparation of purified cathepsin with pH optimum 5.0 was
ioslated [658] from bovine brain in Orekhovich's laboratory. The
cathepsin was thermostable, was inhibited considerably by phosphate ions but not by ions of the heavy metals, and was slightly
activated by cysteine. The preparation was homogeneous on electrophoresis in starch gel and in the analytical ultracentrifuge.

Workers in Palladin's laboratory, using modern methods of
fractionating protein mixtures from bovine brain, obtained [332-
334, 531, 536, 1478] a highly purified cathepsin preparation with
pH optimum for the hydrolysis of hemoglobin at 3.5 and of serum
albumin at 4.1. The enzyme activity of this preparation was 2000
times higher than in the original tissue. This enzyme is thermo-
labile and stable at pH 5.5-8.5. Its activity is unaffected by the
ions of many bivalent metals, cysteine, glutathione, ascorbic acid,
EDTA, p-chloromercuribenzoate, or soy trypsin inhibitor, but it
is completely inhibited in urea solutions. The enzyme hydrolyzes
mainly the peptide bonds that are hydrolyzed by splenic cathepsin D.

Marks and Lajtha [1350] separated the acid (pH optimum 3.5)
and neutral (pH optimum 7.6) proteinases from rat brain, partially
purified these enzymes, and studied some of their properties.
They determined the Michaelis constant for various protein substrates, the pH optima, temperature dependence, response to inhibitors and activators, specificity with respect to hydrolysis of
synthetic substrates, and certain other parameters. The preparations proved to be heterogeneous on electrophoresis. Each evidently consists of several acid or neutral peptide hydrolases re-

spectively. The preparation of acid proteinase consists chiefly
of an enzyme similar in its specificity to cathepsin D, although
cathepsin of types B and C are also present. The neutral proteinase preparation contains small quantities of an enzyme similar
to the Ca^{++}-activated proteinase isolated by Guroff [1085].

Nerve tissue, like other tissues, contains natural inhibitors
of intracellular peptide hydrolases [809, 995, 1117, 1349, 1362,
1365]. A thermostable trypsin inhibitor, firmly bound with the
membranes, has been found [862] in human and bovine brain tissue.

The activity of the proteolytic enzymes [769] and the ability
of these enzymes to hydrolyze brain proteins [322, 541] increases
during phylogeny and runs parallel to the increase in metabolic
activity of the tissue proteins in phylogeny. It is interesting that
the sensitivity of total tissue proteins to proteolysis decreases in
ontogeny [76], probably on account of a decrease in the propensity
of these protein structures to undergo renewal with age. Considerable changes in the levels of activity of the tissue peptide
hydrolases are found in various physiological and pathological
states [164, 167, 514, 661, 1037, 1702], and species differences are
found in the activity of acid [532, 769] and neutral [1085, 1473]
peptide hydrolases in brain tissue.

Numerous investigations have shown that proteolytic enzyme
sytems are present in all parts of the CNS, in different types of
cells and their organelles [153, 769, 861, 936, 1137, 1213, 1248,
1349, 1492, 1495], and also in peripheral nerve tissues, their individual structures, and isolated protein fractions [743, 745, 1161,
1316, 1496]. As a rule the activity of most proteolytic enzymes
is higher in parts of the nervous system and in cells [153, 769,
1248, 1492, 1494-1496] that perform more complex functions and
have higher rates of protein metabolism [461, 543, 638, 1036, 1250,
1255, 1524, 1733, 1738]. One interesting exception is that one of
the neutral peptide hydrolases is most active in the white matter
of the brain [769], i.e., in a tissue with a low intensity of protein
renewal. These differences in the activity of individual proteolytic
enzymes in the structures of nerve tissue probably indicate differences in their functional role. Whereas some participate in
the intravital degradation of protein molecules, others may perform quite different functions, possibly specific for nerve tissue.

Histochemical investigations have shown [1495] that peptidase
activity is higher in the bodies of neurons than in their axons.
This may account for the low peptidase activity in the corpus

callosum, which contains no nerve cell bodies. The corpus callosum also differs from other parts of the brain in its low level of
free amino acids [1539]. Histochemical investigations of the layer-
by-layer distribution of dipeptidases in the cortex and subcortical
white matter of the frontal lobe revealed [1492, 1494] a higher
level of activity in the perikaryon of the neurons and glial cells.

Experiments in Palladin's laboratory showed [27, 532, 1473]
that phylogenetically and functionally different parts of the nervous system differ from each other significantly in the level of
activity of their neutral (pH 7.2, substrate protamine) and acid
(pH 3.5, substrate hemoglobin) proteinases.

The highest neutral proteinase activity was found in the white
matter of the cerebral hemispheres of the rabbits, and the lowest
in the cerebral cortex (Fig. 21, mean values from 12 to 13 experiments). The levels of neutral proteinase activity in the other five
parts of the brain tested are very similar and occupy an intermediate position between those for the gray and white matter of
the hemispheres. Neutral proteinase activity in spinal cord tissue
is very low and is close to the activity found in the cerebral cortex.

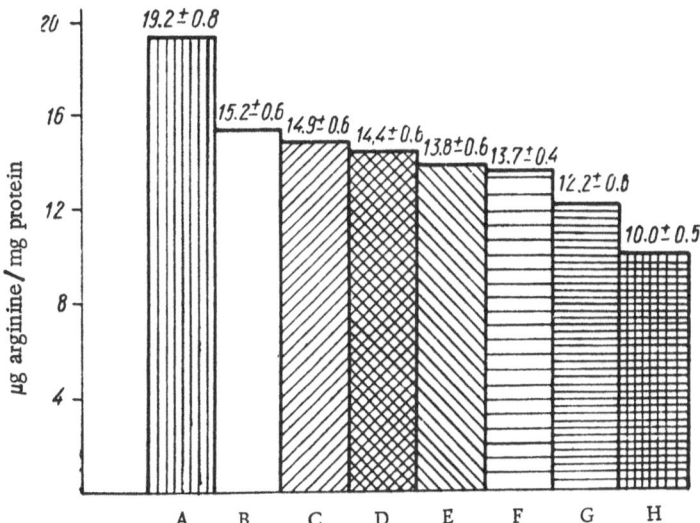

Fig. 21. Neutral proteinase activity in different parts of the rabbit
nervous system: A) white matter of cerebral hemispheres, B) diencephalon, C) mesencephalon, D) medulla, E) cerebellum, F) pons,
G) spinal cord, H) gray matter of cerebral hemisphere.

The distribution of acid proteinase in the nervous system differs substantially from the regional distribution of neutral proteinase [532]. In bovine brain, for instance, the highest acid proteinase activity is found in the gray matter of the cerebral hemispheres and cerebellum. Activity of this enzyme in the white matter of these parts of the brain, calculated per wet and dry weight of tissue, is 2.5 and 5 times less, respectively, than in the gray matter. The cathepsin activity in the cerebral cortex is a little higher than in the gray matter of the cerebellum. Acid proteinase activity in the tissue of the medulla is much lower than in the gray matter but higher than in the white matter of the cerebral hemispheres. Very low acid proteinase activity is found in spinal cord tissue. In the rabbit, cathepsin activity is practically equal in the tissues of the cerebellum and the gray matter of the cerebral hemispheres and is much higher than in the tissue of the white matter of the cerebral hemispheres.

Investigations in Palladin's laboratory [27, 33, 38, 172, 532, 1473] showed that acid and neutral proteinases are found in all subcellular structures of adult rabbit brain tissue tested, but that they are distributed differently among them (Fig. 22).

Neutral proteinase is most active in structures of the microsomal fraction. The activity of this enzyme in the fractions of nerve endings and purified mitochondria and in the soluble cytoplasmic fraction, calculated per weight of protein, is practically the same and is only 30% lower than in the microsomal fraction. Very low proteolytic activity is found in the pure nuclei obtained by Avdeev and Palladin's method [4] and, in particular, in purified myelin isolated by Autilio's method [779]. It will be clear from Fig. 23 that myelin preparations containing traces of membranes of the endoplasmic reticulum as impurities are distinguished by comparatively high neutral proteinase activity [172].

The highest acid proteinase activity was found (Fig. 22) in the mitochondrial fraction, which accounts for about two-thirds of the total activity of this enzyme in the original homogenate, and in the fraction of nerve endings [38, 532]. Particularly high proteolytic activity was found in the fractions of mitochondrial granules sedimented with the usual centrifugal force, while the fractions of heavier and lighter mitochondrial structures have much lower specific acid proteinase activity [38]. Cathepsin activity is much lower in the microsomal fraction and lower still in the purified nuclear fraction compared to the mitochondrial fraction. The

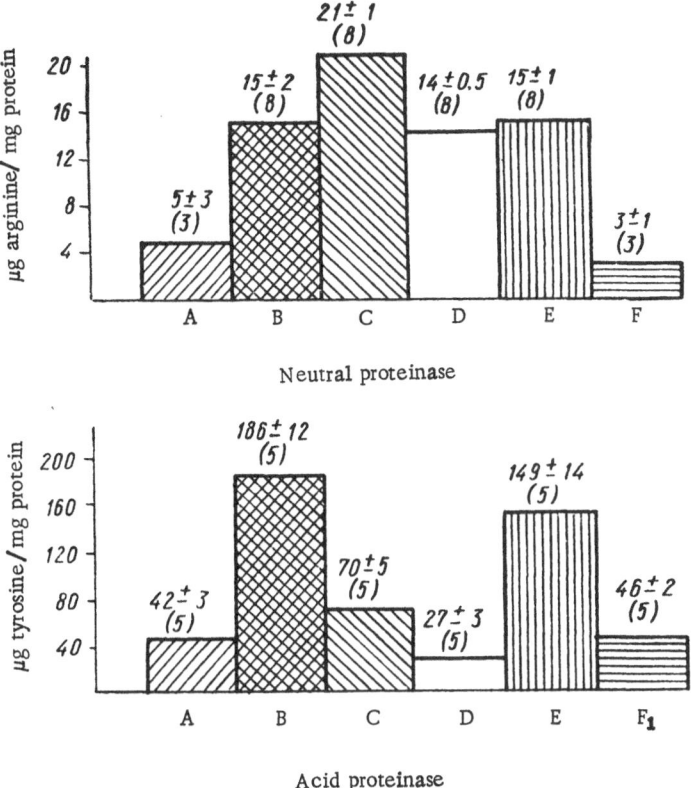

Fig. 22. Activity of neutral and acid proteinases in subcellular fractions of rabbit brain tissue: A) purified nuclei [4], A₁) nuclear, B)
purified mitochondria [33, 1769], C) microsomal, D) soluble, E) nerve
endings [33, 1769], F) purified myelin [779], F₁) myelin [33, 1769]
(citations of papers describing methods of obtaining corresponding fractions are given).

activity of the enzyme is very low in the soluble cytoplasmic
fraction.

Mitochondrial neutral and acid proteinases are localized
chiefly in the fraction of internal membranes [172].

Results of fundamental importance to the understanding of
the role of intracellular proteinases were obtained by determining
the neutral proteinase activity in the soluble and insoluble proteins
of tissue homogenates which had been treated with ultrasound and
with solutions of Triton X-100, sodium deoxycholate, and sodium

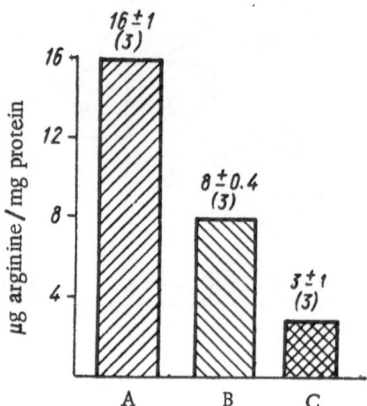

Fig. 23. Neutral proteinase activity in
myelin preparations obtained by different methods from the white matter of the
cat brain: A) myelin obtained by Whittaker's method [1769] as modified by Belik
et al. [33], B) by Autilio's method [779],
unpurified preparation, C) by Autilio's method [779], purified preparation.

dodecylsulfate. These results [172] are conclusive evidence that
neutral proteinase exists in the tissues of the nervous system in
two forms, i.e., in a soluble form or bound to the tissue structures.
More than half of the tissue's activity is firmly bound to the membrane structures of the cell. Neither ultrasound nor detergent
treatment can convert this enzyme activity into a soluble form.

Most of the acid proteinase of nerve tissue is also bound to
the membranes of subcellular structures. The enzyme activity
is not converted to a soluble form by homogenization of the
brain tissue manually or mechanically, by subjecting the homogenate to prolonged multistage fractionation procedures, or by
washing the mitochondrial fractions obtained by differential centrifugation of the fractions of myelin, nerve endings, and purified
mitochondria [38, 532] obtained by Whittaker's method [1769].
For instance, the soluble cytoplasmic fraction contains not more
than 7% of the total tissue enzyme activity, and practically all the
enzyme activity in the original unwashed fractions remains in
the washed structures of the myelin, nerve endings, and mitochondria.

The Lysosomes and Their Role in
Intracellular Protein Catabolism

The biological role of intracellular peptide hydrolases with acid pH optima is closely linked with special cell organelles known as lysosomes. De Duve et al. [965, 966] first isolated these intracellular particles from liver tissue by differential centrifugation and described their biochemical, and later their morphological properties. Subsequently lysosomal granules were found in the cells of all tissues investigated [772, 785, 835, 868, 966, 1581, 1714, 1753] and, in particular, in tissue of the nervous system [202, 348, 349, 424, 951, 1145, 1166, 1211-1213, 1600-1602, 1714]. These lysosomes can be detected within a cell by fluorescence microscopy and with the aid of tetrazolium salts [1211, 1212].

The history of the discovery of lysosomes and the basic facts concerning their morphology, their structural and functional polymorphism, their chemical composition, and certain other properties are adequately described in the literature [202, 203, 235, 424, 557, 965, 966]. In this section only a brief account will therefore be given of the information essential to understanding the role of these organelles and the enzymes contained in them in the functions of the cell.

The distinguishing feature of the lysosomal granules of any cell is that they contain a complex set of hydrolytic enzymes with an acid pH optimum. Enzymes of other classes have not yet been found in lysosomes. Already about 30 acid hydrolases have been found in lysosomes. Included among those identified are enzymes that can hydrolyze biologically important cell components such as proteins (cathepsins, collagenase), phosphoproteins (phosphoprotein phosphatase), nucleic acids (ribonuclease and deoxyribonuclease), phosphoric esters (acid phosphatase), mucopolysaccharides and their derivatives (hyaluronidase, β-glucuronidase, β-galactosidase, α-mannosidase, arylsulfatases A and B), etc. It is entirely possible that other acid hydrolases, essential for degrading other chemical components of the cell to low-molecular-weight metabolites that can be transported and assimilated by the intracellular systems, will be found in lysosomes in the future.

In intact lysosomes the enzymes are separated from substrates present in the cytoplasm by a single lipoprotein membrane. This membrane is a delicate apparatus preventing the uncontrolled hydrolysis of intracellular components by the lysosomal hydrolases.

There is reason to suppose [1213, 1581] that the lysosomal enzymes, or at least most of them, do not form stable chemical complexes with the structures of the lysosomes. These enzymes are
found in the lysosomal matrix either in the free state or adsorbed
on the lipoprotein membrane. The strength of the bond linking
the various hydrolases with the structures of the lysosomes evidently varies, for exposure of lysosomal granules to the same
procedure is followed by the conversion of different quantities of
each enzyme into the free form [773, 1581, 1600, 1602, 1714].

An essential condition for the hydrolysis of the appropriate
substrates in the cell by lysosomal enzymes is that either these
substances enter the lysosomal granules or the enzymes leave
the granules and enter the cytoplasm through a damaged membrane.
Disturbance of the integrity of the lysosomal membranes and
liberation of the enzymes can be produced in vitro by mechanical
injury to these organelles by homogenization, by alternately freezing and thawing the tissue, by treatment with hypotonic medium
(osmotic shock), by x rays, raising the temperature, treatment
with detergents, carbon tetrachloride, and vitamin A, an excess
of oxygen in the medium or anaerobiosis, a change in pH to the
acid side, etc. [202, 424, 759, 771, 773, 948, 966, 1145, 1213, 1581,
1602, 1753, 1762, 1777].

The opposite effect, i.e., the stabilization of lysosomal membranes and prevention of the release of the hydrolases, can be
secured by the addition of cortisone, hydrocortisone, and predni-
solone [759, 966, 1762] or of gangliosides [1600] to the medium.
Neither N-acetylnuraminic acid nor serum albumin has any
stabilizing action on lysosomal membranes [1600]. Calcium ions
protect isolated liver nuclei against the hydrolytic action of lysosomal enzymes, although under suitable conditions these enzymes
hydrolyze proteins, nucleic acids, and other components in a whole
tissue homogenate as well as in the mitochondrial and microsomal fractions intensively [1582].

Disturbance of the integrity of the lysosomal membranes in
vivo and liberation of hydrolases into the cytoplasm have been
observed as a result of exposure to high temperature, viruses,
microorganisms, toxins, certain convulsants, vitamin A, detergents, and other substances [424, 966, 1145, 1601, 1713, 1753, 1762].
A similar effect is observed during increased physical exertion
and fatigue [1037], and in pathological states such as starvation
[523, 966], hemorrhagic shock [835], experimental encephalo-

malacia induced by avitaminosis E [1166], hereditary muscular dystrophy, and dystrophy due to vitamin E insufficiency [966, 1666, 1758, 1801].

Activation of the hydrolytic enzymes of lysosomes also arises in the course of natural physiological wastage of the cells ("physiological regeneration," ontogenetic involution of the thymus and other organs), or their death in pathological processes such as necrosis of injured tissues, destruction of tumor cells, etc. [202, 424, 434, 966, 1079, 1601]. A high degree of activation of the lysosomal hydrolases is observed in spontaneous metamorphosis of amphibian organs [1234, 1574], in inflammatory processes [1312], and tissue regeneration [337, 868, 1462, 1463]. Manina has recently shown [348, 349] intensive formation of lysosomal granules in the cytoplasm of neurons and glial cells after x-ray injury and administration of certain drugs. Cortisone preparations, it is interesting to note, stabilize the lysosomal membranes of the liver in vivo just as efficiently as in vitro [1761, 1762].

The factors listed above, or at least some of them, that disturb the integrity of the lysosomal membranes may thereby act as biochemical trigger mechanisms essential for the lysosomal hydrolases to exhibit their activity in the intact or injured cell. There is a distinct possibility that neurohumoral factors play a definite role in the regulation of lysosome functions [424, 966].

Until recently the biological role of lysosomes in the cell was a matter for conjecture. As a result of the considerable volume of experimental material that has now accumulated [202, 424, 524, 966], the important role of the lysosomal enzymes is now recognized in the hydrolysis of biologically important cell components, in the processes of development and regression (fertilization, early embryonic development, metamorphosis, involution), and also in functions connected with secretion, immunity, detoxication, and the absorption of thrombi. Lysosomes evidently have a particularly important role in the activity of the intracellular structures of the brain, for the well-developed BBB substantially limits the flow of many substances into the metabolic reserves of the brain cells and their organelles. Under these circumstances the utilization of endogenous substrates, which may be available in the structural components in the large numbers of cells dying during postembryonic development, may become particularly important in biosynthetic reactions. It has been suggested that lysosomes may play an important role in the formation of axons and dendrites [806].

The Possible Biological Role of
Intracellular Peptide Hydrolases

To understand the biological role of intracellular endo- and exopeptidases the following questions must be answered. Are the conditions in the tissues suitable for proteolytic enzymes to manifest their activity and how can the processes of proteolysis be regulated? What biochemical reactions are catalyzed by intracellular peptide hydrolases and in what physiological functions or pathological processes do they participate? Is there any definite relationship between the intensity of protein metabolism and the activity of the enzyme systems of proteolysis in tissues, or their structures, with different functions, in different cells and their intracellular microstructures?

Considering the paucity of the available data, we cannot give a final answer to these questions. Such facts as we do possess which shed light on the biological role of intracellular systems of proteolysis in the functions of nerve and other tissues will be described in this section.

Physicochemical and Biological Optima of Enzyme Action. When the possible biological role of intracellular proteinases is assessed, the fact that the pH optima of many of them lie beyond the physiological conditions of the tissue environment often causes surprise. Misgivings regarding the ability of these enzyme systems to function in the tissues arise chiefly because sufficient attention is not always paid to the conditions actually existing in the tissues, and correct comparisons are not always made with conditions artificially created in vitro when the physicochemical properties and kinetics of action of isolated enzymes are studied. Due regard is not always given to changes in the environmental conditions taking place in tissues and their individual structural formations in certain physiological or pathological states. If all the features of action of the intracellular enzymes and all possible functionally determined variations in the composition and physicochemical properties of the intracellular medium were known, the apparent incongruities between the conditions for optimal activity of the enzymes (pH, temperature, etc.) in vitro (physicochemical) and those existing in reality (biological) would be less striking.

Unfortunately we do not yet know many of the conditions that determine the harmony between the action of proteolytic and other enzymes and the functions of any particular cell. The most important differences in the conditions of action of enzymes in vivo and in vitro will be described briefly below. Knowledge of these differences and allowance for them are essential for understanding the biological role, particularly of those proteolytic enzymes whose physicochemical optima appear at first glance not to correspond to conditions existing in the cell.

Differences in the Reaction of the Medium in Tissues, Cells, Organelles, and Different Parts of the Organelles. The reaction of the medium in different tissues, cells, organelles, and in different parts of the membranous structures under normal conditions may not only be neutral, but also acid or alkaline [154, 268, 345, 564]. The acidity of the medium in macrophages and lymphocytes is very low — the pH is about 3.0 or even less [1565], whereas the pH of fertilized frog's eggs is 8.5 [564]. Appreciable variations have also been observed in the reaction of the intracellular medium toward the acid and alkaline side during changes in the functional state of the nervous system and, in particular, in various pathological states [268, 270, 514, 564, 813]. The reaction of the medium in the cytoplasm is usually one pH unit lower than in the cell nucleus [564]. The surface layer of the cytoplasm of the starfish egg has a pH of about 5.0, whereas in areas more distant from the surface the pH of the cytoplasm reaches 7.6 [564]. A gradient of 1-1.5 pH units may even arise between the outer and inner surfaces of the mitochondrial membrane if an increased Ca^{++} concentration is present in the medium [750, 890, 891].

During a seizure, the reaction of the medium in some parts of the rat brain shifts toward the acid side by more than 0.5 pH unit [514]. In the glandular tissues of the nonlactating mammary gland the pH is close to neutral but during the period of lactation it falls to 4.6 [270]. In brain and muscle cells the reaction of the medium is close to neutral and it falls very little throughout the period of postnatal ontogeny. However, during physical exertion the pH of brain tissue may fall by 0.5 pH unit [813]. The pH of the protoplasm of cells falls sharply after their death [124, 564].

The disparity between the biological and physicochemical optima of enzymes may also be due largely to the fact that many enzymatic reactions in the intact cell take place on partition surfaces, whereas in vitro they take place in the aqueous medium. These differences may substantially modify the conditions of enzyme action, for the pH optima of enzymatic reactions on partition

surfaces and in the surrounding solution may differ by two or more
pH units [345]. As an example to illustrate this statement, the differences in the pH optima of glutamate dehydrogenase in intact
and disintegrated liver mitochondria (pH 8.0 and 6.5, respectively)
may also be cited. Even the pH at the surface of the protein molecule itself, i.e., where the enzyme–substrate complex is formed,
may differ from the pH of the surrounding medium [930].

Differences between the Action of Enzymes in vivo and in vitro.
An enzyme may exhibit its maximal activity in vivo and in vitro
at different pH values, different temperatures, and so on [1368,
1415, 1416]. This can be attributed primarily to differences in the
chemical composition of the media compared – the artificially
created medium in vitro and the intracellular medium as it actually
exists. Differences in the concentrations of ions, coenzymes, activators, and inhibitors may significantly affect both the rate of
formation and breakdown of enzyme–substrate complexes and
interaction of enzyme and substrate with the components of the
medium. Not only the velocity of the catalyzed reactions, but also
the sensitivity of the tissue proteins to the proteinases, may depend on the composition of the medium. For instance, substantial
changes in the pH optima of enzymes have been observed in experiments in vitro under the influence of various anions [1368, 1415],
and in vivo in animals receiving different diets [269, 270]. Glutathione, present in the tissues in very high concentrations, has
been shown [1294] to increase the sensitivity of protein substrates
to catheptic hydrolysis considerably. The ability of tissue proteins to undergo degradation is also modified by the action of certain pathological factors [323, 434, 438]. In different media proteolytic enzymes may even catalyze reactions in completely opposite directions, i.e., hydrolysis in an acid medium and synthesis
of peptide bonds in a neutral medium.

When differences between the biological and physicochemical
optima of action of a given enzyme are evaluated, due regard must
be paid to the fact that in vitro, especially with highly purified enzyme
preparations, the optimal conditions are optimal only for the isolated enzyme functioning independently of other enzymes or of
their action in a complex chain of consecutive reactions. In the
living cell, on the other hand, highly complex polyenzyme systems
are the rule. In these systems each enzyme may have its own pH,
temperature, or other optimum. Depending on the changing condi-

tions of the tissue medium, any link in the chain of reactions may become the limiting factor and thus put a brake on the overall reaction. On the other hand, when the system of enzymes acts together in vivo the products of the preceding reactions act as substrates for enzymes of the next reaction, so that the velocities of the individual reactions are automatically regulated and the efficiency of the system as a whole is high. The fact that the efficiency of action of a series of proteolytic enzymes (trypsin, chymotrypsin, and carboxypeptidase) was found to be higher in vitro than the sum of their individual efficiencies is therefore not surprising [12, 689, 690, 692]. Another important feature is that in vitro the pH and other optima for proteolytic enzymes are usually determined on denatured exogenous protein substrates, which differ in their chemical nature from the native cell proteins.

By contrast with conditions in vitro, much of the potential activity of most, if not all, enzymes remains unused in vivo [607, 757, 1250]. It has been calculated, on the basis of the mean half-life of total brain protein and the rate of proteolysis under optimal conditions of incubation [1250], that only one-tenth of the catheptic enzymes present in the tissue is necessary for maintaining the dynamic protein balance of the brain. The corresponding calculations have been undertaken for cathepsins of other organs [757]. If, therefore, the intracellular peptide hydrolases functioned in vivo at the same rates observed at the physicochemical optima in vitro, all the tissue proteins would very soon be degraded to free amino acids. Of course such uncontrolled hydrolysis of the tissue substrates does not take place in vivo. This is prevented by the mechanisms of intracellular regulation of the action of proteolytic systems.

Since many amino acids in vitro inhibit the hydrolysis of proteins by trypsin and chymotrypsin [691], the possibility that tissue free amino acids may also help regulate the activity of the intracellular proteinases by a feedback mechanism can be postulated. Many other mechanisms for the regulation of activity of the intracellular endo- and exopeptidases undoubtedly exist. Some possibilities have already been mentioned.

Substrate Differences. Another circumstance of great importance to the determination of the possible biological role of proteolytic enzymes may be mentioned. The optimal conditions differ for the degradation of exogenous and endogenous substrates and also for individual exogenous and individual endogenous substrates. For example, purified brain cathepsin hydrolyzes hemoglobin optimally at pH 3.0-3.1 but serum albumin at pH 4.1 [332]. Splenic cathepsin C hydrolyzes these substrates optimally at pH 5.0 and 4.0, respectively [1660]. Cases of preferential hydrolysis of endogenous over exogenous proteins by peptide hydrolases of muscle [1219, 1783] and lens [1117, 1202, 1740] and of exogenous substrates over endogenous by proteinases of the brain [1197, 1349, 1350] were mentioned earlier. Examples of differences in the intensity of hydrolysis of different exogenous proteins by the same enzyme systems of the spleen [1034, 1035], liver [1389], and brain [49, 332, 1349] and also of the hydrolysis of different endogenous brain proteins by trypsin [917] are also known.

Some proteins of nerve tissue (neurokeratins, one of the proteins of the proteolytic complex) are known not to be hydrolyzed at all by proteolytic enzymes [1019, 1273, 1692]. Such proteins are also found [1078] in the mitochondrial membranes of cardiac muscle. Also, the action of pathological factors may alter the sensitivity of tissue proteins to proteolytic enzymes [323, 434]. Changes in the physicochemical properties of proteins which may affect their susceptibility to attack by intracellular peptide hydrolases are also found during changes in the functional states of the organism. For instance, excitation and inhibition of nervous activity are accompanied by changes in the macrostructure of the protein molecules, their adsorption properties, solubility, isoelectric point, and composition of reactive groups [238-240, 356-358, 401, 514, 1698, 1699, 1702].

The many differences in the pH of the medium in the tissues, cells, and their organelles, the variation in the pH optimum of the enzymes depending on environmental conditions, and the unequal sensitivity of proteins to proteolysis, in conjunction with the lysosomal mechanisms of regulating intracellular hydrolytic activity, are evidence that the use of intracellular proteolytic enzymes with pH optima far from neutral is perfectly possible in the living organisms. The presence of comparatively high temperature optima for some enzymes [28, 1350], much higher than body temperature, must likewise not cause surprise.

Biochemical Reactions Catalyzed by Intracellular Peptide Hydrolases and Their Possible Biological Role. The following biochemical reactions, which can be catalyzed by digestive and intracellular peptide hydrolases, have been described in the literature.

Hydrolysis of Peptide Bonds. The ability of intracellular proteloytic enzymes to hydrolyze peptide bonds has been proven by observations on tissue autolysis, and by studies in which the criterion of proteolysis was the intensity of liberation of radio-active amino acids from proteins previously labeled in vivo [1215, 1482, 1483, 1613, 1635, 1636]. It is now undisputed that tissue proteins can be degraded to free amino acids with the aid of intracellular peptide hydrolases. However, the mechanisms of proteolysis have received inadequate study. This makes it much more difficult to elucidate the biological role of the intracellular proteinases in the maintenance of the dynamic state of the tissue proteins.

The question of the energy requirement for proteolysis has not yet been settled. Some evidence seems to suggest that the hydrolysis of peptide bonds by proteolytic enzymes can take place without the consumption of energy. Intensive hydrolysis of tissue proteins during tissue autolysis takes place long after the energy-producing processes have ceased. Proteolytic activity in liver homogenates in acid and neutral media is unchanged or even increased somewhat under anaerobic conditions [757, 782]. Dinitrophenol has no effect on the intensity of autolysis in liver homogenates [1635]. However, there is other evidence to show that proteolysis is dependent on sources of energy. Degradation of protein molecules in tissue slices is inhibited [912, 1613, 1635, 1636] by cyanide, dinitrophenol, fluorophenylalanine, and organophosphorus inhibitors, as well as under anaerobic conditions, but is stimulated [1482, 1483] by coenzyme A and ATP, i.e., by the same substances that inhibit or activate protein synthesis under the same conditions.

We must therefore consider the possibility that protein degradation in animal cells follows a more complex pathway than during autolysis. Protein catabolism in living cells might take place without the expenditure of energy, but it is interlinked with and controlled by the anabolic phase of protein metabolism, which does require the expenditure of energy. The suggestion has been made [1250, 1482, 1613, 1635, 1744] that at least two mechanisms of protein degradation exist in the tissues: a simple hydrolytic mechanism, not requiring energy, and a more complex mechanism in which certain steps (the transport of metabolic products, the formation of intermediate compounds, etc.) require the expenditure of energy.

It has been shown [1482] that during hydrolysis of proteins by proteinases, compounds of nucleotide nature and containing amino

acids are formed. This fact is of considerable interest in explaining the role of intracellular proteinases in protein renewal, for
the same compounds are also formed during protein biosynthesis.
It is possible that, through these and similar activated compounds
of amino acids with nucleic acids, the processes of protein synthesis and protein breakdown are interlinked [1482, 1651, 1652, 1744].

On the basis of all the experimental evidence available it can
now be taken as established that in some cases intracellular peptide hydrolases participate in certain biochemical conversions,
physiological functions, and pathological processes, of which the
most important are:

1. Catabolism of protein molecules during their renewal in
   vivo;
2. Degradation of protein structures in cells in the course of
   physiological wear and tear;
3. Certain specific structural changes in protein molecules:
   e.g., enzymic activation of proenzymes, inactivation of insulin, the formation of fibrin monomer from fibrinogen,
   etc. [39, 40, 808, 1264, 1389, 1712];
4. Hydrolysis of protein structures during metamorphosis
   of organs in amphibians [1574], involution of the uterus,
   the thymus, and the mammary gland in mammals [1079,
   1783], and in the specific structural changes in the proteins
   of muscles and the reproductive organs in fishes during
   spawning [1610];
5. The destruction of protein molecules and their complexes
   with other biologically important compounds in such pathological states as muscular dystrophies of varied origin
   [832, 1095, 1096, 1649, 1650, 1666, 1758, 1801], the destruction of tumor cells [332, 434, 1011], tissue necrosis caused
   by mechanical, ischemic, thermal, and other factors [810,
   1417, 1700], demyelination of nerve tissue from various
   causes [814, 1166, 1497, 1692], degeneration of nerves after
   their section or other treatment [743, 746, 1316, 1496],
   damage to the tissues by ionizing radiation [169, 518, 1777],
   cataract [947], and inflammation [1312, 1697];
6. Proteolysis of protein molecules connected with ammonia
   formation in nerve tissue [125, 262, 277, 1756];
7. Reversible structural changes in protein molecules observed during excitation [1698, 1699, 1702];

8. Regeneration of injured tissue structures [337, 433, 1462, 1463];
9. The formation of physiologically active peptides (such as bradykinin) in certain active forms of diseases of the CNS [894].

In conclusion, it must be admitted that the biochemical mechanisms whereby intracellular proteinases participate in many of these functions are not yet clear.

Hydrolysis of Amide and Ester Bonds. With respect to the amidase and esterase activity of the proteolytic enzymes only those data will be given that have a bearing on the explanation of the metabolic conversions of proteins in vivo.

Increased functional activity of nerve tissue is accompanied by an increase in intensity of ammonia formation, and a decrease in functional activity is accompanied by a corresponding decrease in ammonia formation [125, 261, 813]. Proteins are important sources of ammonia formation in nerve tissue [125, 277, 1756]. The content of amide groups in brain proteins, like the total content of tissue proteins, falls (on account of deamination and proteolysis) during an increase, and rises (on account of amination and protein resynthesis) with a decrease in functional activity of the brain [125, 277, 813]. This suggests that the hydrolytic and amidase activities of proteolytic enzyme systems may play an essential role in the formation and elimination of ammonia, processes closely linked with the function of the nervous system [128, 514, 663, 664], and also in the reversible structural reorganization of the proteins [125, 131, 273, 356–358, 635] of nerve tissue.

The esterase activity of the proteolytic enzymes may also play an important role in protein metabolism. It was shown recently [569, 1518] that the ester bonds in the aminoacyl tRNAs, intermediate products in protein biosynthesis, are split by trypsin and chymotrypsin just as readily as the corresponding bonds in esters of the amino acids and aliphatic alcohols, phenols, adenosine, and certain other compounds. The necessary data to answer the question whether intracellular proteinases with esterase activity participate in certain stages of protein biosynthesis and, if they do so participate, the extent of their role in this process, are not yet available. Experimental studies of these problems would undoubtedly help show whether intracellular proteinases do in fact participate in the biosynthesis and renewal of tissue proteins in vivo.

Transfer Reactions. The hypothesis that peptide bonds may be formed with the aid of transamination and transpeptidation reactions catalyzed by proteolytic enzymes, first discovered by Bergmann et al. [822, 823], is now regarded as proven. The synthesis of peptide bonds in reactions of this type has been confirmed by numerous investigations [54, 55, 171, 692, 1173, 1175, 1377, 1435, 1442, 1741] in which both digestive and intracellular peptide hydrolases were used. Dipeptides and tripeptides have now been successfully synthesized from intermediate products of protein biosynthesis (amino acids conjugated with tRNA) by means of a transpeptidation reaction catalyzed by trypsin [246, 1518]. However, we do not know whether intracellular proteinases carry out such condensations in vivo resulting in protein biosynthesis from aminoacyl tRNAs.

The tissues of the brain, liver, and kidneys contain an enzyme system that ruptures the peptide bond between glutamic acid and cysteine in glutathione [1010]. This catalyzed degradation of glutathione is not simple hydrolysis but a transfer reaction, for it takes place intensively only in the presence of certain substances (acceptors of the glutamine group), primarily amino acids and peptides.

Transpeptidase activity is evidently not a characteristic feature of all proteolytic enzymes. Two pepsin preparations with identical hydrolytic activity relative to hemoglobin have been obtained from pepsinogen [1435, 1436]. One of these enzymes has transpeptidase activity, the other does not. Acid proteinases with closely similar substrate specificity for hydrolysis of peptide bonds, but differing in transferase activity, are present in tissues of the lungs [931] and pituitary gland [983]. The lung proteinase catalyzes peptide synthesis from amino acid esters at pH 6.8, while the pituitary enzyme has no polymerase activity whatsoever.

Because of the thermodynamic conditions in the cell the synthesis of peptide bonds from free amino acids by spontaneous hydrolysis is virtually impossible in vivo. Transpeptidase reactions, on the other hand, like other transfer reactions, have low activation energies [59, 135, 340], for it is not free amino acids but their activated forms, consisting of peptides, esters, and amides, that participate in the formation of peptide bonds [340, 1303]. The experimental data cited above [1641] are interesting in this respect. They suggest the existence of a transpeptidase pathway of protein synthesis in the mitochondria, intracellular or-

ganelles with well-developed energy-generating systems. The
intracellular medium as a whole, it is important to note, is compatible with the synthesis of peptide bonds by transfer reactions,
for the proteolytic enzymes exhibit maximal transferase activity
at physiological pH values, i.e., in the pH range of 6.0-8.0 [931,
1173, 1175, 1377, 1435, 1442], far from the pH optimum for hydrolytic action by most cathepsins.

However, because of the necessity of providing a strictly
specific sequence of amino acids in the various protein molecules synthesized, it is unlikely that transferase activity of the intracellular
proteinases, which have wide ranges of specificity, is used independently as a physiological mechanism of protein biosynthesis.
There are probably more solid grounds for assuming that the
transpeptidase activity of the intracellular proteinases may be
used to lengthen polypeptide chains and to carry out various struc-
tral changes in protein molecules in vivo. For example, intracellular proteinases may be active in activation of proenzymes, poly-
merization of structural proteins, formation of hormones from
peptide fragments synthesized previously on template nucleic acids,
and also for reassembly of intermediate products of protein catabolism, etc. [135, 366, 1744].

Synthesis of Peptide Bonds during Plastein Formation. Detailed information on the role of proteolytic enzymes in the formation of plasteins in vitro will be found in a number of publications
[178, 282, 342, 681]. All that needs to be said here, therefore, is that
research has shown that proteolytic enzymes can participate, in
principle, in the resynthesis of peptide bonds. However, since
this resynthesis can take place only in unphysiological conditions
(high concentrations of enzymes and products of protein digestion,
strongly acid reaction of the medium), and since the proteinlike
substances formed under these circumstances lack the properties
of natural proteins, such as native structure and specificity, enzymatic, hormonal, antigenic, etc., it is difficult to imagine how
the plastein-forming action of the proteolytic enzymes could be
utilized in vivo.

Connection of the Intracellular Peptide Hydrolases with the Functions of Tissues and Their
Protein Metabolism. Many references have been made to
literature confirming the existence of a connection between the
intensity of protein metabolism and the activity of proteolytic
systems in structures of the nervous system subserving different

functions. This relation is expressed by the general rule that the activity of the intracellular proteinases is higher in structures with a higher intensity of protein metabolism. The same dependence of activity of proteolytic systems on the level of metabolic activity of their proteins is observed in different tissues [28, 985, 1096, 1141, 1349, 1416, 1418, 1473, 1564, 1696]. Whereas catheptic activity in brain tissues is four or five times less than in the liver and kidneys, the incorporation of various labeled amino acids into brain proteins is from four to ten times lower [1564].

Other examples are: The activity of certain proteolytic enzymes in immature reticulocytes, which are capable of synthesizing protein, is from two to eight times higher than in mature erythrocytes, which have lost their biosynthetic functions [985]; after ligation of the submandibular gland the secretion of saliva ceases and, at the same time, the activity of the cathepsins is reduced by half [1564]. It is interesting to note that the proteolytic activity of blood serum is very low [28, 1473], although serum proteins are renewed at a high rate in vivo. However, in this case it must be remembered that the catabolism of the serum proteins is an extravascular process [774, 1062, 1237, 1318].

The change in the relative velocities of synthesis and breakdown of proteins evidently serves as a unique type of mechanism for the regulation of protein metabolism in the tissues. For instance, an increase (or decrease) in the protein content in the adult organism in a state of dynamic equilibrium can be induced both by an increase (or decrease) in the intensity of synthesis and by a decrease (or increase) in the rate breakdown. Some workers have stated [1096, 1452, 1529], for example, that the growth and regeneration of tissues are due not so much to the intensification of protein synthesis as to a delay in protein breakdown.

The following facts are also relevant to the explanation of the biological role of proteolytic enzymes. In brain, the ribosomes, the cellular machinery for protein synthesis, have no endopeptidase activity and contain only traces of exopeptidases, whereas in the fraction of endoplasmic membranes, although endopeptidases are completely absent, high exopeptidase activity is present [936]. It has been suggested that membrane exopeptidases participate both in protein degradation and in transport functions. Experiments in Palladin's laboratory [172] have shown no activity for acid (hemoglobin substrate) or neutral (protamine substrate) proteinases in washed ribosomes. Practically all the activity of the acid and neutral proteinases of the postmitochondrial fraction of rabbit brain, from which the ribosomes were isolated, is concentrated in the fraction of membranes of the endoplasmic reticulum.

Data showing that the functional state of an organism and its tissues is reflected in the intensity of protein biosynthesis were given in previous sections of this book. Since the anabolic and catabolic phases of protein metabolism are interconnected and interdependent, the intensity of the catabolic phase must also be connected with the functional activity of the tissues. The data illustrating this point will now be discussed.

In cats deprived of photic stimulation for several days the intensity of proteolysis in the visual centers and tracts was reduced but was unchanged in parts of the brain not connected with visual function. Proteolytic activity returned to normal 15 min after the resumption of exposure to light [167]. A similar dependence of the intensity of proteolysis on the functional state of the auditory centers was found in dogs deprived of acoustic stimulation for 3 days [661].

A close connection between the activity of the intracellular endo- and exopeptidases and the functions of central and peripheral nerves in animals of different species has also been demonstrated by biochemical and histochemical methods. Wallerian degeneration of nerves is accompanied by demyelination and by increased activity of acid and neutral proteinases [746, 1316, 1496]. Experiments in vitro have shown that in the process of demyelination, proteolytic enzymes rupture the bonds between lipids and proteins sensitive to trypsin and hydrolyze the proteins. Activation of proteolytic enzymes is also observed in brain tissue throughout the period of development of experimental allergic encephalomyelitis, which is also accompanied by demyelination. Toward the end of the disease, when the demyelination process comes to an end, the activity of the proteinases in the brain tissue returns to normal [814].

Other findings confirm the important role of intracellular proteinases in the development of demyelination [912, 1497]. Organophosphorus inhibitors of acetylcholinesterase not only inhibit the activity of this enzyme, but also induce demyelination and increase the activity of intracellular proteinases in the tissues of the spinal cord and sciatic nerve. In brain tissue, on the other hand, these poisons strongly inhibit acetylcholinesterase but do not induce demyelination and do not change the activity of the proteolytic systems [1497].

The activity of acid and neutral proteinases in the tissue of the central and peripheral nervous systems also rises after fatiguing physical exertion [164, 813, 1037, 1161], during electrical stimulation of nerve tissue [262, 514, 1699], during the action of ionizing radiation [518], and during stimulation of nervous activity by various pharmacological agents [514, 1702]. Inhibitory processes in the tissues of the nervous system, on the other hand, are accompanied by a decrease in the activity of proteolytic enzymes [514, 1702]. It is also interesting to note that many anesthetics also inhibit the activity of the brain peptide hydrolases in vitro [1698].

Another interesting relationship of the intracellular proteinases to organ function was revealed in the lens. In the normal lens active enzymes with proteolytic and esterase action in the pH range of 7.2-7.5 were found [947, 1117, 1740]. During the development of senile cataract both these activities disappeared [947]. If cataract was produced experimentally, the cathepsin activity in the lens also decreased considerably [1803]. These enzymes evidently play an important role in maintaining the transparency of the lens medium.

Proteolytic enzyme activity in the muscles rises appreciably in response to changes in their function, for example, in the dystrophies: hereditary; alimentary, due to vitamin E deficiency [832, 1666, 1758, 1801], and after denervation [1095, 1096, 1649, 1650]. The activity of these enzymes is gradually restored after reinnervation [1095]. It is notable that in rapidly contracting muscles (latissimus dorsi posterior) the activity of the exo- and endopeptidases rises much more after denervation than in more slowly contracting muscles (latissimus dorsi anterior). Under these conditions, incidentally, hypertrophy occurs to a large extent only in the latter muscles [1096]. Proteolysis in the muscles also increases appreciably during the performance of work either of short or long duration. The increase is greater in the latter case and, particularly, in untrained muscles [558]. Acid proteinase activity in inflamed synovial tissue of the knee joint is increased fourfold [1312]. These facts clearly indicate that proteolytic enzymes participate in the regulation of protein metabolism during changes in the function of an organ.

Tissue endo- and exopeptidases play an important biological role in the utilization of tissue protein reserves as sources of free amino acids for the biosynthesis of specific proteins. The following cases will serve as examples of the use of amino acids of some proteins for the biosynthesis of others: the comparatively high stability of the protein concentration in the brain tissue during prolonged protein insufficiency coupled with a marked decrease in the protein content in other tissues [83, 676, 760, 1281, 1344]; the intensive formation of milk proteins in the isolated lactating mammary gland [242]; the 10- to 20-fold increase in weight of the ovaries in the salmon before egg-laying combined with a simultaneous 40% decrease in the weight of the muscles [1610]. Characteristically the activity of proteolytic enzymes in the muscles of fishes is 5 to 10 times higher than in mammalian muscles [1608, 1610], and it is four or more times higher in the lactating and involuting mammary gland than during pregnancy [1079].

All of these data, and more could be cited [522], are evidence that endogenous tissue proteins under certain physiological and pathological conditions may be an important reserve of amino acids for the synthesis of new proteins in vivo. The leading role in the utilization of this important source of amino acids is played by intracellular endo- and exopeptidases.

There are many other example that could be quoted as evidence of the role of intracellular proteinases in physiological and pathophysiological processes in different organs. But enough examples have already been discussed, and we conclude that all the known facts suggest that the enzymes of protein catabolism in the tissues are linked in some way or other with the functions of those organs and with the general trend and intensity of protein metabolism in them. The mechanisms of this link are not clear, for the mechanisms regulating the activity of the intracellular proteinases are also unknown. The identification of these mechanisms must help to reveal the biological role of the proteolytic enzyme systems in the tissues of the body, including the nervous system.

On the basis of the writers' experimental results and their comparison with data in the literature, a role for intracellular proteinases can be postulated in the renewal of the tissue proteins in the body and possibly in other functions of the cell.

The fact that neutral proteinase activity in the organs correlates in general with the intensity of renewal of their proteins indicates that intracellular proteinases with a neutral pH optimum participate in the renewal of tissue proteins by degrading them to free amino acids. Tissue proteolysis as a source of free amino acids for the biosynthesis of new protein molecules plays a particularly important role in nerve tissue, partially separated as it is from other sources of free amino acids by the powerful BBB, and containing a large proportion of permanent cells that function throughout the life of the organism.

A fact that merits special attention when the data on the subcellular localization of neutral and acid proteinases are examined is that both these enzyme systems are contained in all subcellular structures, although the levels of their activity vary considerably in different parts of the cell. Just as the enzyme systems of protein synthesis are more active in some structural components of the cell than in others, the activity of protein-degrading enzymes is also distributed irregularly between the various parts of the cell.

The differences observed in the concentration of neutral and acid proteinases in different parts of the nervous system and its subcellular fractions not only point to the existence of two different enzyme systems of proteolysis in brain tissue but are also evidence of their different functional roles. Acid proteinases evidently hydrolyze protein structures mainly in dying cells, whereas neutral proteolytic systems exhibit their activity in functioning cells. Whereas the role of acid proteinases is particularly important in pathological processes, the functions of the neutral proteinases are more characteristic of normal cells.

Another possibility is that the proteolytic enzymes of cells, or at least some of them, participate not only in the renewal of proteins, but also in other, possibly specific, functions of nerve tissue. The argument on which this hypothesis is based is as follows. If the intracellular proteinases participated only in the catabolic phase of protein metabolism, their activity in the gray matter of the brain would be higher than in the white; but neutral proteinase is almost twice as active in the white matter of the cerebral hemispheres. Correlation between the intensity of incorporation of labeled amino acids into proteins and the level of proteolytic activity is not observed in all subcellular structures of brain tissue. For example, in nuclei amino acids are incorporated rapidly into proteins although the neutral proteinase activity in nuclei is comparatively low. In the mitochondria, on the other hand, neutral proteinase activity is quite high but mitochondrial proteins are renewed comparatively slowly. Although the neutral proteinase activity that we found in myelin is negligible, its level does not correlate with the very low intensity of renewal of the myelin proteins. The high activity of the various peptide hydrolases in the nerve endings fraction is evidence of their participation in synaptic functions [1347].

The results indicating that most intracellular proteinases are located in the structural components of the cell and are firmly bound with the membranes are of great interest. They suggest that protein catabolism in the cell takes place chiefly on its membranous structures.

Conclusions

The facts described in the preceding chapters on the brain proteins and their metabolism show the changes that have taken place in this field during recent years. Whereas only a few years ago the brain proteins had received very little study, despite recognition of their important role in the functions of the nervous system, research in this field has now expanded greatly. This expansion occurred as a result of the introduction and perfection of new techniques (the labeling of compounds with radioactive carbon and other isotopes, chromatography, disc electrophoresis, differential centrifugation), as a result of the rejection of the aspirations of "pure" neurophysiology to regard electrical potentials as the sole basis of all the complex phenomena of activity of nerve cells, and also as a result of the wide application of biochemical methods to metabolic processes not only at the level of tissues, but also at the molecular level.

Although the isolation and study of proteins specific for nerve tissue, such as protein S-100, and the investigation of their distribution among the various structures of the brain, their metabolism, and their functional role are essentially only just beginning, the results already obtained are extremely promising.

Much evidence has been obtained supporting the view that the proteins of the brain are the basis of all its complex functions, and participate in all aspects of mental activity of the brain and its dysfunctions, and that protein metabolism in the brain is closely connected with various aspects of behavior (instinct, learning, memory, etc.).

Recent investigations give reason to expect progress in the very near future in our knowledge of the functions of nerve tissue

and in the development of ways of overcoming disturbances of the
mental activity of the brain. Mental disorders such as homo-
cystinuria and arginine-succinaturia, in whose development specific brain proteins participate, are now being studied. The effect
of a protein-deficient diet on the mental development of children
is firmly established. Schizophrenia is very probably linked with
protein metabolism. Disorders of behavior connected with al-
coholism and narcotic poisoning may be due, according to some
authors, to damage to protein structures of the brain. Very slight
structural changes in the protein molecule are known to have important consequences for the organism: for example, sickle-cell
anemia is due to the substitution of one amino acid residue in the
hemoglobin molecule. In precisely the same way, if one amino
acid is replaced by another in the insulin molecule, it loses its
antidiabetic activity.

In the light of the latest research, a bright and fruitful future
for the biochemistry of the nervous system and particularly for
the study of the protein metabolism of the brain can be expected.
All that is required is further sustained and dedicated work by
scientists in the various specialities — neurochemists, neuro-
physiologists, psychologists — for the protein metabolism of the
nervous system lies at the basis of practically all its functions.

The authors of this book will be satisfied if their work is help-
ful to research workers and, in particular, to those just beginning
to study the problem of protein metabolism in nerve tissue. We
trust this book will provide a solid basis for their work and equip
them with new ideas for future investigations.

Bibliography

1. Avdeev, V. G. Ukr. Biokhim. Zh., 1968, 40:558-561.
2. Avdeev, V. G., and Belik, Ya. V. Ukr. Biokhim. Zh., 1967, 39:584-589.
3. Avdeev, V. G., and Belik, Ya. V. In: Metabolism of the Cell Nucleus and Nucleo-cytoplasmic Relations [in Russian], Naukova Dumka, Kiev, 1970, pp. 82-84.
4. Avdeev, V. G., and Palladin, A. V. Ukr. Biokhim. Zh., 1967, 39:18-24.
5. Avtandilov, G. G. In: Tissue—Blood Barriers [in Russian], Izd. AN SSSR, Moscow, 1961, pp. 375-380.
6. Agranoff, B. W. In: Molecules and Cells, Vol. 4 [Russian translation], Mir, Moscow, 1969, pp. 107-118.
7. Azhipa, Ya. I., Kayushin, L. P., and Nikishkin, E. I. In: Problems in Neurochemistry [in Russian], Nauka, Moscow, 1966, pp. 209-214.
8. Aleksandrovskaya, M. M. In: Problems in the Study of Tissue—Blood Barriers [in Russian], Nauka, Moscow, 1965, pp. 288-292.
9. Aleksidze, N. G. Zh. Évolyuts. Biokhim. i Fiziol., 1970, 6:381-383.
10. Antonov, V. K. Izvest. Akad. Nauk SSSR, Ser. Biol., 1970:823-837.
11. Aprikyan, G, V., and Paronyan, Zh. A. In: Problems in Brain Biochemistry, Vol. 3 [in Russian], Izd. AN Arm. SSR, Erevan, 1967, pp. 67-82.
12. Afanas'ev, P. V., and Mosolov, V. V. Dokl. Akad. Nauk SSSR, 1955, 100:507-510.
13. Acidosis, In: Great Medical Encyclopedia, Vol. 1 [in Russian], Gos. Izd. Med. Lit., Moscow, 1957, pp. 1246-1253.
14. Acs, G., Balazs, R., and Straub, F. B. Ukr. Biokhim. Zh., 1953, 25:17-27.
15. Babii, T. P., Skvirskaya, E. B., and Kovalenko, M. I. Ukr. Biokhim. Zh., 1965, 37:33-42.
16. Babskaya, Yu. E., and Pogosova, A. V. Ukr. Biokhim. Zh., 1969, 41:52-55.
17. Baev, A. A. In: Amino Acid Metabolism [in Russian], Metsniereba, Tbilisi, 1967, pp. 39-54.
18. Balashova, E. K., et al. Biokhimiya, 1958, 23:674-682.
19. Banshchikov, V. M., and Stolyarov, G. V. Zh. Nevropat. i Psikhiat., 1961, 61:934-943.

20.   Baranov, M. N., and Pevzner, L. Z. Uspekhi Sovr. Biol., 1964, 58:221-241.
21.   Barmina, O. N. In: Biochemistry of the Brain [in Russian], Gor'kii, 1941,
        pp. 180-190.
22.   Barmina, O. N. Third All-Union Conference on Biochemistry of the Nervous
        System [in Russian], Izd. AN Arm. SSR, Erevan, 1963, pp. 573-580.
23.   Barmina, O. N. In: Problems in Biochemistry of the Brain [in Russian], Volgo-
        Vyatskoe Izd., Gor'kii, 1966, pp. 38-42.
24.   Barmina, O. N., and Gorodisskaya, G. Ya. In: Problems in Biochemistry of
        the Brain [in Russian], Volgo-Vyatskoe Izd., Gor'kii, 1966, pp. 5-13.
25.   Belik, Ya. V. Third All-Union Conference on Biochemistry of the Nervous
        System [in Russian], Izd. AN Arm. SSR, Erevan, 1963, pp. 39-45.
26.   Belik, Ya. V. Ukr. Biokhim. Zh., 1970, 42:386-400.
27.   Belik, Ya. V., et al. Uk. Biokhim. Zh., 1968, 40:543-548.
28.   Belik, Ya. V., Grinenko, O. G., and Smerchinskaya, L. S. Ukr. Biokhim. Zh.,
        1968, 40:532-537.
29.   Belik, Ya. V., and Krachko, L. S. Ukr. Biokhim. Zh., 1959, 31:322-329.
30.   Belik, Ya. V., and Krachko, L. S., Ukr. Biokhim. Zh., 1961, 33:684-692.
31.   Belik, Ya. V., Palladin, A. V., and Smerchinskaya, L. S. Zh. Évolyuts. Bio-
        khim. i Fiziol., 1967, 3:394-401.
32.   Belik, Ya. V., Smerchinskaya, L. S., and Glovatskaya, O. P. Ukr. Biokhim.
        Zh., 1969, 41:3-10.
33.   Belik, Ya. V., Terletskaya, Ya. T., and Smerchinskaya, L. S. Proceedings of
        the 4th All-Union Conference on Biochemistry of the Nervous System [in Rus-
        sian], Tartu, 1969, pp. 214-224.
34.   Belik, Ya. V., Terletskaya, Ya. T., and Tyulenev, V. I. Ukr. Biokhim. Zh.,
        1965, 37:839-849.
35.   Belik, Ya. V., Terletskaya, Ya. T., and Tyulenev, V. I. Ukr. Biokhim. Zh.,
        1966, 38:343-348.
36.   Belik, Ya. V., and Tyulenev, V. I. In: Problems in Neurochemistry [in Rus-
        sian], Nauka, Moscow, 1966, pp. 18-27.
37.   Belik, Ya. V., and Tyulenev, V. I. Zh. Évolyuts. Biokhim. i Fiziol., 1966,
        2:333-338.
38.   Belik, Ya. V., and Tyulenev, V. I. Ukr. Biokhim. Zh., 1967, 39:343-351.
39.   Belik, Ya. V., and Khodorova, E. L. The Biochemistry of Blood Clotting [in
        Russian], Izd. AN Ukr. SSR, Kiev, 1957.
40.   Belitser, V. A., Khodorova, E. L., and Varetskaya, T. V. Ukr. Biokhim. Zh.,
        1961, 33:753-778.
41.   Belitser, V. A., Khodorova, E. L., and Varetskaya, T. V. In: Chemical Basis
        of Living Processes [in Russian], Medgiz, Moscow, 1962, pp. 146-155.
42.   Proteins, Vol. 3, Parts I-II. The Biochemistry of Proteins [Russian translation],
        Moscow, 1958-1959.
43.   Berdyshev, G. D. Uspekhi Sovr. Biol., 1970, 70:376-396.
44.   Beritov, I. S. Structure and Functions of the Cerebral Cortex [in Russian],
        Nauka, Moscow, 1969.
45.   The Biosynthesis of Protein and Nucleic Acids [in Russian], Nauka, Moscow,
        1965.

46. Blinkov, S. M., and Glezer, I. I. The Human Brain in Figures and Tables [in Russian], Meditsina, Leningrad, 1964. (English translation, Plenum Press, New York, 1968.)

47. Block, R. J., and Bolling, D. The Amino Acid Composition of Proteins and Food Products [Russian translation], IL, Moscow, 1949.

48. Blokhina, V. D. Uspekhi Sovr. Briol., 1963, 55:34-44.

49. Blyum, É., Yakovchuk, A. I., and Yarmoshkevich, A. I. Byull. Éksperim. Biol. i Med., 1936, 1:16-17.

50. Bogdanova, E. V., and Tolkacheva, G. M. Zh. Évolyuts. Biokhim. i Fiziol., 1969, 4:37-41.

51. Bogolepov, N. N., and Dovedova, E. L. Tsitologiya, 1969, 11:189-200.

52. Bogolov, V. P. Nauch. Trudy Ryazan. Med. Inst., 1964, 18:262-267.

53. Borisov, I. N. Uspekhi Sovr. Biol., 1966, 62:222-236.

54. Botvinnik, M. M., et al. Zh. Obshch. Khimii, 1961, 31:3234-3242.

55. Botvinnik, M. M., and Ostoslavskaya, V. I. Dokl. Akad. Nauk SSSR, 1958, 123:285-288.

56. Braines, A. S. In: Problems in Bionics [in Russian], Nauka, Moscow, 1967, p. 240.

57. Braun, A. D., et al. Tsitologiya, 1965, 7:494-500; 1967, 9:112-116.

58. Braun, A. D., and Nesvetaeva, N. M. Tsitologiya, 1967, 9:112-116.

59. Braunshtein, A. E. In: Amino Acid Metabolism [in Russian], Metsniereba, Tbilisi, 1967, pp. 11-23.

60. Brachet, J. In: Current Problems in Cytology [Russian translation], Moscow, 1955, pp. 11-38.

61. Brachet, J. Biochemical Embryology [Russian translation], IL, Moscow, 1961.

62. Bresler, S. E. Introduction to Molecular Biology [in Russian], Izd. AN SSSR, Moscow, 1963.

63. Brik, I. L. Vestn. Leningrad. Gos. Univ., 1955, 10(4):57-66.

64. Brodskii, V. Ya. Dokl. Akad. Nauk SSSR, 1960, 130:189-192.

65. Brodskii, V. Ya. Tsitologiya, 1961, 3:312-326.

66. Brodskii, V. Ya. Zh. Obshch. Biol., 1964, 25:39-50.

67. Brodskii, V. Ya. Cell Nutrition [in Russian], Nauka, Moscow, 1966.

68. Broun, R. G. Vestn. Leningrad. Gos. Univ., 1955, 10(4):67-75.

69. Broun, R. G. In: The Nervous System, Vol. 2 [in Russian], Leningrad University Press, 1960, pp. 11-14.

70. Broun, R. G., and Goncharova, V. P. Ukr. Biokhim. Zh., 1962, 34:734-740.

71. Brumberg, V. A. Tsitologiya, 1968, 10:1193-1197.

72. Brumberg, V. A., and Pevzner, L. Z. Tsitologiya, 1968, 10:1452-1459.

73. Budanova, A. M. Byull. Éksperim. Biol. i Med., 1936, 1(2):114-115.

74. Buikis, I. M. Izvest. Akad. Nauk Latv. SSR, 1966, 10:121-130.

75. Bulankin, I. N., et al. Trudy Nauch.-Issled. Inst. Biol. Khar'kov. Univ., 1954, 21:87-98.

76. Bulankin, I. N., and Blyumina, M. A. Trudy Nauch.-Issled. Inst. Biol. Khar'kov. Univ., 1947, 12:61-73.

77. Bulankin, I. N., Novikova, N. M., and Parina, E. V. Ukr. Biokhim. Zh., 1953, 25:147-156.

78.   Bulankin, I. N., and Parina, E. V. In: Current Problems in Biochemistry. 1.
      The Biochemistry of Proteins [in Russian], Moscow, 1959, pp. 205-216.
79.   Bulankin, I. N., and Parina, E. V. In: Problems in Gerontology and Geriatrics
      [in Russian], Medgiz, Leningrad, 1962, pp. 9-13.
80.   Bulankin, I.N., and Parina, E. V. Trudy. Nauch.-Issled. Inst. Biol. Biol. Fak.
      Khar'kov. Univ., 1962, 33-34:285-315.
81.   Bulankin, I. N., Parina, E. V., and Golovko, N. I. Dokl. Akad. Nauk SSSR,
      1960, 134:1461-1463.
82.   Bulankin, I. N., Parina, E. V., and Golovko, N. I. Trudy Nauch.-Issled. Inst.
      Biol. Biol. Fak. Khar'kov. Univ., 1962, 33-34:21-26.
83.   Bulankin, I. N., Parina, E. V., and Sergienko, E. F. In: Age Changes in Metab-
      olism and Reactivity of the Organism [in Russian], Izd. AN Ukr. SSR, Kiev,
      1951, 27-35.
84.   Bunyatyan, G. Kh. Third All-Union Conference on Biochemistry of the Ner-
      vous System [in Russian], Izd. AN Arm. SSR, Erevan, 1963, pp. 133-151.
85.   Bunyatyan, G. Kh. In: The Role of Gamma-aminobutyric Acid in the Ac-
      tivity of the Nervous System [in Russian], Leningrad University Press, 1964,
      pp. 9-27.
86.   Bunyatyan, G. Kh. In: Problems in Neurochemistry [in Russian], Nauka, Mos-
      cow, 1966, pp. 148-157.
87.   Bunyatyan, G. Kh., and Aprikyan, G. V. In: Problems in Biochemistry, Vol. 2
      [in Russian], Izd. AN Arm. SSR, Erevan, 1961, pp. 5-15.
88.   Bunyatyan, G. Kh., and Arutyunyan, A. V. Dokl. Akad. Nauk Arm. SSR, 1965,
      40:209-215.
89.   Bunyatyan, G. Kh., and Oganesyan, V. S. In: Problems in Biochemistry,
      Vol. 2 [in Russian], Izd. AN Arm.SSR, Erevan, 1961, pp. 17-28.
90.   Busch, H. Histones and Other Nuclear Proteins [Russian translation], Mir,
      Moscow, 1967. (English edition: Academic Press, New York, 1965.)
91.   Varypaeva, I. S. Uch. Zap. Gor'k. Med. Inst., 1957, 1:35-44.
92.   Veksler, Ya. I., and Gershenovich, Z. S. Biokhimiya, 1965, 30:449-456.
93.   Verbolovich, P. A. Myoglobin and Its Role in the Physiology and Pathology
      of Animals and Man [in Russian], Medgiz, Moscow, 1961.
94.   Veremienko, K. M. Ukr. Biokhim. Zh., 1963, 35:294-297.
95.   Veremienko, K. M., and Belitser, V. A. Dop. Akad. Nauk URSR, 1963, 4:501-504.
96.   Verzhbinskaya, N. A., In: Problems in Biochemistry of the Nervous System
      [in Russian], Izd. AN Ukr.SSR, Kiev, 1957, pp. 187-199.
97.   Verzhbinskaya, N. A. In: Tissue—Blood Barriers [in Russian], Izd. AN SSSR,
      Moscow, 1961, pp. 146-157.
98.   Verzhbinskaya, N. A., and Volkova, R. I. Dokl. Akad. Nauk SSSR, 1958,
      118:135-138.
99.   Vertaimer, N. Author's Abstract of Dissertation for the Degree of Candidate
      of Sciences, Kiev State University, 1954.
100. Vilenchik, M. M. Uspekhi Sovr. Biol., 1970, 69:380-397.
101. Vinnikov, Ya. A., Zhinkin, I. L., and Shchukolyukov, S. A. Fiziol. Zh. SSSR,
      1964, 50:1329-1334.
102. Vinogradov, A. G., and Ékman, N. V. In: The Nervous System, Vol. 10 [in
      Russian], Leningrad University Press, 1969, pp. 84-88.

103. Vladimirov, G. E. Functional Biochemistry of the Brain [in Russian], Izd. AN SSSR, Moscow, 1954.
104. Vladimirov, G. E. In: Problems in Biochemistry of the Nervous System [in Russian], Izd. AN Ukr. SSR, Kiev, 1957, pp. 247-257.
105. Vladimirov, G. E. In: The Study of the Animal Organism [in Russian], Izd. AN SSSR, Moscow, 1957, pp. 5-9.
106. Vladimirov, G. E. In: Current Problems in Biochemistry. I. Biochemistry of Proteins [in Russian], Institute of Biological and Medical Chemistry, Academy of Medical Sciences of the USSR, Moscow, 1959, pp. 114-125.
107. Vladimirov, G. E., et al. In: The Nervous System, Vol. 2 [in Russian], Leningrad University Press, 1960, pp. 3-10.
108. Vladimirov, G. E., Ivanova, T. N., and Pravdina, N. I. Biokhimiya, 1956, 21:155-162.
109. Vladimirov, G. E., Ivanova, T. N., and Pravdina, N. I. In: Problems in Biochemistry of the Nervous System [in Russian], Izd. AN Ukr. SSR, Kiev, 1957, pp. 61-68.
110. Vladimirov, G. E., Myul'berg, A. A., and Sytinskii, I. A. Vopr. Med. Khimii, 1961, 7:65-70.
111. Vladimirov, G. E., and Panteleeva, N. S. Functional Biochemistry. Selected Chapters [in Russian], Leningrad University Press, 1965.
112. Vladimirov, G. E., and Sytinskii, I. A. Uspekhi Sovr. Biol., 1961, 51:3-20.
113. Vladimirov, G. E., and Urinson, A. P. Biokhimiya, 1957, 22:709-714.
114. Vladimirova, E. A. Third All-Union Conference on Biochemistry of the Nervous System [in Russian], Izd. AN Arm. SSR, Erevan, 1963, pp. 207-221.
115. Vladimirova, E. A. Zh. Vyssh. Nerv. Deyat., 1968, 18:399-406.
116. Vladimirova, E. A. Proceedings of the 4th All-Union Conference on Biochemistry of the Nervous System [in Russian], Tartu State University Press, 1969, pp. 73-85.
117. Vladimirova, E. A., Gordon, B. G., and Nilova, N. S. Ukr. Biokhim. Zh., 1965, 37:538-545.
118. Voitkevich, A. A. In: Current Problems in Endocrinology, Vol. 1 [in Russian], Medgiz, Moscow, 1960, pp. 48-93.
119. Voitkevich, A. A., and Dedov, I. I. Dokl. Akad. Nauk SSSR, 1968, 182:197-200.
120. Voitkevich, A. A., and Dedov, I. I. Tsitologiya, 1969, 11:1234-1240.
121. Volzhina, N. S. In: The Problem of Brain Development and the Effect of Harmful Factors on It [in Russian], Medgiz, Moscow, 1960, pp. 215-221.
122. Voronin, L. G., et al. Zh. Vyssh. Nerv. Deyat., 1967, 17:553-555.
123. Voronina, N. P., and Veksler, Ya. I. Ukr. Biokhim. Zh., 1967, 39:25-28.
124. Vorontsova, M. A., and Liozner, L. D. Physiological Regeneration [in Russian], Sovetskaya Nauka, Moscow, 1955.
125. Vrba, R., Uspeki Sovr. Biol., 1956, 41:321-352.
126. Gavrilova, T. N. Zh. Nevropat. i Psikhiat., 1967, 67:1014-1019.
127. Gavrilova, T. N. Tsitologiya, 1967, 9:68-72.
128. Gaevskaya, M. S. Biochemistry of the Brain during Dying and Resuscitation [in Russian], Medgiz, Moscow, 1963.
129. Gaevskaya, M. S., et al. In: Problems in Neurochemistry [in Russian], Nauka, Moscow-Leningrad, 1966, pp. 88-96.

130. Gaevskaya, M. S., and Nosova, E. A. Ukr. Biokhim. Zh., 1961, 33:407-419.
131. Gaevskaya, M. S., Nosova, E. A., and Slez, L. M. Ukr. Biokhim. Zh., 1965, 37:691-696.
132. Galoyan, A. A. Dokl. Akad. Nauk Arm. SSR, 1964, 38:305-308.
133. Galoyan, A. A., and Srapionyan, R. M. Biol. Zh. Armenii, 1966, 19(9):15-20.
134. Gannushkina, I. V. Zh. Nevropat. i Psikhiat., 1958, 58:1025-1030.
135. Haurowitz, F. Chemistry and Function of Proteins [Russian translation], Mir, Moscow, 1965. (English edition: Academic Press, New York, 1963.)
136. Haurowitz. F. Immunochemistry and the Biosynthesis of Antibodies [Russian translation], Mir, Moscow, 1969. (English edition: John Wiley, New York, 1967.)
137. Haurowitz, F., and Crampton, C. In: Current Problems in Cytology [Russian translation], IL, Moscow, 1955, pp. 39-50.
138. Gachev, É. Lab. Delo, 1958, (2):8-11.
139. Gvozdev, V. A. Biokhimiya, 1960, 25:920-930.
140. Gvozdev, V. A., and Khesin, R. B. Dokl. Akad. Nauk SSSR, 1960, 134:1226-1228.
141. Geinisman, Yu. A., Larina, V. N., and Mats, V. N. Tsitologiya, 1970, 12:1028-1038.
142. Gel'man, N. S. Uepekhi Sovr. Biol., 1967, 64:379-398.
143. Gel'man, N. S. Uspekhi Sovr. Biol., 1969, 68:3-18.
144. The Blood—Brain Barrier [in Russian], Moscow, 1935.
145. Georgiev, G. P. Uspekhi Sovr. Biol., 1962, 54:285-308.
146. Georgievskaya-Petrun'kina, A. M. Russkii Fiziol. Zh., 1921, 4:268-269.
147. Gershenovich, Z. S., et al. Third All-Union Conference on Biochemistry of the Nervous System [in Russian], Izd. Akad. Nauk Arm. SSR, Erevan, 1963, pp. 91-101.
148. Gershenovich, Z. S., and Krichevskaya, A. A. Biokhimiya, 1960, 25:310-317.
149. Gershenovich, Z. S., Krichevskaya, A. A., and Kheruvimova, V. A. In: Problems in Neurochemistry [in Russian], Nauka, Moscow, 1966, pp. 75-87.
150. Gershenovich, Z. S., Krichevskaya, A. A., and Shumskaya, V. I. Dokl. Akad. Nauk SSSR, 1965, 162:1415-1417.
151. Gershenovich, Z. S., Sinichkin, A. A., and Rumbesht, L. M. Dokl. Akad. Nauk SSSR, 1969, 186:474-476.
152. Gershenovich, Z. S., and Émirbekov, É. Z. Ukr. Biokhim. Zh., 1968, 40:270-273.
153. Gershtein, L. M. Tsitologiya, 1966, 8:639-642.
154. Giese, A. C. Cell Physiology [Russian translation], IL, Moscow (1959). (English 1st edition, Saunders, Philadelphia, 1957.)
155. Tissue—Blood Barriers [in Russian], Izd. AN SSSR, Moscow, 1961.
156. Tissue—Blood Barriers and Ionizing Radiation [in Russian], Izd. AN SSSR, Moscow, 1963.
157. Histology [in Russian], Medgiz, Moscow, 1963.
158. Gladkii, A. P. Arkh. Anat., Gistol., i Émbriol., 1958, 35(1):59-62.
159. Glebov, R. N. Uspekhi Sovr. Biol., 1970, 70:26-40.
160. Glezer, I. I., Mitashova, N. I., and Pevzner, V. I. In: Problems in the Study of Tissue—Blood Barriers [in Russian], Nauka, Moscow, 1965, pp. 310-323.

161.   Golubitskaya, R. I. Pratsi N.-D. Zool.-Biol. Inst. Khar'k. Derzh. Univ., 1938, 5:81-101.
162.   Goncharenko, E. N., and Utevskaya, L. B. In: Tissue—Blood Barriers and Ionizing Radiation [in Russian], Izd. AN SSSR, Moscow, 1963, pp. 52-59.
163.   Goncharova, V. P., and Broun, R. G. Ukr. Biokhim. Zh., 1964, 36:126-131.
164.   Gorachkova, M., Jakoubek, B., and Gutmann, E. V. In: Nervous Mechanisms of Motor Activity [in Russian], Nauka, Moscow, 1966, pp. 15-19.
165.   Gordienko, É. A. Zh. Évolyuts. Biokhim. i Fiziol., 1967, 3:35-39.
166.   Gordienko, É. A. Zh. Évolyuts. Biokhim. i Fiziol., 1970, 6:335-337.
167.   Gorodiskaya, G. Nauk. Zap. Ukr. Biokhim. Inst., 1926, 1:105-113.
168.   Grabar, P. Biokhimiya, 1957, 22:49-59.
169.   Graevskaya, B. M., and Shchedrina, R. N. Radiobiologiya, 1963, 3:168-173.
170.   Gracheva, N. D. Autoradiography of Nucleic Acid and Protein Synthesis in the Nervous System [in Russian], Nauka, Leningrad, 1968.
171.   Green, N., and Neurath, H. In: The Proteins [Russian translation], Vol. 3, Part II, IL, Moscow, 1959, pp. 7-165.
172.   Grinenko, O. G. Proceedings of the Ukrainian Biochemical Congress [in Ukrainian], Naukova Dumka, Kiev, 1971, pp. 72-73.
173.   Gromova, E. A., Serotonin and Its Role in the Organism [in Russian], Meditsina, Moscow, 1966.
174.   Gromova, L. G. Trudy Omsk. Med. Inst., 1962, 35:141-146.
175.   Gubin, G. D., and Petrova, R. M. Tsitologiya, 1967, 9:116-120.
176.   Gulidova, G. P. Byull. Éksperim. Biol. i Med., 1965, 59(4):42-45.
177.   Gulidova, G. P. Author's Abstract of Candidate's Dissertation, Brain Institute, Academy of Medical Sciences of the USSR, Moscow, 1966.
178.   Gulyi, M. F. Protein Biosynthesis [in Russian], Izd. Akad. Nauk Ukr. SSR, Kiev, 1963.
179.   Gutnikov, Z. Results of a Study of the Chemical Composition of the Human Brain, Dissertation, Khar'kov, 1893.
180.   Gaito, J. Molecular Psychology [Russian translation], Mir, Moscow, 1969. (English edition: C. C. Thomas, Springfield, Ill., 1966.)
181.   Davydova, S. Ya. Tsitologiya, 1967, 9:1248-1264.
182.   Davydova, S. Ya., and Konikova, A. S. Dokl. Akad. Nauk SSSR, 1950, 73:349-350.
183.   Danilevskii, A. Fiziologicheskii Sbornik, 1892, 2:141-166.
184.   Danilevskii, A. Fiziologicheskii Sbornik, 1891, 2:167-188.
185.   Danilevskii, A. Ya. Russkii Fiziol. Zh., 1919, 2:128-144.
186.   Daudova, G. M., and Ipat'eva, N. V. Zh. Évolyuts. Biokhim. i Fiziol., 1968, 4:348-353.
187.   Dounce, A. L. In: Current Problems in Cytology [Russian translation], IL, Moscow, 1955, pp. 51-71.
188.   Dounce, A. L. In: Nucleic Acids [Russian translation], Moscow, 1957, pp. 51-101.
189.   Debabov, V. G., and Rebentish. B. A. Uspekhi Sovr. Biol., 1970, 70:147-165.
190.   Demin, N. N. Third All-Union Conference on Biochemistry of the Nervous System [in Russian], Izd. Akad. Nauk Arm. SSR, Erevan, 1963, pp. 551-560.

191. Demin, N. N. In: Problems in Neurochemistry [in Russian], Nauka, Moscow, 1966, pp. 197-208.
192. Deryabina, T. I. Uch. Zap. Gor'k. Med. Inst., 1957, (1):45-54.
193. Desenko, V. F., Chermnikh, N. M., and Gerashchenko, T. A. Ukr. Biokhim. Zh., 1968, 40:331-335.
194. Diasamidze, G. A. Vopr. Med. Khimii, 1970, 16:244-250.
195. Diasamidze, G. A., and Kometiani, P. A. Uspekhi Sovr. Biol., 1970, 69:364-379.
196. Dixon, M., and Webb, E. C. Enzymes, Longmans, London, 1964.
197. Dmitriev, N. I. In: The Development of the Animal Brain [in Russian], Nauka, Moscow, 1969, pp. 123-144.
198. Dobrynina, V. I. Fiziol. Zh. SSSR, 1940, 29:220-224.
199. Dobrynina, V. I. Proceedings of the 6th Scientific Conference on Age Morphology, Physiology, and Biochemistry [in Russian], Izd. Akad. Ped. Nauk SSSR, Moscow, 1963, pp. 322-323.
200. Dovedova, E. L., and Bogolepov, N. N. Tsitologiya, 1966, 8:469-474.
201. Dusheiko, A. A. Ukr. Biokhim. Zh., 1960, 32:823-831.
202. De Duve, C. In: Structural Components of the Cell [Russian translation], IL, Moscow, 1962, pp. 128-172.
203. De Duve, C., and Muller, M. In: The Functional Biochemistry of Cell Structures [Russian translation], Moscow, 1970, pp. 322-336.
204. Dyumbaum, V. I. Vestn. Leningrad. Gos. Univ., Ser. Biol., 1966, 3(1):75-84.
205. Elaev, N. R. Biokhimiya, 1964, 29:413-419.
206. Elaev, N. R. Biokhimiya, 1966, 31:234-240.
207. Elaev, N. R. Tsitologiya, 1966, 8:45-54.
208. Elaev, N. R., and Mashanskii, V. F. Ukr. Biokhim. Zh., 1966, 38:461-468.
209. Elaev, N. R., and Rykhlik, I. Biokhimiya, 1963, 28:1047-1052.
210. Eliseeva, Yu. E., and Orekhovich, V. N. Dokl. Akad. Nauk SSSR, 1963, 153:954-956.
211. Ermolaeva, L. P. Biokhimiya, 1966, 31:936-942.
212. Ermolaeva, L. P. In: Structure and Functions of the Cell Nucleus [in Russian], Nauka, Moscow, 1967, pp. 122-125.
213. Ermolaeva, L. P., and Zbarskii, I. B. Vopr. Med. Khimii, 1968, 14:262-266.
214. Zhabotinskii, Yu. M. Arkh. Anat., Gistol., i Émbriol., 1958, 35(3):19-28.
215. Zhabotinskii, Yu. M. The Normal and Pathological Morphology of the Neuron [in Russian], Meditsina, Leningrad, 1965.
216. Zhabotinskii, Yu. M. Vestn. Akad. Med. Nauk SSSR, 1966, (1):50-62.
217. Zhinkin, L. N. The Renewal of Cells in the Organism [in Russian], Leningrad, 1962.
218. Zhinkin, L. N. In: Textbook of Cytology, Vol. 2 [in Russian], Nauka, Moscow, 1966, pp. 550-574.
219. Zhukova, T. P. In: Problems in Development of the Brain and the Effect of Harmful Factors on It [in Russian], Medgiz, Moscow, 1960, pp. 93-123.
220. Zaets, T. L. Vopr. Med. Khimii, 1969, 15:43-47.
221. Zaiko, N. N. Izvest. Akad. Nauk SSSR, Ser. Biol., 1958, pp. 698-711.
222. Zaiko, N. N. In: Tissue—Blood Barriers [in Russian], Izd. AN SSSR, Moscow, 1961, pp. 179-187.

223. Zaiko, N. N., Gaude, V., and Mints, S. M. In: Tissue—Blood Barriers [in Russian], Izd. AN SSSR, Moscow, 1961, pp. 239-246.
224. Zaletaeva, T. A., and Sukhova, Z. I. Tsitologiya, 1969, 11:1323-1326.
225. Zakharov, N. V., and Orlyanskaya, R. L. Vopr. Med. Khimii, 1960, 6:249-253.
226. Zakharov, N. V., and Orlyanskaya, R. L. Trudy Mosk. Obl. N.-I. Klin. Inst., 1961, (4):243-250.
227. Zakharov, N. V., and Sharkov, B. M. Byull. Éksperim. Biol. i Med., 1968, 66(12):42-44.
228. Zbarskii, I. B. Uspekhi Sovr. Biol., 1962, 54:265-284.
229. Zbarskii, I. B. In: Structure and Function of the Cell Nucleus [in Russian], Nauka, Moscow, 1967, pp. 111-121.
230. Zbarskii, I. B. Uspekhi Sovr. Biol., 1969, 67:323-341.
231. Zbarskii, I. B., and Georgiev, G. P. Biokhimiya, 1959, 24:192-199.
232. Zbarskii, I. B., and Samarina, O. P. Biokhimiya, 1962, 27:557-564.
233. Zererov, E. G. Biokhimiya, 1960, 25:727-734.
234. Zelenin, A. V. Uspekhi Sovr. Biol., 1962, 53:364-374.
235. Zotikov, L. A., and Pinchuk, V. G. Tsitologiya, 1969, 11:1205-1220.
236. Zryakov, O. N. Author's Abstract of Dissertation for the Degree of Candidate of Sciences, Institute of Biochemistry, Academy of Sciences of the Ukrainian SSR, Kiev, 1968.
237. Zubkova, S. R., Rapoport, S. Ya., and Shchors, N. V. In: The Development and Regulation of the Tissue—Blood Barriers [in Russian], Nauka, Moscow, 1967, pp. 174-181.
238. Ivanenko, E. F., and Dunaeva, V. F. Third All-Union Conference on Biochemistry of the Nervous System [in Russian], Izd. Akad. Nauk Arm. SSR, Erevan, 1963, pp. 109-120.
239. Ivanenko, E. F., and Dunaeva, V. F. Ukr. Biokhim. Zh., 1964, 36:72-79.
240. Ivanenko, E. F., and Dunaeva, V. F. Ukr. Biokhim. Zh., 1964, 36:183-189.
241. Ivanov, I. I. Uspekhi Sovr. Biol., 1966, 32:943-951.
242. Ivanov, I. I., et al. Biokhimiya, 1967, 32:943-951.
243. Ilkov, A., and Nikolov, T. Vopr. Med. Khimii, 1959, 5:388-392.
244. Kazakova, T. B. Uspekhi Sovr. Biol., 1965, 60:198-214.
245. Kazakova, T. B. In: Mechanisms of Integration of Cell Metabolism [in Russian], Nauka, Leningrad, 1967, pp. 157-205.
246. Kazakova, T. B., and Markosyan, K. A. In: Mitochondria. Structure and Functions [in Russian], Nauka, Moscow, 1966, pp. 42-44.
247. Kazakova, T. B., and Neifakh, S. A. In: Chemistry and Metabolism of Carbohydrates [in Russian], Nauka, Moscow, 1965, pp. 272-280.
248. Kalabukhov, N. I. Hibernation in Animals [in Russian], Khar'kov University Press, 1956.
249. Kapitonova, G. V., and Dish, T. N. In: The Development and Regulation of Tissue—Blood Barriers [in Russian], Nauka, Moscow, 1967, pp. 186-191.
250. Kaplanskii, S. Ya. Acid—Base Balance in the Organism and Its Regulation [in Russian], Medgiz, Moscow, 1940.
251. Kaplanskii, S. Ya., and Akopyan, Zh. I. Biokhimiya, 1966, 31:265-269.
252. Kaplanskii, S. Ya., and Akopyan, Zh. I. Ukr. Biokhim. Zh., 1967, 29:34-37.
253. Kasavina, B. S., and Umanskaya, M. B. Biokhimiya, 1958, 23:587-591.

254.   Kassil', G. N. The Blood—Brain Barrier [in Russian], Izd. AN SSSR, Moscow, 1963.

255.   Kafiani, K. A. In: Enzymes [in Russian], Nauka, Moscow, 1964, pp. 269-311.

256.   Kashkin, K. P., et al. Vopr. Med. Khimii, 1969, 15:235-243.

257.   Kedrovskii, B. V. The Cytology of Protein Syntheses in the Animal Cell [in Russian], Izd. AN SSSR, Moscow, 1959.

258.   Kirsenko, O. V. In: Protoplasmic Membranes and Their Functional Role [in Russian], Naukova Dumka, Kiev, 1965, pp. 180-192.

259.   Kirsenko, O. V., Palladin, A. V., Rozhmanova, O. M., and Eismont, S. S. Ukr. Biokhim. Zh., 1963, 35:807-815.

260.   Klein, E. É. Uspekhi Sovr. Biol., 1956, 41:161-176.

261.   Klein, E. É., Iordanishvili, G. S., and Gvaliya, N. V. Third All-Union Conference on Biochemistry of the Nervous System [in Russian], Izd. Akad. Nauk Arm. SSR, Erevan, 1963, pp. 193-205.

262.   Klein, E. É., Kurtskhaliya, É. G., and Gvaliya, N. V. Soobshch. Akad. Nauk Gruz. SSR, 1966, 44:331-336.

263.   Klein, E. É., Kurtskhaliya, É. G., and Chogovadze, I. S. Ukr. Biokhim. Zh., 1969, 41:512-518.

264.   Klimenko, A. I. In: The Molecular and Functional Basis of Ontogeny [in Russian], Meditsina, Moscow, 1970, pp. 89-109.

265.   Klosovskii, B. N. The Circulation of the Blood in the Brain [in Russian], Medgiz, Moscow, 1951.

266.   Klosovskii, B. N. In: Problems in the Study of Tissue—Blood Barriers [in Russian], Nauka, Moscow, 1965, pp. 277-287.

267.   Klosovskii, B. N., and Kosmarskaya, E. N. Active and Inhibitory States of the Brain [in Russian], Medgiz, Moscow, 1961.

268.   Koval'skii, V. V. Zh. Obshch. Biol., 1965, 26:14-35.

269.   Koval'skii, V. V., and Kapner, R. B. Dokl. Akad. Nauk SSSR, 1957, 112:905-908.

270.   Koval'skii, V. V., and Shumkova, I. A. Dokl. Akad. Nauk SSSR, 1963, 152:1467-1470.

271.   Kogan, A. B. Zh. Évolyuts. Biokhim. i Fiziol., 1969, 5:3-9.

272.   Kozlov, N. B., and Shashkevich, I. K. Ukr. Biokhim. Zh., 1969, 41:519-522.

273.   Kometiani, P. A., Ukr. Biokhim. Zh., 1965, 37:721-733.

274.   Kometiani, P. A. In: Amino Acid Metabolism [in Russian], Metsniereba, Tbilisi, 1967, pp. 99-121.

275.   Kometiani, P. A. Proceedings of the 4th All-Union Conference on Biochemistry of the Nervous System [in Russian], Tartu (1969), pp. 22-45.

276.   Kometiani, P. A. Biokhimiya, 1970, 35:394-403.

277.   Kometiani, P. A., et al., In: Problems in Biochemistry of the Nervous and Muscular Systems [in Russian], Metsniereba, Tbilisi, 1965, pp. 41-63.

278.   Komissarenko, V. P. In: Current Problems in Endocrinology [in Russian], Vol. 1, Medgiz, Moscow, 1960, pp. 30-47.

279.   Komissarchik, Ya. Yu., Mashanskii, V. F., and Mosevich, T. N. In: Protoplasmic Membranes and Their Functional Role [in Russian], Naukova Dumka, Kiev, 1965, p. 5-12.

280. Konikova, A. S. Ukr. Biokhim. Zh., 1950, 22:3-10.
281. Konikova, A. S., et al. In: Investigations into the Use of Radioactive Isotopes in Medicine [in Russian], Medgiz, Moscow, 1955, pp. 255-258.
282. Konikova, A. S., and Kritsman, M. G. Pathways of Protein Synthesis [in Russian], Meditsina, Moscow, 1965.
283. Konikova, A. S., Kritsman, M. G., and Samarina, O. P. Dokl. Akad. Nauk SSSR, 1956, 109:593-596.
284. Komienko, V. M., and Gorbacheva, A. P. In: The Molecular Biology of Aging [in Russian], Naukova Dumka, Kiev, 1969, pp. 103-105.
285. Komienko, V. M., Trubach, M. I., and Malyarova, O. E. In: The Molecular Biology of Aging [in Russian], Naukova Dumka, Kiev, 1969, pp. 100-103.
286. Kosmarskaya, E. N. In: The Problem of Brain Development and the Effect of Harmful Factors on It [in Russian], Medgiz, Moscow, 1960, pp. 34-40, 209-215.
287. Kravchinskii, E. M. Ukr. Biokhim. Zh., 1959, 31:665-670.
288. Kravchinskii, E. M., and Silich, T. P. Ukr. Biokhim. Zh., 1957, 29:25-32.
289. Kreps, E. M., Ukr. Biokhim. Zh., 1965, 37:734-741.
290. Kreps, E. M. In: The Functional Evolution of the Nervous System [in Russian], Nauka, Moscow, 1965, pp. 30-40.
291. Kreps, E. M. In: The Biochemistry and Function of the Nervous System [in Russian], Nauka, Leningrad, 1967, pp. 134-146.
292. Kreps, E. M., et al., Zh. Vyssh. Nerv. Deyat., 1952, 2:46-57.
293. Kreps, E. M., et al. Third All-Union Conference on Biochemistry of the Nervous System [in Russian], Izd. Akad. Nauk Arm. SSR, Erevan, 1963, pp. 355-365.
294. Kreps, E. M., et al. In: The Evolution of Functions [in Russian], Nauka, Moscow, 1964, pp. 211-218.
295. Kreps, E. M., et al. In: Problems in Neurochemistry [in Russian], Nauka, Moscow, 1966, 124-136.
296. Kreps, E. M., and Verzhbinskaya, N. A. Izvest. Akad. Nauk SSSR, Ser. Biol., 1959, pp. 855-864.
297. Kreps, E. M., Smirnov, A. A., and Chetverikov, D. A. In: The Biochemistry of the Nervous System [in Russian], Izd. Akad. Nauk Ukr. SSR, Kiev, 1954, pp. 125-138.
298. Kruglova, É. É. Zh. Évolyuts. Biokhim. i Fiziol., 1970, 6:464-467.
299. Kudinov, S. A., and Krisa, N. K. Ukr. Biokhim. Zh., 1967, 39:571-576.
300. Kudinov, S. A., and Polyakova, N. M. Ukr. Biokhim. Zh., 1966, 38:455-460.
301. Kudinov, S. A., and Polyakova, N. M. Proceedings of the 4th All-Union Conference on Biochemistry of the Nervous System [in Russian], Tartu, 1969, pp. 239-250.
302. Kudinov, S. A., and Polyakova, N. M. Ukr. Biokhim. Zh., 1969, 41:227-232.
303. Kudinov, S. A., Polyakova, N. M., and Krisa, N. K. Ukr. Biokhim. Zh., 1969, 41:501-506.
304. Kudinov, S. A., and Sokhina, A. M. Ukr. Biokhim. Zh., 1969, 39:115-118.
305. Kudryashov, B. A. Problems in Blood Clotting and Thrombosis [in Russian], Vysshaya Shkola, Moscow, 1960.

306. Kurokhtina, T. P. Biokhimiya, 1954, 19:16-18.
307. Kurskii, M. D., and Zryakov, O. M. Ukr. Biokhim. Zh., 1964, 36:679-684.
308. Kushko, V. M., and Panchenko, L. F. Patol. Fiziol. i Éksperim. Terapiya, 1959, 3:22-27.
309. Laguchev, S. S. Uspekhi Sovr. Biol., 1963, 56:274-283.
310. Laguchev, S. S., et al. Tsitologiya, 1962, 4:381-390.
311. Lang, K. In: Current Problems in Cytology [Russian translation], IL, Moscow, 1955, pp. 149-165.
312. Lebedeva, E. M., Maslova, M. N., and Rozengart, V. I. Dokl. Akad. Nauk SSSR, 1955, 102:563-566.
313. Lebedinskii, A. V., and Nakhil'nitskaya, Z. N. The Effect of Ionizing Radiations on the Nervous System [in Russian], Atomizdat, Moscow, 1960.
314. Levinson, L. B. Dokl. Akad. Nauk SSSR, 1952, 83:745-748.
315. Levitina, M. V. Uspekhi Sovr. Biol., 1970, 69:113-125.
316. Levyant, M. I., Levchuk, T. P., and Orekhovich, V. N. Biokhimiya, 1959, 24:177-180.
317. Lehninger, A. In: Horizons in Biochemistry [Russian translation], Mir, Moscow, 1964, pp. 326-337.
318. Lehninger, A. The Mitochondrion [Russian translation], Mir, Moscow, 1966. (English edition: W. A. Benjamin, New York, 1964.)
319. Lents, A. Izvest. Voen.-Med. Akad., 1913, 27:541-554, 735-747, 875-890.
320. Lents, A. K. Russkii Fiziol. Zh., 1919, 2:145-169.
321. Leslie, I. In: Nucleic Acids [Russian translation], IL, Moscow, 1957, pp. 7-50.
322. Li Chao-t'e. Dokl. Akad. Nauk SSSR, 1958, 120:650-653.
323. Libinzon, R. E., and Tseveleva, I. A. Biokhimiya, 1959, 24:263-266.
324. Lindberg, O., and Ernster, L. In: Problems in Cytophysiology [Russian translation], IL, Moscow, 1957, pp. 111-225.
325. Liozner, L. D., and Sidorova, V. F. Byull. Éksperim. Biol. i Med., 1959, 48(12):93-96.
326. Lisovskaya, N. P. Biokhimiya, 1954, 19:626-637.
327. Lisovskaya, N. P. In: Problems in Biochemistry of the Nervous System [in Russian], Izd. Akad. Nauk Ukr. SSR, Kiev, 1957, pp. 76-82.
328. Lisovskaya, N. P. Uspekhi Biol. Khimii, 1967, 8:93-116.
329. Lisovskaya, N. P., and Livanova, N. B. Biokhimiya, 1959, 24:799-810.
330. Lisovskaya, N. P., and Livanova, N. B. Phosphoproteins [in Russian], Izd. AN SSSR, Moscow, 1960.
331. Litovchenko, G. P. Pratsi N.-D. Zool.-Biol. Inst. Khar'k. Univ., 1938, 5:102-137.
332. Lishko, V. K. Ukr. Biokhim. Zh., 1963, 35:874-880.
333. Lishko, V. K. Ukr. Biokhim. Zh., 1964, 36:565-573.
334. Lishko, V. K. Ukr. Biokhim. Zh., 1965, 37:163-168.
335. Logunov, V. V. Trudy Voronezh. Med. Inst., 1960, 36:103-105.
336. Lukash, A. I., and Gershenovich, Z. S. In: Nucleic Acids [in Russian], Meditsina, Moscow, 1966, pp. 295-298.
337. Lukashevich, E. F. Vopr. Med. Khimii, 1968, 14:510-513.
338. Lutsenko, N. G., and Promyslov, M. Sh. Vopr. Med. Khimii, 1963, 9:60-63.

339. L'vova, S. P. Dokl. Akad. Nauk SSSR, 1968, 179:1225-1226.
340. Meister, A. Biochemistry of the Amino Acids [Russian translation], IL, Moscow, 1961. (English 2nd edition: Academic Press, New York, 1965.)
341. Maistrakh, E. V. Hypothermia and Anabiosis [in Russian], Nauka, Moscow, 1964.
342. Makarov, K. S. Biokhimiya, 1967, 32:45-49.
343. Makarov, P. V. Arkh. Anat. Gistol., i Émbriol., 1966, 51(10):92-103.
344. McIlwain, H. Biochemistry and the Central Nervous System [Russian translation], IL, Moscow, 1962. (English edition: Little Brown and Co., Boston, 1957.)
345. McLaren, A., and Babcock, K. In: Structural Components of the Cell [Russian translation], IL, Moscow, 1962, pp. 36-57.
346. Manina, A. A. Dokl. Akad. Nauk SSSR, 1958, 122:504-507.
347. Manina, A. A. In: Mechanisms of Activity of the Central Neuron [in Russian], Nauka, Moscow, 1966, pp. 34-57.
348. Manina, A. A. Dokl. Akad. Nauk SSSR, 1967, 177:1457-1459.
349. Manina, A. A. Byull. Éksperim. Biol. i Med., 1969, 68(12):91-93.
350. Manukyan, K. G. Dokl. Akad. Nauk SSSR, 1955, 101:1083-1088.
351. Manukyan, K. G. Dokl. Akad. Nauk SSSR, 1955, 102:567-570.
352. Manukyan, K. G., Smirnov, A. A., and Chirkovskaya, E. V. Biokhimiya, 1963, 28:246-252.
353. Markova, I. V. Uspekhi Sovr. Biol., 1962, 53:347-363.
354. Markosyan, A. A. The Physiology of Blood Clotting [in Russian], Meditsina, Moscow, 1966.
355. Martinson, É. É., and Lind, A. Ya. Proceedings of the 1st Biochemical Conference of the Baltic Republics and White Russia [in Russian], Tartu, 1961, pp. 62-64.
356. Martinson, É. É., and Tyakhepyl'd, L. Ya. Biokhimiya, 1961, 26:984-992.
357. Martinson, É. É., and Tyakhepyl'd, L. Ya. Proceedings of the 1st Biochemical Conference of the Baltic Republics and White Russia [in Russian], Tartu, 1961, pp. 26-42.
358. Martinson, É. É., and Tyakhepyl'd, L. Ya. In: Problems in Neurochemistry [in Russian], Nauka, Moscow, 1966, pp. 46-52.
359. Maslova, M. N. In: The Role of Gamma-aminobutyric Acid in the Activity of the Nervous System [in Russian], Leningrad University Press, 1964, pp. 49-58.
360. Maslova, M. N., et al. In: Problems in Neurochemistry [in Russian], Nauka, Moscow, 1966, pp. 40-45.
361. Maslova, M. N., et al. Vopr. Med. Khimii, 1966, 12:350-355.
362. Maslova, M. N., and Rozengart, V. I. Third All-Union Conference on Biochemistry of the Nervous System [in Russian], Izd. Akad. Nauk Arm. SSR, Erevan, 1963, pp. 153-162.
363. Maslova, M. N., and Sytinskii, I. A. In: Amino Acid Metabolism [in Russian], Metsniereba, Tbilisi, 1967, pp. 130-133.
364. Maslova, M. N., and Khaurina, R. A. In: Evolutionary Neurophysiology and Neurochemistry [in Russian], Nauka, Leningrad, 1967, pp. 186-191.
365. Makhno, P. M. Trudy Mosk. Vet. Akad., 1962, 43:190-196.

366. Medvedev, Zh. A. Protein Biosynthesis and Problems in Ontogeny [in Russian], Gos. Izd. Med. Lit., Moscow, 1963.
367. Medvedev, Zh. A. Uspekhi Sovr. Biol., 1965, 59:333-353.
368. Medvedeva, N. B. Medichnyi Zhurn., 1940, 10:793-798.
369. Meerson, F. Z. The Connection between Physiological Function and the Genetic Apparatus of the Cell [in Russian], Izd. Inst. Norm. Patol. Fiziol. AMN SSSR, Moscow, 1963.
370. Mezesh, V., and Zakhar, F. Ukr. Biokhim. Zh., 1967, 39:237-241.
371. Meibaum, V. V. Biokhimiya, 1945, 10:353-359.
372. Mel'nichenko, A. V., Genis, E. D., and Khomutovskii, O. A. In: Problems in the Study of Tissue—Blood Barriers [in Russian], Nauka, Moscow, 1965, pp. 324-331.
373. Mendetskii, Yu. Trudy Mosk. Vet. Akad., 1960, 30:131-135.
374. Mendetskii, Yu., and Ruzhitskii, B. Trudy Mosk. Vet. Akad., 1960, 30:128-130.
375. Mendetskii, Yu., and Ruzhitskii, B. Ukr. Biokhim. Zh., 1962, 34:655-665.
376. Mechanisms of Activity of the Central Neuron [in Russian], Nauka, Moscow, 1966.
377. Minaev, P. F. The Effect of Ionizing Radiation on the Central Nervous System [in Russian], Izd. AN SSSR, Moscow, 1962.
378. Minaev, P. F., Chukhrova, A. I., and Antonova, A. M. Third All-Union Conference on Biochemistry of the Nervous System [in Russian], Izd. Akad. Nauk Arm. SSR, Erevan, 1963, pp. 561-571.
379. Miropol'skii, S. V. Trudy Leningrad. Obshch. Estestvoisp., 1950, 69(5):122-142.
380. Mirsky, A. E. Proceedings of the 5th International Biochemical Congress, Symposium II [in Russian], Izd. AN SSSR, Moscow, 1962, pp. 81-90.
381. Mirsky, A. E., and Osawa, S. In: Functional Morphology of the Cell [Russian translation], IL, Moscow, 1963, pp. 9-68.
382. Mitev, I. P. Ukr. Biokhim. Zh., 1958, 30:643-651.
383. Mitina, L. V. Scientific Reports of the 2nd Moscow Medical Institute [in Russian], Moscow, 1957, pp. 53-58.
384. Mikhailova, R. I. In: Problems in Biochemistry of the Brain [in Russian], Volgo-Vyatskoe Izd., Gor'kii, 1966, pp. 116-122.
385. Mishchenko, V. P. Author's Abstract of Candidate's Dissertation, Khar'kov State University, 1968.
386. Mishchenko, V. P. In: Molecular and Functional Bases of Ontogeny [in Russian], Nauka, Moscow, 1970, pp. 58-71.
387. Mogilevskii, A. Ya. Uspekhi Sovr. Biol., 1960, 50:322-337.
388. Monakhov, N. K. Biokhimiya, 1964, 29:955-963.
389. Monakhov, N. K. Molekul. Biol., 1967, 1:114-122.
390. Morales, M. In: Current Problems in Biophysics [Russian translation], Vol. 2, IL, Moscow, 1961, pp. 152-161.
391. Mosolov, A. N., and Ershov, F. I. Izvest. Sib. Otdel. Akad. Nauk SSSR, Ser. Biol.-Med. Nauk, 1965, 8(2):139-147.
392. Mchedlishvili, G. I. Function of the Vascular Mechanisms of the Brain [in Russian], Nauka, Leningrad, 1968.
393. Myul'berg, A. A., and Goryukhina, O. A. Ukr. Biokhim. Zh., 1966, 38:328-333.

394. Myul'berg, A. A., Sytinskii, I. A., and Chaika, T. V. Vopr. Med. Khimii, 1962, 8:58-64.
395. Nagornyi, A. V. The Problem of Aging and Longevity [in Russian], Khar'kov University Press, 1940.
396. Nagornyi, A. V. Trudy N.-I. Inst. Biol. Khar'k. Univ., 1947, 12:19-37, 39-60.
397. Nagornyi, A. V. Aging and the Prolongation of Life [in Russian], Sovetskaya Nauka, Moscow, 1950.
398. Nagornyi, A. V. In: Age Changes in Metabolism and Reactivity of the Organism [in Russian], Izd. Akad. Nauk Ukr. SSR, Kiev, 1951, pp. 5-16.
399. Nagornyi, A. V., Nikitin, V. N., and Bulankin, I. N. The Problem of Aging and Longevity [in Russian], Medgiz, Moscow, 1963.
400. Narepekha, O. M. In: The Biological Action of Radiation [in Russian], Vol. 2, L'vov University Press, 1963, pp. 84-88.
401. Nasonov, D. N. The Local Reaction of Protoplasm and Spreading Excitation [in Russian], Izd. AN SSSR, Moscow, 1962.
402. Nguyen Thi Thin and Sytinskii, I. A. Zh. Obshch. Biol., 1964, 25:389-390.
403. Nguyen Thi Thin and Sytinskii, I. A. Ukr. Biokhim. Zh., 1964, 36:67-71.
404. Nezlin, R. S. The Biochemistry of Antibodies [in Russian], Nauka, Moscow, 1966.
405. Neifakh, A. A. The Problem of Relations between Nucleus and Cytoplasm in Development [in Russian], Izd. Inst. Morfol. Zhivotnykh im. A. N. Severtsova AN SSSR, Moscow, 1962.
406. Neifakh, A. A. In: Mitochondria. Structure and Functions [in Russian], Nauka, Moscow, 1966, pp. 29-42.
407. Neifakh, S. A. In: Mechanisms of Integration of Cell Metabolism [in Russian], Nauka, Leningrad, 1967, pp. 9-65.
408. Nechaeva, G. A. Biokhimiya, 1957, 22:546-553.
409. Nechaeva, G. A. Dokl. Akad. Nauk SSSR, 1963, 152:225-227.
410. Nechaeva, G. A., Sadikova, N. V., and Skvortsevich, V. A. In: Problems in Biochemistry of the Nervous System [in Russian], Izd. Akad. Nauk Ukr. SSR, Kiev, 1957, pp. 31-39.
411. Nechiporenko, Z. Yu. Ukr. Biokhim. Zh., 1946, 18:77-86.
412. Nechiporenko, Z. Yu. Ukr. Biokhim. Zh., 1955, 27:146-160.
413. Nikitin, V. N. Fiziol. Zh. SSSR, 1941, 30:619-626.
414. Nikitin, V. N. Trudy N.-I. Inst. Khar'k. Univ., 1947, 12:95-107.
415. Nikitin, V. N. Trudy N.-I. Inst. Biol. Biol. Fak. Khar'k. Univ., 1962, 33-34:243-284.
416. Nikitin, V. N., et al. In: Age Changes in Metabolism and Reactivity of the Organism [in Russian], Izd. Akad. Nauk Ukr. SSR, Kiev, 1951, pp. 17-26.
417. Nikitin, V. N., and Golubitskaya, R. I. Trudy N.-I. Inst. Biol. Khar'k. Univ., 1954, 21:113-126, 143-151.
418. Nikitin, V. N., Dryuchina, L. A., and Semenova, Z. L. Ukr. Biokhim. Zh., 1949, 21:296-305.
419. Nikitin, V. N., Novikova, N. M., and Tsikalo, A. P. Trudy N.-I. Inst. Biol. Biol. Fak. Khar'k. Univ., 1960, 29:81-94.
420. Nikol'skii, N. N. In: Textbook of Cytology [in Russian], Nauka, Moscow, 1965, pp. 491-556.

421.   Nikulin, V. I. Éksperim. Khirurgiya, 1957, 1:55-60.
422.   Nilova, N. S. Dokl. Akad. Nauk SSSR, 1963, 150:1161-1163.
423.   Nilova, N. S. Ukr. Biokhim. Zh., 1963, 35:220-226.
424.   Novikoff, A. B. In: Functional Morphology of the Cell [Russian translation],
       IL, Moscow, 1963, pp. 113-158.
425.   Novikova, N. M. In: The Molecular and Functional Bases of Ontogeny [in
       Russian], Meditsina, Moscow, 1970, pp. 110-124.
426.   Nomenclature of Enzymes. Recommendations of the International Biochemical
       Union [Russian translation], Izd. VINITI, Moscow, 1966.
427.   Nosova, E. A. Ukr. Biokhim. Zh., 1969, 41:288-291.
428.   Ochs, S. Fundamentals of Neurophysiology [Russian translation], Mir, Moscow,
       1969.
429.   Olenov, S. N., Korochkin, L. I., and Demin, D. V. Uspekhi Sovr. Biol., 1968,
       65:245-266.
430.   Allfrey, V. G. Proceedings of the 5th International Biochemical Congress.
       Symposium II [Russian translation], Izd. AN SSSR, Moscow, 1962, pp. 137-152.
431.   Allfrey, V. G., and Mirsky, A. E. In: Structural Components of the Cell
       [Russian translation], IL, Moscow, 1962, pp. 211-243.
432.   Allfrey, V. G., Mirsky, A. E., and Osawa, S. In: The Chemical Basis of
       Heredity [Russian translation], IL, Moscow, 1960, pp. 162-182.
433.   Orekhovich, V. N. Byull. Éksperim. Biol. i Med., 1937, 3(2):194-196.
434.   Orekhovich, V. N. Biokhimiya, 1940, 5:331-338.
435.   Orekhovich, V. N., et al., Dokl. Akad. Nauk SSSR, 1950, 71:105-107.
436.   Orekhovich, V. N., Kurokhtina, T. P., and Buyanova, N. D. Biokhimiya, 1953,
       18:706-708.
437.   Orekhovich, V. N., Levyant, M. I., and Levchuk-Kurokhtina, T. P. Biokhimiya,
       1954, 19:610-615.
438.   Orekhovich, V. N., and Sokolova, T. P. Dokl. Akad. Nauk SSSR, 1940, 28:748-
       750.
439.   Pavlov, I. P. Lectures on the Work of the Cerebral Hemispheres [in Russian],
       Izd. Akad. Med. Nauk SSSR, Moscow, 1952.
440.   Pavlov, I. P. Lectures in Physiology. Complete Collected Works [in Russian],
       Vol. 5, Izd. AN SSSR, Moscow, 1952.
441.   Palladin, A. V. Fiziol. Zh. SSSR, 1947, 33:727-736.
442.   Palladin, A. V. Ukr. Biokhim. Zh., 1947, 19:293-305.
443.   Palladin, A. V. Vestn. Akad. Nauk SSSR, 1952, (10):37-62.
444.   Palladin, A. V. The Biochemistry of the Brain [in Russian], Izd. AN SSSR,
       Moscow, 1955.
445.   Palladin, A. V. Izvest. Akad. Nauk SSSR, Ser. Biol., 1956, pp. 11-22.
446.   Palladin, A. V. In: Current Problems in Biochemistry [in Russian], Vol. 1,
       Izd. Inst. Biol. Med. Khimii AMN SSSR, Moscow, 1959, pp. 102-113.
447.   Palladin, A. V. Ukr. Biokhim. Zh., 1959, 31:765-779.
448.   Palladin, A. V. Ukr. Biokhim. Zh., 1961, 33:602-621.
449.   Palladin, A. V. Ukr. Biokhim. Zh., 1962, 34:621-632.
450.   Palladin, A. V. Third All-Union Conference on Biochemistry of the Nervous
       System [in Russian], Izd. Akad. Nauk Arm. SSR, Erevan, 1963, pp. 9-23.

451. Palladin, A. V. Problems in Biochemistry of the Nervous System [in Russian], Naukova Dumka, Kiev, 1965.
452. Palladin, A. V. In: Problems in Neurochemistry [in Russian], Nauka, Moscow, 1966, pp. 5-17.
453. Palladin, A. V. Proceedings of the 4th All-Union Conference on Biochemistry of the Nervous System [in Russian], Tartu, 1969, pp. 9-21.
454. Palladin, A. V., et al. In: Problems in Biochemistry of the Nervous System [in Russian], Izd. Akad. Nauk Ukr. SSR, Kiev, 1957, pp. 9-30.
455. Palladin, A. V., et al. Proceedings of the 1st Conference on Cytochemistry and Histochemistry [in Russian], Moscow, 1960, pp. 77-80.
456. Palladin, A. V., and Belik, Ya. V. Abstracts of Proceedings of the 5th All-Union Conference on Neurochemistry [in Russian], Tbilisi, 1968, pp. 50-52.
457. Palladin, A. V., and Belik, Ya. V. In: The Molecular and Functional Bases of Ontogeny [in Russian], Meditsina, Moscow, 1970, pp. 35-57.
458. Palladin, A. V., and Belik, Ya. V. The 5th All-Union Conference on Neurochemistry [in Russian], Metsniereba, Tbilisi, 1970, pp. 73-86.
459. Palladin, A. V., Belik, Ya. V., and Krachko, L. S. Biokhimiya, 1957, 22:359-368.
460. Palladin, A. V., Belik, Ya. V., and Krachko, L. S. Dokl. Akad. Nauk SSSR, 1959, 127:702-705.
461. Palladin, A. V., and Vertaimer, N. Dokl. Akad. Nauk SSSR, 1955, 102:319-321.
462. Palladin, A. V., and Gulyi, M. F. Ukr. Biokhim. Zh., 1935, 7:73-89.
463. Palladin, A. V., and Kirsenko, O. V. Biokhimiya, 1961, 26:385-390.
464. Palladin, A. V., and Kirsenko, O. V. In: In Memory of Academician Ignat Emanuilov [in Bulgarian], Izv. na Bolgarsk. Akad. na Naukite, Sofia, Bulgaria, 1969, pp. 193-202.
465. Palladin, A. V., Kirsenko, O. V., and Vavilova, G. A. Biokhimiya, 1970, 35:404-411.
466. Palladin, A. V., and Kudinov, S. A. Ukr. Biokhim. Zh., 1964, 36:548-558.
467. Palladin, A. V., and Polyakova, N. M. Dokl. Akad. Nauk SSSR, 1956, 107:568-570.
468. Palladin, A. V., and Polyakova, N. M. Ukr. Biokhim. Zh., 1959, 31:307-313.
469. Palladin, A. V., and Polyakov, N. M. Byull. Pol'skoi Akad. Nauk., Ser. Biol., 1959, 7(2):47-52.
470. Palladin, A. V., Polyakova, N. M., and Gotovtseva, O. P. Ukr. Biokhim. Zh., 1958, 30:323-332.
471. Palladin, A. V., Polyakova, N. M., and Malysheva, M. K. Dokl. Akad. Nauk SSSR, 1960, 134:1236-1239.
472. Palladin, A. V., Polyakova, N. M., and Malysheva, M. K. In: Biochemistry and Function of the Nervous System [in Russian], Nauka, Leningrad, 1967, pp. 104-109.
473. Palladin, A. V., Polyakova, N. M., and Silich, T. P. Fiziol. Zh. SSSR, 1957, 43:611-618.
474. Palladin, A. V., and Rashba, O. Ya. Ukr. Biokhim. Zh., 1934, 7(2):51-71; (3-4):85-116.
475. Palladin, A. V., Rashba, O. Ya., and Gel'man, R. M. Ukr. Biokhim. Zh., 1935, 8:5-26; 8:27-46.

476. Palladin, A. V., Rashba, O. Ya., and Gel'man, R. M. Ukr. Biokhim. Zh., 1936, 9:169-192.

477. Palladin, A. V., Rashba, O. Ya., and Shtutman, Ts. M. Ukr. Biokhim. Zh., 1951, 22:265-274.

478. Palladin, A. V., Terletskaya, Ya. T., and Belik, Ya. V. Zh. Évolyuts. Biokhim. i Fiziol., 1969, 5:140-144.

479. Palladin, A. V., Terletskaya, Ya. T., and Kozulina, O. P. Ukr. Biokhim. Zh., 1970, 42:144-154.

480. Palladin, A. V., Shtutman, Ts. M., and Rashba, O. Ya. Ukr. Biokhim. Zh., 1951, 23:170-177.

481. Panchenko, L. F. Scientific Reports of the 2nd Moscow Medical Institute [in Russian], Vol. 6, Moscow, 1957, pp. 41-48.

482. Panchenko, L. F. Farmakol. i Toksikol., 1958, 21:53-57.

483. Panchenko, L. F. Fiziol. Zh. SSSR, 1958, 44:243-248.

484. Panchenko, L. F. Author's Abstract of Doctoral Dissertation, 2nd Moscow Medical Institute, 1966.

485. Panchenko, L. F., and Archakov, A. I. Byull. Éksperim. Biol. i Med., 1965, 59(5):51-53.

486. Parina, E. V. Proceedings of the 7th Scientific Conference on Age Morphology, Physiology, and Biochemistry [in Russian], Moscow, 1965, pp. 411-412.

487. Parina, E. V. In: Leading Problems in Age Physiology and Biochemistry [in Russian], Meditsina, Moscow, 1966, pp. 150-167.

488. Parina, E. V. Age and Protein Metabolism [in Russian], Khar'kov University Press, 1967.

489. Parina, E. V. In: The Molecular and Functional Bases of Ontogeny [in Russian], Meditsina, Moscow, 1970, pp. 125-149.

490. Parina, E. V., and Derevyanko, G. I. In: The Molecular Biology of Aging [in Russian], Naukova Dumka, Kiev, 1969, pp. 66-69.

491. Parina, E. V., and Mishchenko, V. P. Zh. Évolyuts. Biokhim. i Fiziol., 1966, 2:439-445.

492. Paronyan, Zh. A. Author's Abstract of Dissertation for the Degree of Candidate of Sciences, Institute of Biochemistry, Academy of Sciences of the Armenian SSR, 1968.

493. Pas'ko, S. G. Dokl. Akad. Nauk SSSR, 1953, 91:1211-1212.

494. Patrikeeva, M. V. Dokl. Akad. Nauk SSSR, 1964, 154:1235-1237.

495. Pashkova, A. A., and Popova, L. Ya. In: The Molecular and Functional Bases of Ontogeny [in Russian], Meditsina, Moscow, 1970, pp. 158-190.

496. Pevzner, L. Z. Ukr. Biokhim. Zh., 1963, 35:448-477.

497. Pevzner, L. Z. Vopr. Med. Khimii, 1969, 15:211-223.

498. Pevzner, L. Z. Author's Abstract of Doctoral Dissertation, I. P. Pavlov Institute of Physiology, Academy of Sciences of the USSR, Leningrad, 1969.

499. Pevzner, L. Z. Uspekhi Sovr. Biol., 1969, 68:340-360.

500. Pevzner, L. Z., Koval', V. A., and Kuchin, A. A. Tsitologiya, 1964, 6:216-219.

501. Petrashkaite, S. K. Biokhimiya, 1965, 30:551-558.

502. Petrov, V. S. Arkh. Anat., Gistol., i Émbriol., 1968, 54(3):77-85.

503. Petrov, I. R., and Gubler, E. V. Artificial Hypothermia [in Russian], Medgiz, Leningrad, 1961.

504. Pigareva, Z. D. In: Problems in Biochemistry of the Nervous System [in Russian], Izd. Akad. Nauk Ukr. SSR, Kiev, 1957, pp. 217-227.
505. Pigareva, Z. D. Zh. Évolyuts. Biokhim. i Fiziol., 1965, 1:413-418.
506. Pigareva, Z. D. Zh. Nevropat. i Psikhiat., 1966, 66:1716-1721.
507. Pigareva, Z. D. In: Development of the Brain in Animals [in Russian], Nauka, Leningrad, 1969, pp. 172-191.
508. Pigareva, Z. D., et al. In: Mitochondria. Structure and Functions [in Russian], Nauka, Moscow, 1966, pp. 25-27.
509. Pigareva, Z. D., et al. In: Abiogenesis and the Initial Stages of Evolution of Life [in Russian], Nauka, Moscow, 1968, pp. 159-168.
510. Pinus, E. A., and Metlitskaya, A. Z. Uspekhi Sovr. Biol., 1970, 70:3-25.
511. Pinchuk, V. G. Tsitologiya, 1964, 6:3-12.
512. Pogodaev, B. F. Byull. Éksperim. Biol. i Med., 1961, 52(9):56-59.
513. Pogodaev, K. I. Ukr. Biokhim. Zh., 1957, 29:428-436.
514. Pogadaev, K. I. The Biochemistry of the Epileptic Fit [in Russian], Meditsina, Moscow, 1964.
515. Pogodaev, K. I. In: Problems in Neurochemistry [in Russian], Nauka, Moscow-Leningrad, 1966, pp. 34-39.
516. Pogodaev, K. I., et al. Ukr. Biokhim. Zh., 1960, 32:808-822.
517. Pogodaev, K. I., and Mekhedova, A. Ya. Problems in Biochemistry of the Nervous System [in Russian], Izd. Akad. Nauk Ukr.SSR, Kiev, 1957, pp. 40-47.
518. Pogodaev, K. I., Osipova, M. S., and Kunaeva, Z. I. Trudy Inst. Vyssh. Nerv. Deyat. AN SSSR. Ser. Fiziol., 1960, 4:236-243.
519. Pogodaev, K. I., and Turova, N. F. Ukr. Biokhim. Zh., 1967, 39:29-33.
520. Pogodaev, K. I., and Turova, N. F. The Biochemistry of the Brain in Atherosclerosis [in Russian], Meditsina, Moscow, 1969.
521. Pogorelova, T. N. Dokl. Akad. Nauk SSSR, 1966, 167:1421-1422.
522. Pogosova, A. V., Rapoport, É. A., and Zelenina, V. P. Dokl. Akad. Nauk SSSR, 1964, 154:1206-1209.
523. Pokrovskii, A. A., and Tutel'yan, V. A. Biokhimiya, 1968, 33:809-816.
524. Pokrovskii, A. A., and Tutel'yan, V. A. Uspekhi Sovr. Biol., 1969, 68:318-339.
525. Polenov, A. L. Hypothalamic Neurosecretion [in Russian], Nauka, Leningrad, 1968.
526. Polyakova, N. M. Dokl. Akad. Nauk SSSR, 1956, 109:1174-1175.
527. Polyakova, N. M. Ukr. Biokhim. Zh., 1956, 28:286-295.
528. Polyakova, N. M. Ukr. Biokhim. Zh., 1959, 31:314-321.
529. Polyakova, N. M. Ukr. Biokhim. Zh., 1960, 32:120-148.
530. Polyakova, N., M. Author's Abstract of Doctoral Dissertation, Leningrad State University, 1962.
531. Polyakova, N. M. The 3rd All-Union Conference on Biochemistry of the Nervous System [in Russian], Izd. Akad. Nauk Arm. SSR, Erevan, 1963, pp. 25-38.
532. Polyakova, N. M., Belik, Ya. V., and Tsaryuk, L. A. Ukr. Biokhim. Zh., 1960, 32:623-635.
533. Polyakova, N. M., and Gotovtseva, O. P. Ukr. Biokhim. Zh., 1957, 29:400-408.
534. Polyakova, N. M., and Kabak, K. S. Dokl. Akad. Nauk SSSR, 1958, 122:275-277.
535. Polyakova, N. M., and Lishko, V. K. Ukr. Biokhim. Zh., 1962, 34:10-22.

536.   Polyakova, N. M., and Lishko, V. K. Ukr. Biokhim. Zh., 1962, 34:208-216.
537.   Polyakova, N. M., and Malysheva, M. K. Ukr. Biokhim. Zh., 1961, 33:713-731.
538.   Polyakova, N. M., and Malysheva, M. K. Dokl. Akad. Nauk SSSR, 1962,
        144:1394-1397.
539.   Pomazanskaya, L. F. Zh. Évolyuts. Biokhim. i Fiziol., 1970, 6:477-484.
540.   Portugalov, V. V. Zh. Nevropat. i Psikhiat., 1958, 58:641-649.
541.   Portugalov, V. V. In: Current Problems in Biochemistry. 1. Biochemistry of
        Proteins [in Russian], Moscow, 1959, pp. 126-131.
542.   Portugalov, V. V., Krasnov, I. B., and Dovedova, E. L. Byull. Éksperim. Biol.
        i Med., 1965, 60(11):103-106.
543.   Portugalov, V. V., Tsvetkova, I. V., and Yakovlev, V. A. Tsitologiya, 1959,
        1:422-430.
544.   The Problem of Brain Development and the Effect of Harmful Factors on It [in
        Russian], Medgiz, Moscow, 1960.
545.   Problems in the Study of Tissue–Blood Barriers [in Russian], Nauka, Moscow,
        1965.
546.   Promyslov, M. Sh. In: Biochemistry of the Nervous System [in Russian], Izd.
        Akad. Nauk Ukr. SSR, Kiev, 1954, pp. 179-189.
547.   Promyslov, M. Sh. Dokl. Akad. Nauk SSSR, 1956, 110:417-419.
548.   Prokhorova, M. I. In: The Nervous System [in Russian], Leningrad University
        Press, 1960, pp. 24-32.
549.   Prokhorova, M. I., et al. In: The Physiology and Biochemistry of the Nervous
        System and Muscles [in Russian], Leningrad University Press, 1957, pp. 272-286.
550.   The Development and Regulation of Tissue–Blood Barriers [in Russian], Nauka,
        Moscow, 1967.
551.   Rashba, O. Ya., and Shtutman, Ts. M. Ukr. Biokhim. Zh., 1951, 23:89-102.
552.   Reva, A. D., et al. Abstracts of Section Proceedings of the 2nd All-Union Bio-
        chemical Congress. Section 17 [in Russian], Fan, Tashkent, 1969, pp. 102-103.
553.   Reva, A. D., Starodub, M. F., and Kukharenko, R. I. Ukr. Biokhim. Zh., 1967,
        39:60-65.
554.   Reznikov, K. Yu. Uspekhi Sovr. Biol., 1966, 62:237-247.
555.   Rybak, V. I., Reva, A. D., and Nazarenko, V. I. Ukr. Biokhim. Zh., 1969,
        41:121-125.
556.   Rich, A. In: The Functional Biochemistry of Cell Structures [Russian transla-
        tion], Nauka, Moscow, 1970, pp. 218-231.
557.   De Robertis, E., Nowinski, W. W., and Saez, F. A. Cell Biology [Russian transla-
        tion], Mir, Moscow, 1967. (English 4th edition: Saunders, Philadelphia, 1966.)
558.   Rogozkin, V. A. Vopr. Med. Khimii, 1959, 5:358-361.
559.   Rozhmanova, O. M., and Kirsenko, O. V. Proceedings of the 4th All-Union
        Conference on Biochemistry of the Nervous System [in Russian], Tartu Univer-
        sity Press, 1969, pp. 267-277.
560.   Rozengart, V. I., and Maslova, M. N. Dokl. Akad. Nauk SSSR, 1956, 109:1176-
        1179.
561.   Rozengart, V. I., and Maslova, M. N. Biokhimiya, 1957, 22:947-953.
562.   Rozengart, V. I., and Maslova, M. N. Vopr. Med. Khimii, 1963, 9:3-15.
563.   The Role of Gamma-aminobutyric Acid in the Activity of the Nervous System
        [in Russian], Leningrad University Press, 1964.

564. Rubinshtein, D. L. General Physiology [in Russian], Medgiz, Moscow, 1947.
565. Rudakov, V. V. Uspekhi Sovr. Biol., 1966, 61:32-38.
566. Ruzhitskii, B. Trudy Mosk. Vet. Akad., 1960, 30:136-139.
567. Rushchak, M., et al. Ukr. Biokhim. Zh., 1964, 36:584-592.
568. Ryzhkov, V. L. Izvest. Akad. Nauk SSSR, Ser. Biol., 1965, pp. 533-541.
569. Rykhlik, I., Dancheva, K. I., and Tsergova, M. In: Nucleic Acids [in Russian], Meditsina, Moscow, 1966, pp. 131-135.
570. Saakov, B. A. Hypothermia [in Russian], Gos. Med. Izd. Ukr. SSR, Kiev, 1957.
571. Savitskii, I. V., and Zelinskii, V. G. Ukr. Biokhim. Zh., 1964, 36:14-21.
572. Savich, K. V., and Yakovlev, V. A. Vopr. Med. Khimii, 1957, 3:121-128.
573. Savluchinskaya, L. G., and Rozanov, A. Ya. Ukr. Biokhim. Zh., 1969, 41:448-451.
574. Sadikova, N. V., and Kudryashova, G. K. In: The Nervous System [in Russian], Vol. 5, Leningrad University Press, 1964, pp. 16-21.
575. Sadikova, N. V., and Skvortsevich, V. A. Vopr. Med. Khimii, 1956, 2:128-132.
576. Salganik, R. I. Biokhimiya, 1954, 19:641-644.
577. Salganik, R. I. Vopr. Med. Khimii, 1956, 2:424-427.
578. Samarina, O. P. Biokhimiya, 1961, 26:61-69.
579. Samarina, O. P., and Georgiev, G. P. Dokl. Akad. Nauk SSSR, 1960, 133:694-697.
580. Saudargene, D. S., and Pevzner, L. Z. Tsitologiya, 1969, 11:1275-1286.
581. Svanidze, I. K. Zh. Obshch. Biol., 1967, 28:697-707.
582. Sviridov, S. M., and Polyakova, E. V. Dokl. Akad. Nauk SSSR, 1969, 187:925-927.
583. Free-Radical Processes in Biology Systems [in Russian], Nauka, Moscow, 1966.
584. Free Radicals in Biological Systems [Russian translation], IL, Moscow, 1963.
585. Sergienko, E. F., and Timkovitskaya, A. M. Trudy N.-I. Inst. Biol. Khar'k. Univ., 1954, 21:81-86.
586. Silakova, A. I. In: Problems in the Biochemistry of Muscles [in Russian], Izd. Akad. Nauk Ukr. SSR, Kiev, 1954, pp. 221-243.
587. Silich, T. P. Ukr. Biokhim. Zh., 1957, 29:166-172.
588. Sisakyan, N. M. Uspekhi Sovr. Biol., 1961, 51:129-152.
589. Sisakyan, N. M. Ukr. Biokhim. Zh., 1965, 37:640-649.
590. Sisakyan, N. M., et al. Biokhimiya, 1963, 28:326-333.
591. Skachkova, A. S. In: Research into the Use of Radioactive Isotopes in Medicine [in Russian], Medgiz, Moscow, 1955, pp. 53-57.
592. Skvirskaya, E. B., and Babii, T. P. Ukr. Biokhim. Zh., 1961, 33:647-656.
593. Skvirskaya, E. B., and Silich, T. P. In: Biochemistry of the Nervous System [in Russian], Izd. Akad. Nauk Ukr. SSR, Kiev, 1954, pp. 36-46.
594. Skvirskaya, E. B., and Silich, T. P. Ukr. Biokhim. Zh., 1955, 27:385-393.
595. Skvirskaya, E. B., and Silich, T. P. Ukr. Biokhim. Zh., 1957, 29:33-41.
596. Skvirskaya, E. B., and Chepnoga, O. P. Dokl. Akad. Nauk SSSR, 1953, 92:1007-1010.
597. Slovtsov, B. I. Russkii Fiziol. Zh., 1921, 3:31-32.
598. Slovtsov, B. I., and Georgievskaya, A. M. Russkii Fiziol. Zh., 1921, 4:35-51.
599. Smerchinskaya, L. S. Third All-Union Conference on Biochemistry of the Nervous System [in Russian], Izd. Akad. Nauk Arm. SSR, 1963, pp. 47-54.

600. Smerchinskaya, L. S. Ukr. Biokhim. Zh., 1964, 36:355-366.
601. Smerchinskaya, L. S. Author's Abstract of Candidate's Dissertation, Institute of Biochemistry, Academy of Sciences of the Ukrainian SSR, Kiev, 1965.
602. Smerchinskaya, L. S., and Avdeev, V. G. Ukr. Biokhim. Zh., 1964, 36:836-837.
603. Smerchinskaya, L. S., Belik, Ya. V., and Berezhnyi, G. A. Ukr. Biokhim. Zh., 1972, 44:3-9.
604. Smirnov, A. A. Dokl. Akad. Nauk SSSR, 1955, 105:185-187.
605. Smirnov, A. A., and Chetverikov, D. A. Dokl. Akad. Nauk SSSR, 1953, 90:631-633.
606. Smirnova-Zamkova, A. I., and Mel'nichenko, A. V. In: Tissue—Blood Barriers [in Russian], Izd. AN SSSR, Moscow, 1961, pp. 355-360.
607. Sopin, E. F. Visnik Kiivs. Univ., Ser. Biol., 1958, 1(1):149-161.
608. Sorks, T. L. Zh. Vses. Khim. Obshch. im. D. I. Mendeleeva, 1964, 9:381-394.
609. Spirin, A. S. Uspekhi Sovr. Biol., 1969, 68:454-470.
610. Spirin, A. S., et al. Uspekhi Biol. Khimii, 1963, 5:3-60.
611. Spirin, A. S., and Gavrilova, L. P. The Ribosome [in Russian], Nauka, Moscow, 1968.
612. Strizhova-Salova, N. I., and Baluev, S. I. Ukr. Biokhim. Zh., 1968, 40:261-264.
613. Sukhomlinov, B. F., Oleinik, Ya. V., and Pakosh, M. P. In: The Biological Action of Radiation [in Russian], Vol. 2, L'vov University Press, 1963, p. 26-39.
614. Sukhomlinov, B. F., and Fornyak, N. M. Ukr. Biokhim. Zh., 1965, 37:315-323.
615. Sytinskii, I. A. In: The Role of Gamma-aminobutyric Acid in the Activity of the Nervous System [in Russian], Leningrad University Press, 1964, pp. 36-48.
616. Sytinskii, I. A., Zh. Évolyuts. Biokhim. i Fiziol., 1970, 6:162-171.
617. Taranova, N. P. Uspekhi Sovr. Biol., 1968, 66:339-351.
618. Tashmukhamedov, B. A., and Lisovskaya, N. P. Biofizika, 1965, 10:699-701.
619. Terletskaya, Ya. T. Ukr. Biokhim. Zh., 1963, 35:542-548.
620. Terletskaya, Ya. T. Author's Abstract of Dissertation for the Degree of Candidate of Sciences, Institute of Biochemistry, Academy of Sciences of the Ukrainian SSR, Kiev, 1964.
621. Terletskaya, Ya. T., Belik, Ya. V., and Smerchinskaya, L. S. In: Mitochondria. Biochemistry and Morphology [in Russian], Nauka, Moscow, 1967, pp. 57-61.
622. Terletskaya, Ya. T., Palladin, A. V., and Pisarevich, O. V. Ukr. Biokhim. Zh., 1963, 35:737-746.
623. Timkin, V. N., et al. Zh. Vyssh. Nervn. Deyat., 1970, 20:185-190.
624. Todorov, I. N., Galkin, A. P., and D'yachenko, A. G. In: Problems in Experimental and Clinical Radiology [in Russian], Vol. 4, Zdorov'ya, Khar'kov, 1968, pp. 19-26.
625. Toropova, G. P. Vopr. Pitaniya, 1955, 14(4):12-14.
626. Troshin, A. S. Problems in Cellular Permeability [in Russian], Izd. AN SSSR, Moscow, 1956.
627. Troshin, A. S. Tsitologiya, 1963, 5:601-604.
628. Troshin, A. S. In: Protoplasmic Membranes and Their Functional Role [in Russian], Naukova Dumka, Kiev, 1965, pp. 125-133.
629. Turpaev, T. M. The Mediator Function of Acetylcholine and the Nature of the Cholinergic Receptor [in Russian], Izd. AN SSSR, Moscow, 1962.

630. Turpaev, T. M., and Nistratova, S. N. In: Protoplasmic Membranes and Their Functional Role [in Russian], Naukova Dumka, Kiev, 1965, pp. 197-210.

631. Tustanovskii, A. A. Biokhimiya, 1938, 3:218-230.

632. Tyulenev, V. I. Ukr. Biokhim. Zh., 1967, 39:11-17.

633. Tyulenev, V. I., and Belik, Ya. V. Ukr. Biokhim. Zh., 1967, 39:231-236; 577-583.

634. Tyulenev, V. I., and Belik, Ya. V. Zh. Évolyuts. Biochim. i Fiziol., 1970, 6:25-29.

635. Tyakhepyl'd, L. Ya. Vopr. Med. Khimii, 1962, 8:264-270.

636. Tyakhepyl'd, L. Ya. Uch. Zap. Tartu. Univ., 1964, 163:128-134; 142-148.

637. Tyakhepyl'd, L. Ya. Proceedings of the 4th All-Union Conference on Biochemistry of the Nervous System [in Russian], Tartu University Press, 1969, pp. 61-72.

638. Upirvitskaya, A. K. In: The Biochemistry of the Brain [in Russian], Gor'kii Medical Institute Press, 1941, pp. 35-47.

639. Usacheva, N. T., and Milova, G. N. Vopr. Pitaniya, 1964, 23(6):17-21.

640. Uspenskii, V. I. Histamine [in Russian], Medgiz, Moscow, 1963.

641. Utevskii, A. M. In: Current Problems in Endocrinology [in Russian], Vol. 1, Medgiz, Moscow, 1960, pp. 5-58.

642. Utevskii, A. M., and Baru, A. M. Zh. Vses. Khim. Obshch. im. D. I. Mendeleeva, 1964, 9:374-380.

643. Utina, I. A. Biofizika, 1960, 5:626-627.

644. Webb, L. Enzyme Inhibitors and Metabolism [Russian translation], Mir, Moscow, 1966.

645. Fedorov, O. M. Ukr. Biokhim. Zh., 1963, 35:520-527.

646. Fedorov, O. M. Ukr. Biokhim. Zh., 1966, 38:128-131.

647. Fedorov, O. M., and Palladin, A. V. Ukr. Biokhim. Zh., 1963, 35:690-699.

648. Fedorova, N. A., Goryukhina, O. A., and Tolchenova, G. A. Vestn. Leningrad. Gos. Univ. Ser. Biol., 1966, (4):156-158.

649. Fedorova, N. A., Komkova, A. I., and Zarubailo, T. T. Third All-Union Conference on Biochemistry of the Nervous System [in Russian], Izd. Akad. Nauk Arm. SSR, Erevan, 1963, pp. 77-80.

650. Ferdman, D. L. Uspekhi Sovr. Biol., 1936, 5:431-450.

651. Ferdman, D. L. Ukr. Biokhim. Zh., 1941, 17:95-104.

652. Ferdman, D. L. Uspekhi Biol. Khimii, 1950, 1:216-241.

653. Ferdman, D. L. In: Amino Acid Metabolism [in Russian], Metsniereba, Tbilisi, 1967, pp. 77-84.

654. Ferdman, D. L., and Silakova, A. I. Biokhimiya, 1957, 22:283-294.

655. Ferdman, D. L., and Sopin, E. F. Nauk. Zap. Kiivs. Univ., 1957, 16(20):71-76.

656. Ferdman, D. L., Frenkel', S. R., and Silakova, A. I. Biokhimiya, 1942, 7:43-58.

657. Fernandes-Moran, G. In: Current Problems in Biophysics [Russian translation], Vol. 2, IL, Moscow, 1961, pp. 58-68.

658. Firfarova, K. F., Morozkin, A. D., and Orekhovich, V. N. Biokhimiya, 1964, 29:673-679.

659. Fleisher, G. V. Russkii Vrach, 1908, 7:395-397.

660.   Folch-Pi, J. In: Biochemistry and Function of the Nervous System [Russian translation], Nauka, Leningrad, 1967, pp. 123-133.

661.   Fomin, S. Nauk. Zap. Ukr. Biokhim. Inst., 1928, 3:143-146.

662.   Fomichenko, K. V. Third All-Union Conference on Biochemistry of the Nervous System [in Russian], Izd. Akad. Nauk Arm. SSR, Erevan, 1963, pp. 581-588.

663.   Frenkel', S. R. In: Problems in Neurochemistry [in Russian], Nauka, Moscow, 1966, pp. 97-102.

664.   Frenkel', S. R., and Gordienko, E. A. Third All-Union Conference on Biochemistry of the Nervous System [in Russian], Izd. Akad. Nauk Arm. SSR, Erevan, 1963, pp. 223-235.

665.   Fridman-Pogosova, A. V. Dokl. Akad. Nauk SSSR, 1955, 102:1227-1229.

666.   Haggis, G. H., et al. Introduction to Molecular Biology, Longmans, London, 1964.

667.   Khaikina, B. I. Dokl. Akad. Nauk SSSR, 1956, 111:1061-1063.

668.   Khaikina, B. I., and Krachko, L. S. Ukr. Biokhim. Zh., 1957, 29:10-19.

669.   Khachatryan, G. S. Trudy Erevan. Med. Inst., 1963, 13:97-105.

670.   Kheruvimova, V. A. Dokl. Akad. Nauk SSSR, 1961, 136:968-970.

671.   Khesin, R. B. The Biochemistry of Cytoplasm [in Russian], Izd. AN SSSR, Moscow, 1960.

672.   Khesin, R. B., Gvozdev, V. A., and Astaurova, O. B. Biokhimiya, 1961, 26:807-816.

673.   Hydén, H. In: The Functional Morphology of the Cell [in Russian], IL, Moscow, 1963, pp. 185-260.

674.   Hydén, H., and Lange, P. V. Zh. Évolyuts. Biochim. i Fiziol., 1969, 5:145-157.

675.   Hodge, A. In: Current Problems in Biophysics [Russian translation], Vol. 2, IL, Moscow, 1961, pp. 142-151.

676.   Khritinina, K. M. Byull. Éksperim. Biol. i Med., 1942, 14(2):28-31.

677.   Khrushchev, G. K., and Brodskii, V. Ya. Uspekhi Sovr. Biol., 1961, pp. 181-207.

678.   Tsanev, R. G., and Markov, G. G. Biokhimiya, 1960, 25:151-159.

679.   Tsaryuk, L. A. Ukr. Biokhim. Zh., 1962, 34:815-824.

680.   Tsaryuk, L. A. Ukr. Biokhim. Zh., 1964, 36:334-342.

681.   Tsiperovich, A. S. Biokhim. Zh., 1941, 17:173-200.

682.   Chagovets, R. V., et al. In: Problems in Biochemistry of the Nervous System [in Russian], Izd. Akad. Nauk Ukr. SSR, Kiev, 1957, pp. 258-267.

683.   Chang Tsin, Biokhimiya, 1960, 25:1099-1104.

684.   Chang Tsin. Vopr. Med. Khimii, 1961, 7:154-158.

685.   Chentsov, Yu. S. In: Structure and Functions of the Cell Nucleus [in Russian], Nauka, Moscow, 1967, pp. 39-50.

686.   Cherkasova, L. S. In: The Biochemistry of Small Doses of Ionizing Radiation [in Russian], Nauka i Tekhnika, Minsk, 1964, pp. 5-24.

687.   Cherkasova, L. S., et al. Ionizing Radiation and Metabolism [in Russian], Izd. Akad. Nauk Belor. SSR, Minsk, 1962.

688.   Chernevskaya, Kh. I. In: Problems in Biochemistry of the Brain [in Russian], Volgo-Vyatskoe Izd., Gor'kii, 1966, pp. 14-21.

689.   Chernikov, M. P. Biokhimiya, 1957, 22:5-13.

690.   Chernikov, M. P. In: Current Problems in Biochemistry. 1. Biochemistry of Proteins [in Russian], Moscow, 1959, pp. 170-177.

691. Chernikov, M. P. Biokhimiya, 1963, 28:285-287.
692. Chernikov, M. P., and Evtikhina, Z. F. Uspekhi Sovr. Biol., 1964, 57:50-70.
693. Chetverikov, D. A., and Gasteva, S. V., Dokl. Akad. Nauk SSR, 1963, 151:718-721.
694. Chikalo, I. I., and Cherevichnaya, E. V. In: Problems in Experimental and Clinical Radiology [in Russian], Vol. 1, Zdorov'ya, Kiev, 1965, pp. 153-157.
695. Chikvaidze, V. N. Third All-Union Conference on Biochemistry of the Nervous System [in Russian], Izd. Akad. Nauk Arm. SSR, Erevan, 1963, pp. 181-189.
696. Shabadash, A. L. Dokl. Akad. Nauk SSSR, 1957, 114:658-661.
697. Shabadash, A. L. Tsitologiya, 1959, 1:15-34.
698. Shabadash, A. L. Radiobiologiya, 1961, 1:212-222.
699. Shabadash, A. L. In: Mitochondria. Structure and Functions [in Russian], Nauka, Moscow, 1966, pp. 5-22.
700. Shabadash, A. L. Proceedings of the 4th All-Union Conference on Biochemistry of the Nervous System [in Russian], Tartu University Press, 1969, pp. 203-213.
701. Shabadash, A. L., Zelikina, T. I., and Agracheva, N. D. Arch. Anat., Gistol., i Émbriol., 1963, 44(2):3-9.
702. Shamirzaev, A. Yu. Med. Zh. Uzbekistana, 1964, 1:60-65.
703. Chantrenne, H. Biosynthesis of Proteins [Russian translation], IL, Moscow, 1963. (English edition: Pergamon Press, London, 1961.)
704. Shapot, V. S. Uspekhi Sovr. Biol., 1952, 34:244-267.
705. Shapot, V. S., and Nemchinskaya, V. L. Dokl. Akad. Nauk SSSR, 1950, 70:465-468.
706. Schweet, R., and Owen, R. In: Current Problems in Biochemistry [Russian translation], IL, Moscow, 1961, pp. 324-343.
707. Sjöstrand, F. In: Functional Morphology of the Cell [Russian translation], IL, Moscow, 1963, pp. 69-85.
708. Shkarin, A. N. The Protein Composition of the Cerebral Cortex Depending on Age and Some Other Physiological Conditions. Dissertation, St. Petersburg, 1902.
709. Shmerling, Zh. G. Izvest. Akad. Nauk SSSR, Ser. Biol., 1970, pp. 847-862.
710. Shnyak, E. I. Dokl. Akad. Nauk SSSR, 1962, 146:734-736.
711. Shtark, M. B. Uspekhi. Sovr. Biol., 1965, 60:384-410.
712. Shtark, M. B., and Danilyuk, V. P. Dokl. Akad. Nauk SSSR, 1963, 151:740-743.
713. Shtern, L. S. In: The Blood—Brain Barrier [in Russian], Moscow, 1935, pp. 532-546.
714. Shtern, L. S. Uspekhi Sovr. Biol., 1958, 45:328-348.
715. Shtern, L. S. The Immediate Nutrient Medium of Organs and Tissues [in Russian], Izd. AN SSSR, Moscow, 1960.
716. Shtern, L. S. In: Development and Regulation of Tissue—Blood Barriers [in Russian], Nauka, Moscow, 1967, pp. 3-10.
717. Shtern, L. S., and Peiro, R. In: The Blood—Brain Barrier [in Russian], Moscow, 1935, pp. 186-187.
718. Shtern, L. S., and Rapoport, Ya. L. In: The Blood—Brain Barriers [in Russian], Moscow, 1935, pp. 188-190.
719. Shtern, L. S., Rapoport, Ya. L., and Lokshina, É. S. In: The Blood—Brain Barrier [in Russian], Moscow, 1935, pp. 191-192.

720. Stich, H. In: Problems in Cytophysiology [Russian translation], IL, Moscow,
      1957, pp. 42-52.
721. Shtutman, Ts. M. Ukr. Biokhim. Zh., 1949, 21:73-76.
722. Shtutman, Ts. M. Ukr. Biokhim. Zh., 1958, 30:852-859.
723. Shubnikova, E. Uspekhi Sovr. Biol., 1946, 21:442-444.
724. Eccles, J. C. The Physiology of Synapses, Springer, Berlin, 1964.
725. Él'piner, I. E. Uspekhi Sovr. Biol., 1966, 61:212-229.
726. Émirbekov, É. Z. Nitrogen Metabolism of the Brain in Hypothermia and Hiber-
      nation [in Russian], Dauchpedgiz, Makhachkala, 1969.
727. Émirbekov, É. Z. Ukr. Biokhim. Zh., 1969, 41:233-235.
728. Émirbekov, É. Z., and L'vova, S. P. Uspekhi Sovr. Biol., 1970, 70:276-285.
729. Éngel'gardt, V. A., and Lisovskaya, N. P. In: The Biochemistry of the Nervous
      System [in Russian], Izd. Akad. Nauk Ukr. SSR, Kiev, 1954, pp. 77-86.
730. Éngel'gardt, V. A., and Lisovskaya, N. P. Biokhimiya, 1955, 20:225-235.
731. Yavich, M. P. Éksperim. Khirurgiya, 1959, 2:55-56.
732. Yasenchak, S. A. Sborn. Nauch. Rabot L'vov. Gos. Med. Inst., 1959, 17:232-233.
733. Yasenchak, S. A. Author's Abstract of Dissertation for the Degree of Can-
      didate of Sciences, Uzhgorod State University, 1963.
734. Abadia-Fenoll, F., Blanco, M., and Fairen, A. Rev. Roum. Neurol., 1970,
      7:19-25.
735. Abadom, P. N., and Scholefield, P. G. Canad. J. Biochem. Physiol., 1962,
      40:1575-1590; 1591-1602; 1603-1618.
736. Abdel-Latif, A. A. Biochim. Biophys. Acta, 1966, 121:403-406.
737. Abdel-Latif, A. A., and Abood, L. G. J. Neurochem., 1964, 11:9-15.
738. Abdel-Latif, A. A., and Abood, L. G. J. Neurochem., 1966, 13:1189-1196.
739. Acs, G., Neidle, A., and Schneiderman, N. Biochim. Biophys. Acta, 1962,
      56:373-374.
740. Acs, G., Neidle, A., and Waelsch, H. Biochim. Biophys. Acta, 1961, 50:403-
      404.
741. Adair, L. B., et al. Proc. Nat. Acad. Sci. (Washington), 1968, 61:606-613.
742. Adair, L. B., Wilson, J. E., and Glassman, E. Proc. Nat. Acad. Sci. (Washing-
      ton), 1968, 61:917-922.
743. Adams, C. W. M., and Bayliss, O. B. J. Histochem. Cytochem., 1961, 9:473-476.
744. Adams, C. W. M., and Bayliss, O. B. J. Histochem. Cytochem., 1968, 16:110-114
745. Adams, C. W. M., and Glenner, G. G. J. Neurochem., 1962, 9:233-239.
746. Adams, C. W. M., and Tugan, N. A. J. Neurochem., 1961, 6:334-341.
747. Adams, D. H. Biochem. J., 1966, 98:636-640.
748. Adams, D. H., and Lim, L. Biochem. J., 1966, 99:261-265.
749. Adams, E., and Smith, E. L. J. Biol. Chem., 1951, 191:651-664.
750. Addanki, S., Canill, F. D., and Sotos, J. F. J. Biol. Chem., 1968, 243:2337-
      2348.
751. Agranoff, B. W. In: Protein Metabolism of the Nervous System, Plenum Press,
      New York, 1970, pp. 553-541.
752. Agranoff, B. W., et al. In: First International Meeting of the International
      Society for Neurochemistry, Strasbourg, 1967, p. 4.
753. Agrawal, H. C., Davis, J. M., and Himwich, W. A. J. Neurochem., 1966,
      13:607-615.

754. Agrawal, H. C., Davis, J. M., and Himwich, W. A. J. Neurochem., 1967, 14:179-181.
755. Agrawal, H. C., Davis, J. M., and Himwich, W. A. J. Neurochem., 1968, 15:529-531; 917-923.
756. Albert, E. Hoppe-Seyler's Z. Physiol. Chem., 1957, 308:189-196.
757. Alberty, K. G. M., and Bartley, W. Biochem. J., 1963, 87:104-114.
758. Allegranza, A., and Marobbio, C. World Neurol., 1962, 3:316-325.
759. Allison, A. C. Nature, 1965, 205:141-143.
760. Allison, J. B., Wannemacher, R. W., and Banks, W. L. Federat. Proc., 1963, 22:1126-1130.
761. De Almeida, D. F., and Pearse, A. G. E. J. Neurochem., 1958, 3:132-138.
762. Altman, J. Science, 1962, 135:1127-1128.
763. Altman, J. Nature, 1963, 199:777-780.
764. Amaducci, L. Federat. Proc., 1960, 19:231.
765. Amaducci, L. J. Neurochem., 1962, 9:153-160.
766. Amino Acid Pools, Elsevier, Amsterdam, 1962.
767. Andre, J., and Marinozzi, V. J. Microsc., 1965, 4:615-626.
768. Ansell, G. B., and Richter, D. Biochem. J., 1954, 57:70-73.
769. Ansell, G. B., and Richter, D. Biochim. Biophys. Acta, 1954, 13:87-91; 92-97.
770. Anson, M. L. J. Gen. Physiol., 1938, 22:79-89.
771. Appelmans, F., Wattiauz, R., and de Duve, C. Biochem. J., 1955, 59:438-445.
772. Aronson, N. N., and Davidson, E. A. J. Biol. Chem., 1965, 240:3222-3224.
773. Artizzu, M., et al. Biochim. Biophys. Acta, 1964, 82:454-462.
774. Athineos, E., Kukral, J. C., and Winzler, R. J. Arch. Biochem. Biophys., 1964, 106:338-342.
775. Austin, L., and Morgan, I. G. J. Neurochem., 1967, 14:377-387.
776. Austin, L., Morgan, I. G., and Bray, J. J. In: Protein Metabolism of the Nervous System, Plenum Press, New York, 1970, pp. 271-287.
777. Autilio, L. Federat. Proc., 1966, pp. 25, 764.
778. Autilio, L. A., Appel, S. H., and Pettis, P. Biochemistry, 1968, 7:2615-2622.
779. Autilio, L. A., Norton, W. T., and Terry, R. D. J. Neurochem., 1964, 11:17-27.
780. Azzi, A., Chappel, J. B., and Robinson, B. H. Biochem. Biophys. Res. Commun., 1967, 29:148-152.
781. Bachelard, H. S. Biochem. J., 1966, 100:131-137.
782. Bailey, B., et al. J. Biol. Chem., 1942, 143:721-728.
783. Bailey, B. F. S., and Heald, P. J. J. Neurochem., 1961, 6:342-349.
784. Bakay, L. In: Metabolism of the Nervous System, Pergamon Press, London, 1957, pp. 136-150.
785. Balasubramaniam, K., and Deiss, W. P. Biochim. Biophys. Acta, 1965, 110:564-575.
786. Balazs, R., et al. J. Neurochem., 1968, 15:1335-1349.
787. Balazs, R., and Gaitonde, M. K. Biochem., J., 1968, 106:1p-2p.
788. Banik, N. L., and Davison, A. N. Biochem. J., 1969, 115:1051-1062.
789. Barkulis, S. S., et al. J. Neurochem., 1960, 5:339-348.
790. Barondes, S. H. Science, 1964, 146:779-781.
791. Barondes, S. H. J. Neurochem., 1966, 13:721-727.
792. Barondes, S. H. J. Neurochem., 1968, 15:343-350; 699-706.

793.   Barondes, S. H., and Cohen, H. D. In: First International Meeting of the In-
        ternational Society for Neurochemistry, Strasbourg, 1967, p. 17.
794.   Barondes, S. H., and Cohen, H. D. Brain Res., 1967, 4:44-51.
795.   Barondes, S. H., and Cohen, H. D. Science, 1968, 160:556-557.
796.   Barron, K. D., Bernsohn, J., and Hess, A. R. J. Histochem. Cytochem., 1963,
        11:139-156.
797.   Bass, N. H., and Hess, H. H. J. Neurochem., 1969, 16:731-750.
798.   Bass, N. H., Netsky, M. G., and Young, E. Neurology, 1969, 19:405-414.
799.   Battistin, L., Grynbaum, A., and Lajtha, A. J. Neurochem., 1969, 16:1459-
        1468.
800.   Baumstark, F. Z. Physiol. Chem., 1885, 9:145.
801.   Bayer, S. M., and McMurray, W. C. J. Neurochem., 1967, 14:696-706.
802.   Beattie, D. S., Basford, R. E., and Koritz, S. B. Biochemistry, 1966, 5:926-930.
803.   Beattie, D. S., Basford, R. E., and Koritz, S. B. Biochemistry, 1967, 6:3099-
        3106.
804.   Beattie, D. S., Basford, R. E., and Koritz, S. B. J. Biol. Chem., 1967, 242:3366-
        3368.
805.   Beaufay, H., Berleur, A.-M., and Doyen, A. Biochem. J., 1957, 66:32p.
806.   Becker, N. H., and Sandbank, U. J. Histochem. Cytochem., 1964, 12:483-485.
807.   Bélanger, L. F., and Migicovsky, B. B. J. Histochem. Cytochem., 1963, 11:734-
        737.
808.   Belitser, V. A., Varetskaja, T. V., and Malneva, G. V. Biochim. Biophys. Acta,
        1968, 154:367-375.
809.   Beloff, A. Biochem. J., 1946, 40:108-115.
810.   Beloff, A., and Peters, R. A. J. Physiol., 1945, 103:461-476.
811.   Benda, Ph., et al. Science, 1968, 161:370.
812.   Benedetta, C. Di, and Brunngraber, E. G. In: Second International Meeting
        of the International Society for Neurochemistry, Milan, 1969, p. 154.
813.   Benetato, Gr., et al. Studii si Cercetari Fiziol. Acad. RPR, 1962, 7:589-599.
814.   Benetato, Gr., Gabrielescu E., and Boros, I. Rev. Roum. Physiol., 1965, 2:379-
        384.
815.   Bennet, E. L., et al. J. Neurochem., 1961, 6:210-218.
816.   Bennet, G. S., and Edelman, G. M. J. Biol. Chem., 1968, 243:6234-6241.
817.   Van den Berg, C. J. J. Neurochem., 1970, 17:973-983.
818.   Van den Berg, C. J., et al. Biochem. J., 1969, 113:281-290.
819.   Van den Berg, C. J., and Van den Velden, J. J. Neurochem., 1970, 17:985-991.
820.   Berger, M., Strecker, H. J., and Waelsch, H. Nature, 1956, 177:1234-1235.
821.   Bergmann, M. Advances Enzymol., 1942, 2:49-68.
822.   Bergmann, M., and Fraenkel-Conrat, H. J. Biol. Chem., 1937, 119:707-720.
823.   Bergmann, M., and Fraenkel-Conrat, H. J. Biol. Chem., 1938, 124:1-6.
824.   Berl, S. In: Progress in Brain Research, Vol. 9, The Developing Brain, Else-
        vier, Amsterdam, 1964, pp. 178-182.
825.   Berl, S., et al. J. Biol. Chem., 1962, 237:2562-2569.
826.   Berl, S., and Frigyesi, T. L. J. Neurochem., 1969, 16:405-415.
827.   Berl, S., Lajtha, A., and Waelsch, H. J. Neurochem., 1961, 7:186-197.
828.   Berl, S., Lajtha, A., and Waelsch, H. In: Chemical Pathology of the Nervous
        System, Pergamon Press, Oxford, 1961, pp. 361-368.

829.   Berl, S., and Purpura, D. P. J. Neurochem., 1963, 10:237-240.
830.   Berl, S., and Purpura, D. P. J. Neurochem., 1966, 13:293-304.
831.   Berl, S., and Waelsch, H. J. Neurochem., 1958, 3:161-169.
832.   Berlinguet, L., and Srivastava, U. Canad. J. Biochem., 1966, 44:613-623.
833.   Bernsohn, J., Barron, K. D., and Hess, A. R. Proc. Soc. Exptl. Biol. (New York), 1961, 107:773-775.
834.   Biesold, D., and Teichgräber, P. In: First International Meeting of the International Society for Neurochemistry, Strasbourg, 1967, p. 26.
835.   Bitensky, L., et al., Nature, 1963, 199:493-494.
836.   Blasberg, R., and Lajtha, A. Arch. Bioch. Biophys., 1965, 112:361-377.
837.   Blasberg, R., and Lajtha, A. Brain Res., 1966, 1:86-104.
838.   Block, R. J., J. Biol. Chem., 1937, 120:467-470.
839.   Blomstrand, C., and Hamberger, A. J. Neurochem., 1969, 16:1401-1407.
840.   Bocci, V. Nature, 1966, 212:826-827.
841.   Bogoch, S. In: Second International Meeting of the International Society for Neurochemistry, Milan, 1969, pp. 97-98.
842.   Bogoch, S. In: Protein Metabolism of the Nervous System, Plenum Press, New York, 1970, pp. 555-569.
843.   Bogoch, S., Rajam, P. C., and Belval, P. C. Nature, 1964, 204:73-75.
844.   Bondy, S. C., and Perry, S. V. J. Neurochem., 1963, 10:593-601.
845.   Bondy, S. C., and Perry, S. V. J. Neurochem., 1963, 10:603-609.
846.   Bondy, S. C., and Waelsch, H. J. Neurochem., 1965, 12:751-756.
847.   Boolj, J. In: Structure and Function of the Cerebral Cortex, Elsevier, Amsterdam, 1960, pp. 358-364.
848.   Borell, U., and Orström, A. Biochem. J., 1947, 41:398-403.
849.   Borst, P. Biochem. J., 1967, 105:37p-38p.
850.   Bossmann, B. H., and Hemsworth, B. A. J. Biol. Chem., 1970, 245:363-371.
851.   Bouma, J. M. W., and Gruber, M. Biochim. Biophys. Acta, 1964, 89:545-547.
852.   Bózkowa, K., Czuryna, A., and Osinska, K. J. Prace i Mater. Nauk. Inst. Matki i Dziecka, 1965, 5:39-52.
853.   Bózkowa, K., Duczynska, N., and Kurzera, St. Prace i Mater. Nauk. Inst. Matki i Dziecka, 1965, 6:43-47.
854.   Bózkowa, K., and Lambert, I. Prace i Mater. Nauk. Inst. Matki i Dziecka, 1965, 6:17-22.
855.   Bradford, H. F., Swanson, P. D., and Gammack, D. B. Biochem. J., 1964, 92:247-254.
856.   Brattgard, S.-O. Acta Radiol., Suppl., 1952, 96:1-80.
857.   Brattgard, S.-O., Edström, J.-E., and Hyden, H. J. Neurochem., 1957, 1:316-325.
858.   Brattgard, S.-O., Hyden, H., and Sjöstrand, J. Nature, 1958, 182:801-802.
859.   Braunitzer et al., Advances Protein Chem., 1964, 19:1-71.
860.   Bray, J. J., and Austin, L. J. Neurochem., 1968, 15:731-740.
861.   Brecher, A. S. J. Neurochem., 1963, 10:1-6.
862.   Brecher, A. S., and Quinn, N. M. Biochem. J., 1967, 102:120-121.
863.   Brierley, J. B. In: Metabolism of the Nervous System, Pergamon Press, London, 1957, pp. 121-135.
864.   Brizzee, K. R., Vogt, J., and Kharetchko, X. In: Progress in Brain Research. Growth and Maturation of the Brain, Elsevier, Amsterdam, 1964, pp. 136-146.

865.   Brodie, T. G., and Halliburton, W. D. J. Physiol., 1904, 31:437-490.
866.   Brody, T. M., and Bain, J. A. J. Biol. Chem., 1952, 195:685-696.
867.   Broome, J. Nature, 1963, 199:179-180.
868.   Brown, W. E., and Hamdy, M. K. Proc. Soc. Exptl. Biol. (New York), 1965,
        119:778-783.
869.   Brunish, R., and Luck, J. M. J. Biol. Chem., 1952, 198:621-628.
870.   Brunngraber, E. G., and Aguilar, V. J. Neurochem., 1962, 9:451-461.
871.   Buchanan, D. L. Arch. Biochem. Biophys., 1961, 94:500-511.
872.   Bulat, M.; and Supek, Z. Nature, 1968, 219:72-73.
873.   Bulman, N., and Campbell, D. N. Proc. Soc. Exptl. Biol. (New York), 1953,
        84:155-159.
874.   Burdman, J. A. J. Neurochem., 1970, 17:1555-1562.
875.   Burdman, J. A., Haglid, K., and Dravid, A. R. J. Neurochem., 1970, 17:669-
        676.
876.   Burdman, J. A. and Journey, L. J. J. Neurochem., 1969, 16:493-500.
877.   Burton, K. Biochem. J., 1956, 62:315:323.
878.   Burton, R. M., et al. Biochem. Biophys. Acta, 1964, 84:441-447.
879.   Burton, R. M., and Gibbons, J. M. Biochim. Biophys. Acta, 1964, 84:220-223.
880.   Du Buy, H. G., Mattern, C. F. T., and Riley, F. L. Biochim. Biophys. Acta,
        1966, 123:298-305.
881.   Calissano, P., Moore, B. W., and Friessen, A. Biochemistry, 1969, 8:4318-4326.
882.   Campbell, M. K., et al. Biochemistry, 1966, 5:1174-1184.
883.   Caravaglios, R., and Chiaverini, P. Experientia, 1956, 12:303-304.
884.   Carnegie, P. R., Bencina, B., and Lamoureux, G. Biochem. J., 1967, 105:559-
        568.
885.   Carnegie, P. R., and Lumsden, C. E. Nature, 1966, 209:1354-1355.
886.   Carver, M. J. J. Neurochem., 1969, 16:113-116.
887.   Carver, M. J., Schain, R. J., and Copenhaver, J. Proc. Soc. Exptl. Biol. (New
        York), 1966, 122:75-78.
888.   Casola, L., Davis, G. A., and Davis, R. E. J. Neurochem., 1969, 16:1037-1041.
889.   Chain, E. B., Larsson, S., and Pocchiari, F. Proc. Roy. Soc., 1960, B, 152:283-
        289.
890.   Chance, B., and Mela, L. J. Biol. Chem., 1966, 241:4588-4599.
891.   Chance, B., and Mela, L. Proc. Nat. Acad. Sci. (Washington), 1966, 55:1243-
        1251.
892.   Chao, L.-P., and Einstein, E. R. J. Biol. Chem., 1968, 243:6050-6055.
893.   Chao, L.-P., and Einstein, E. R. J. Neurochem., 1970, 17:1121-1132.
894.   Chapman, L. F., and Wolff, H. G. Arch. Internal Med., 1959, 103:86-94.
        Cited by: Chem. Abstr., 1959, 53:7354f.
895.   Chen, P. S., Toribara, T. Y., and Warner, H. Analyt. Chem., 1956, 28:1756-
        1758.
896.   Chesler, A., and Himwich, H. E. Amer. J. Physiol., 1944, 141:513-517.
897.   Chevremont, M. Biochem. J., 1962, 85:25p-26p.
898.   Chirigos, M. A., Greengard, P., and Udenfriend, S. J. Biol. Chem., 1960,
        235:2075-2079.
899.   Chitre, V. S., Chopra, S. P., and Talwar, G. P. J. Neurochem., 1964, 11:439-
        448.

900. Chordikian, F., Abood, L. G., and Howard, N. J. Neurochem., 1966, 13:945-954.
901. Christensen, H. N. Fedrat. Proc., 1963, Part I, 22:1110-1114.
902. Christensen, H. N., and Riggs, T. R. J. Biol. Chem., 1956, 220:265-278.
903. Cicero, T. J., et al. Brain Res., 1970, 18:25-34.
904. Clouet, D. H., and Gaitonde, M. K. Biochem. J., 1956, 64:18p.
905. Clouet, D. H., and Gaitonde, M. K. J. Neurochem., 1956, 1:126-133.
906. Clouet, D. H., Gaitonde, M. K., and Richter, D. J. Neurochem., 1957, 1:228-233.
907. Clouet, D. H., and Neidle, A. J. Neurochem., 1970, 17:1069-1074.
908. Clouet, D. H., and Ratner, M. J. Neurochem., 1968, 15:17-23.
909. Clouet, D. H., Ratner, M., and Williams, N. Biochim. Biophys. Acta, 1966, 123:142-150.
910. Clouet, D. H., and Richter, D. J. Neurochem., 1959, 3:219-229.
911. Clouet, D. H., and Waelsch, H. J. Neurochem., 1961, 8:201-215.
912. Clouet, D. H., and Waelsch, H. J. Neurochem., 1963, 10:51-63.
913. Cohen, M. M., and Lin, S. J. Neurochem., 1962, 9:345-352.
914. Cohn, P., Gaitonde, M. K., and Richter, D. J. Physiol. (London), 1954, 126:7p.
915. Cohn, P., and Richter, D. J. Neurochem., 1956, 1:166-172.
916. Cornwell, D. G., and Luck, J. M. Arch. Biochem. Biophys., 1958, 73:391-409.
917. Cotariu, D. Studii si Cercetari Biochim. Acad. RPR, 1962, 5:263-269.
918. Cotman, C. W., and Mahler, H. R. Arch. Biochem. Biophys., 1967, 120:384-396.
919. Cotman, C. W., Moore, W. J., and Mahler, H. R. In: First International Meeting of the International Society for Neurochemistry, Strasbourg, 1967, p. 45.
920. Cravioto, R. O., Massieu, G., and Izquierdo, J. J. Proc. Soc. Exptl. Biol. (New York), 1951, 78:856-858.
921. Cremer, J. E. J. Neurochem., 1964, 11:165-185.
922. Cremer, J. E. Biochem. J., 1967, 104:223-228.
923. Cremer, J. E., et al. J. Neurochem., 1968, 15:1361-1370.
924. Criddle, R. S., et al. Biochem. Biophys. Res. Commun., 1961, 5:75-80.
925. Cummins, I. T., and McIlwain, H. Biochem. J., 1961, 79:330-341.
926. Cutler, R. J. Neurochem., 1970, 17:1017-1027.
927. Dahl, D. R., and Samson, F. E. Amer. J. Physiol., 1959, 196:470-472.
928. Dahlström, A., and Haggendal, J. Acta Physiol. Scand., 1966, 67:278-288.
929. Daniel, R. G., and Waisman, H. A. J. Neurochem., 1969, 16:787-795.
930. Danielli, J. F. Biochem. J., 1941, 35:470-478.
931. Dannenberg, A. M., and Smith, E. L. J. Biol. Chem., 1955, 215:45-54; 55-66.
932. Dannies, P. S., and Levine, L. Biochem. Biophys. Res. Commun., 1969, 37:587-592.
933. Datta, R. K., and Ghosh, J. J. J. Neurochem., 1962, 9:463-464.
934. Datta, R. K., and Ghosh, J. J. J. Neurochem., 1963, 10:363-372; 611-612.
935. Datta, R. K., and Ghosh, J. J. J. Neurochem., 1964, 11:357-366.
936. Datta, R. K., Marks, N., and Lajtha, A. In: First International Meeting of the International Society for Neurochemistry, Strasbourg, 1967, p. 49.
937. Daughaday, W. H., and Lowry, O. H., et al. J. Lab. Clin. Med., 1952, 39:663-665.
938. Davis, B. J. In: Disc Electrophoresis. Cell Research Laboratory. The Mount Sinai Hospital, New York, 1961.

939.   Davis, B. J., and Ornstein, L. In: Disc Electrophoresis. Reprinted by Distillation Product Ind. Rochester, New York, 1962.

940.   Davis, W. E. J. Neurochem., 1970, 17:297-303.

941.   Davison, A. N. Biochem. J., 1961, 78:272-282.

942.   Davison, A. N., and Dobbing, J. Nature, 1961, 191:844-848.

943.   Deibler, G. F., Martenson, R. E., and Kies, M. W. Biochim. Biophys. Acta, 1970, 200:342-352.

944.   Dellweg, H., Gemer, R., and Wacker, A. J. Neurochem., 1968, 15:1109-1119.

945.   Demling, L., Kinzelmeier, H., and Henning, N. Z. Ges. Exptl. Med., 1954, 122:416-430.

946.   Depieds, R., et al. Compt. Rend. Soc. Biol., 1961 (1962), 155; 1966-1970.

947.   Devi, A. Biochim. Biophys. Acta, 1963, 73:155-158.

948.   Dingle, J. T. Biochem. J., 1961, 79:509-512.

949.   Dingman, W., and Spom, M. B. J. Neurochem., 1959, 4:148-153.

950.   Dingman, W., Spom, M. B., and Davies, R. K. J. Neurochem., 1959, 4:154-160.

951.   D'Monte, B., et al. In: Protein Metabolism of the Nervous System. Plenum Press, New York, 1970, pp. 185-217.

952.   Dobbing, J. Physiol. Rev., 1961, 41:130-188.

953.   Dobbing, J. J. Neurochem., 1963, 10:739-742.

954.   Dounce, A. L., and Umana, R. Biochemistry, 1962, 1:811-819.

955.   David, A. R., and Burdman, J. A. J. Neurochem., 1968, 15:25-30.

956.   David, A. R., Himwich, W. A., and Davis, J. M. J. Neurochem., 1965, 12:901-906.

957.   Dravid, A.R., and Jilek, L. J. Neurochem., 1965, 12:837-843.

958.   Droz, B. In: First International Meeting of the International Society for Neurochemistry, Strasbourg, 1967, p. 57.

959.   Droz, B., and Barondes, S. H. Science, 1969, 165:1131-1133.

960.   Droz, B., and Leblond, C. P. Science, 1962, 137:1047-1048.

961.   Droz, B., and Leblond, C. P. J. Comp. Neurol., 1963, 121:325-337.

962.   Droz, B., and Warshawsy, H. J. Histochem. Cytochem., 1963, 11:426-435.

963.   Duggan, A. W., and Johnston, G. A. R. J. Neurochem., 1970, 17:1205-1208.

964.   Dutton, G. R., and Barondes, S. H. J. Neurochem., 1970, 17:913-920.

965.   De Duve, C., et al. Biochem. J., 1955, 60:604-617.

966.   De Duve, C., and Wattiaux, R. Ann. Rev. Physiol., 1966, 28:435-492.

967.   Edel, S., and Poirel, G. Bull. Soc. Chim. Biol., 1966, 48:935-942.

968.   Edlbacher, S., Goldschmidt, E., and Schiappi, V. Hoppe-Seyler's Z. physiol. Chem., 1934, 227:118-123.

969.   Edström, A. J. Neurochem., 1964, 11:309-314.

970.   Edström, A. J. Neurochem., 1966, 13:315-321.

971.   Edström, A., and Sjöstrand, J. J. Neurochem., 1969, 16:67-81.

972.   Edström, J.-E., and Eichner, D. Nature, 1958, 181:619-620.

973.   Edström, R. Intem. Rev. Neurobiol., 1964, 7:153-190.

974.   Edwards, J. L., and Klein, R. E. Amer. J. Pathol., 1961, 38:437-453.

975.   Einstein, E. R., et al. J. Neurochem., 1962, 9:353-361.

976.   Einstein, E. R., et al. Immunochemistry, 1968, 5:567-575.

977.   Einstein, E. R., and Chao, L.-P. In: Protein Metabolism of the Nervous System, Plenum Press, New York, 1970, pp. 643-657.
978.   Einstein, E. R., and Csejtey, J. Trans. Amer. Neurol. Assn., 1966, 91:218.
979.   Einstein, E. R., and Csejtey, J. In: First International Meeting of the International Society for Neurochemistry, Strasbourg, 1967, p. 61.
980.   Einstein, E. R., Dalal, K. B., and Csejtey, J. Brain Res., 1970, 18:35-49.
981.   Eist, H., and Seal, U. S. Amer. J. Psychiatry, 1965, 122:584-586.
982.   Elliott, K. A. C., and van Gelder, N. M. J. Neurochem., 1958, 3:28-40.
983.   Ellis, S. J. Biol. Chem., 1960, 235:1694-1699.
984.   Ellis, S. Biochem. Biophys. Res. Commun., 1963, 12:452-456.
985.   Ellis, D., Sewell, C. E., and Skinner, L. G. Nature, 1956, 177:190-191.
986.   Emanuel, C. F., and Chaikoff, I. L. Biochim. Biophys. Acta, 1957, 24:254-261.
987.   Ende, N., Katayama, Y., and Auditore, J. V. Nature, 1964, 201:1197-1198.
988.   Eng, L. F., et al. Biochemistry, 1968, 7:4455-4465.
989.   Eng, L. F., et al. In: First International Meeting of the International Society for Neurochemistry, Strasbourg, 1967, p. 62.
990.   Ewald, A., and Kühne, W. Verh. Naturhist.-Med. Ver., Heidelberg, N. F., 1877, 1:457-464.
991.   Eylar, E. H., and Thompson, M. Arch. Biochem. Biophys., 1969, 129:468-479.
992.   Eylar, E. H., and Hashim, G. A. Proc. Nat. Acad. Sci. (Washington), 1968, 61:644-650.
993.   Fahn, S., and Côtè, L. J. J. Neurochem., 1968, 15:209-213.
994.   Fanelli, A. R., Antonini, E., and Caputo, A. Advances Protein Chem., 1964, 19:74-222.
995.   Feinstein, R. N., and Ballin, J. C. Proc. Soc. Exptl. Biol. (New York), 1953, 83:10-14.
996.   Felix, K. Advances Protein Chem., 1960, 15:1-16.
997.   Filipowicz, W., et al. Life Sci., 1968, 7:1243-1250.
998.   Finean, J. B. In: Metabolism of the Nervous System, Pergamon Press, London, 1957, 52-57.
999.   Fink, W. Zbl. Allgem. Pathol., 1957, 96:477-483.
1000.   Finkenstaedt, J. T. Proc. Soc. Exptl. Biol. (New York), 1957, 95:302-304.
1001.   Fischer, J., Kalousěk, J., and Lodin, Z. Nature, 1956, 178:1122-1123.
1002.   Fischer, J., Lodin, Z., and Kolousek, J. Nature, 1958, 181:341-342.
1003.   Fiszer, S., and de Robertis, E. Brain Res., 1967, 5:31-44.
1004.   Flangas, A. L., and Bowman, R. E., Science, 1968, 161:1025-1027.
1005.   Flangas, A. L., and Bowman, R. E. J. Neurochem., 1970, 17:1237-1245.
1006.   Fletcher, M. J., and Sanadi, D. R. Biochim. Biophys. Acta, 1961, 51:356-360.
1007.   Flexner, L. B., et al. J. Neurochem., 1965, 12:535-541.
1008.   Flexner, L. B., and Flexner, J. B. Proc. Nat. Acad. Sci. (Washington), 1968, 60:923-927.
1009.   Flexner, L. B., Flexner, J. B., and Roberts, R. B. J. Cell. Comp. Physiol., 1958, 51:385-403.
1010.   Fodor, P., Miller, A., and Waelsch, H. J. Biol. Chem., 1953, 202:551-565.
1011.   Fodor, P. J., Funk, C., and Tomasheisky, Ph. Arch. Bioch. Biophys., 1955, 56:281-289.

1012.   Folbergrova, J. Physiol. Bohemosl., 1961, 10:122-129; 130-138.
1013.   Folbergrova, J. Ceskosl. Fysiol., 1961, 10:13-33.
1014.   Folbergrova, J. Nature, 1962, 194:871-873.
1015.   Folbergrova, J. J. Neurochem., 1963, 10:775-782.
1016.   Folbergrova, J. Physiol. Bohemosl., 1964, 13:21-27.
1017.   Folbergrova, J. J. Neurochem., 1966, 13:553-562.
1018.   Folch-Pi, J., and Le Baron, F. N. Federat. Proc., 1953, 12:203.
1019.   Folch-Pi, J., and Lees, M. J. Biol. Chem., 1951, 191:807-817.
1020.   Folch-Pi, J., and Uzman, L. L. Federat. Proc., 1948, 7:155.
1021.   Folch-Pi, J., Webster, G. R., and Lees, M. Federat. Proc. 1959, 18:228.
1022.   Folch-Pi, J. In: Biochemistry of the Developing Nervous System, Academic
         Press, New York, 1955, 121-133.
1023.   Folch-Pi, J. Protoplasma, 1967, 63:160-164.
1025.   Freeman, K. B., Roodyn, D. B., and Tata, J. R. Biochim. Biophys. Acta, 1963,
         72:129-132.
1026.   Frenster, J. H., Allfrey, V. G., and Mirsky, A. E. Proc. Nat. Acad. Sci.
         (Washington), 1960, 46:432-444.
1027.   Freysz, L., et al. Bull. Soc. Chim. Biol., 1963, 45:1019-1029.
1028.   Freysz, L., Bieth, R., and Mandel, P. Bull. Soc. Chim. Biol., 1966, 48:287-293.
1029.   Friedberg, F., and Greenberg, D. M. J. Biol. Chem., 1947, 168:411-413.
1030.   Friedberg, F., Tarver, H., and Greenberg, D. M. J. Biol. Chem., 1948,
         173:355-361.
1031.   Friede, R. L. Topographic Brain Chemistry, Academic Press, New York, 1966.
1032.   Fries, B. A., and Chaikoff, I. L. J. Biol. Chem., 1941, 141:468-478; 479-485.
1033.   Fruton, J. J. Biol. Chem., 1946, 166:721-738.
1034.   Furlan, M., Naturwissenschaften, 1965, 52:515.
1035.   Furlan, M. Enzymologiya, 1966, 31:9-22.
1036.   Furst, S., Lajtha, A., and Waelsch, H. J. Neurochem., 1958, 2:216-225.
1037.   Gabrielescu, E., and Bordeianu, A. Fisiol. Normala si Patol., 1967, 13:319-329.
1038.   Gaitonde, M. K. Biochem. J., 1961, 80:277-284.
1039.   Gaitonde, M. K. Biochem. J., 1965, 95:803-810.
1040.   Gaitonde, M. K., Dahl, D. R., and Elliott, K. A. C. Biochem. J., 1965,
         94:345-352.
1041.   Gaitonde, M. K., Marchi, S. A., and Richter, D. Proc. Roy. Soc., 1964, B,
         160:124-136.
1042.   Gaitonde, M. K., and Martenson, R. E. J. Neurochem., 1970, 17:551-563.
1043.   Gaitonde, M. K., and Richter, D. Biochem. J., 1955, 59:690-696.
1044.   Gaitonde, M. K., and Richter, D. Proc. Roy. Soc., 1956, B, 145:83-99.
1045.   Gaitonde, M. K., and Richter, D. In: Metabolism of the Nervous System,
         Pergamon Press, New York, 1957, 449-455.
1046.   Gallagher, B. B. J. Neurochem., 1969, 16:701-706.
1047.   Gartside, I. D. Nature, 1968, 220:383-384.
1048.   Geel, S. E., and Timiras, P. S. In: Protein Metabolism of the Nervous
         System, Plenum Press, New York, 1970, 335-353.
1049.   Geffen, L. B., Hunter, C., and Rush, R. A. J. Neurochem., 1969, 16:469-474.
1050.   Geffen, L. B., and Rush, R. A. J. Neurochem., 1968, 15:925-930.

1051. Geiger, A., Horvath, M., and Kawakita, Y. J. Neurochem., 1960, 5:311-322.
1052. Geiger, A., Yamasaki, S., and Lyons, R. Amer. J. Physiol., 1956, 184:239-243.
1053. Geiger, A., Yamasaki, I., and Nebel, L. Federat. Proc., 1957, 16:44.
1054. Gelber, S., et al. J. Neurochem., 1964, 11:221-229.
1055. Van Gelder, N. M., and Elliott, K. A. C. J. Neurochem., 1958, 3:139-143.
1056. Gent, W. L. G., et al. Nature, 1964, 204:553-555.
1057. Gerhardt, W., Clausen, J., and Andersen, H. Acta Neurol. Scand., 1963, 39:31-40.
1058. Gielen, W. Naturwissenschaften, 1966, 53:504-505.
1059. Gilboe, D. D., and Williams, J. N. Proc. Soc. Exptl. Biol. (New York), 1956, 91:535-536.
1060. Giri, K. W. Naturwissenschaften, 1956, 43:36.
1061. Giri, K. W. Naturwissenschaften, 1956, 43:232-233.
1062. Gitlin, D., Klinnenberg, J. R., and Hughes, W. L. Nature, 1958, 181:1065.
1063. Godin, Y., and Mandel, P. J. Neurochem., 1965, 12:455-460.
1064. Gombos, G., et al. C. R. Acad. Sci., 1966, D263:1533-1535.
1065. Gombos, G., et al. In: First International Meeting of the International Society for Neurochemistry, Strasbourg, 1967, p. 86.
1066. Gombos, G., Geiger, A., and Otsuki, S. J. Neurochem., 1963, 10:405-413.
1067. Gombos, G., Vincendon, G., and Uyemura, K. In: Second International Meeting of the International Society for Neurochemistry, Milan, 1969, 191-192.
1068. Gonda, O., and Quastel, J. H. Biochem. J., 1962, 84:394-406.
1069. Gonzalez-Sastre, F., J. Neurochem., 1970, 17:1049-1056.
1070. Gordon, A. H., et al. Collect. Czechosl. Chem. Commun., 1950, 15:1-16.
1071. Gordon, A. H., Keil, B., and Sebesta, K. Nature, 1949, 164-498-499.
1072. Gordon, M. W., et al. J. Neurochem., 1962, 9:477-486.
1073. Gordon, M. W., and Deanin, G. G. J. Biol. Chem., 1968, 243:4222-4226.
1074. Got, K. and Polya, J. B. Enzymologia, 1964, 27:63-76.
1075. Grabar, P., and Williams, C. A. Biochim. Biophys. Acta, 1953, 10:193-194.
1076. Grasso, A., Cicero, T., and Moore, B. W. In: Second International Meeting of the International Society for Neurochemistry, Milan, 1969, 201-202.
1077. Graves, J., and Himwich, H. E. Amer. J. Physiol., 1955, 180:205-208.
1078. Green, D. E., et al. Proc. Nat. Acad. Sci. (Washington), 1968, 60:277-284.
1079. Greenbaum, A. L., and Greenwood, F. C. Biochem. J., 1954, 56:625-631.
1080. Greenberg, D., et al. In: Symposium on Quantiative Biology, 1948, 13:113. Cited by (1044).
1081. Greenberg, D., and Winnick, T. J. Biol. Chem., 1948, 173:199-204.
1082. Gregson, N. A., and Williams, P. L. J. Neurochem., 1969, 16:617-626.
1083. Groves, W. E., Davis, F. C., and Sells, B. H. Analyt. Biochem., 1963, 22:195-210.
1084. Guiditta, A., Dettbarn, W. D., and Brzin, M. Proc. Nat. Acad. Sci. (Washington), 1968, 59:1284-1287.
1085. Guroff, G. J. Biol. Chem., 1964, 239:149-155.
1086. Guroff, G., Hogans, A. F., and Udenfriend, S. J. Neurochem., 1968, 15:489-497.
1087. Guroff, G., King, W., and Udenfriend, S. J. Biol. Chem., 1961, 236:1773-1777.

1088.   Guroff, G., and Udenfriend, S. J. Biol. Chem., 1962, 237:803-806.
1089.   Guroff, G., and Udenfriend, S. In: Amino Acid Pools, Elsevier, Amsterdam, 1962, pp. 545-553.
1090.   Guroff, G., and Udenfriend, S. In: Progress in Brain Research, 9. The Developing Brain, Elsevier, Amsterdam, 1964, pp. 187-197.
1091.   Gutmann, E., et al. Physiol. Bohemosl., 1962, 11:437-442.
1092.   Haber, B. Canad. J. Biochem., 1965, 43:865-876.
1093.   Hadjiolov, A. A., Tencheva, Z. S., and Bojadjieva-Mikhailova, A. G. J. Cell. Biol., 1965, 26:383-393.
1094.   Von Hahn, H. P. Gerontologia, 1966, 12:18-29.
1095.   Hájek, I., Gutmann, E., and Syrový, I. Physiol. Bohemosl., 1965, 14:481-487.
1096.   Hájek, I., Gutmann, E., and Syrový, I. Physiol. Bohemosl., 1966, 15:1-6.
1097.   Häkkinen, H.-M., and Kulonen, E. Biochem. J., 1961, 78:588-593.
1098.   Häkkinen, H.-M., and Kulonen, E. Biochemem. J., 1967, 105:261-269.
1099.   Häkkinen, H.-M., Kulonen, E., and Wallgren, H. Biochem. J., 1963, 88:488-498.
1100.   Haldar, D. Biochem. Biophys. Res. Commun., 1970, 40:129-134.
1101.   Haldar, D., Freeman, K., and Work, T. S. Nature, 1966, 221:9-12.
1102.   Halliburton, W. D. J. Physiol., 1894, 15:90-107.
1103.   Halliburton, W. D. J. Physiol., 1895, 18:306-318.
1104.   Harkness, R. D., et al. Biochem. J., 1954, 56:558-569.
1105.   Härkönen, M., Mustakallio, A., and Niemi, M. J. Neurochem., 1966, 13:269-270.
1106.   Van Harreveld, A., and Kooiman, M. J. Neurochem., 1965, 12:431-439.
1107.   Harris, H., and Watts, J. M. Nature, 1958, 181:1582-1584.
1108.   Hartley, B. S. Nature, 1964, 201:1284-1287.
1109.   Hartley, B. S. In: Sixth International Congress of Biochemistry. Abstr. IV, New York, 1964, pp. 253-254.
1110.   Hashim, G. A., and Eylar, E. H. Arch. Biochem. Biophys., 1969, 129:635-644; 645-654.
1111.   Hatcher, V. B., and MacPherson, C. J. Immunol., 1969, 102:877-883.
1112.   Hatcher, V. B., and MacPherson, C. J. Immunol., 1970, 104:633-640.
1113.   Heald, P. J. Biochem. J., 1957, 66:659-663.
1114.   Heald, P. J. Biochem. J., 1958, 68:580-584.
1115.   Heald, P. J. Biochem. J., 1959, 73:132-141.
1116.   Heath, R. G., and Crupp, I. Arch. Gen. Psychiat., 1967, 16:1-9.
1117.   Van Heiningen, R., and Waley, S. G. Biochem. J., 1962, 86:92-103.
1118.   Heinz, E., and Walsh, P. M. J. Biol. Chem., 1958, 233:1488-1493.
1119.   Hendler, R. W. Nature, 1962, 193:821-823.
1120.   Henshaw, E. C., Bojarski, T. B., and Hiatt, H. H. J. Molec. Biol., 1963, 7:122-129.
1121.   Herman, C. J., and Lapham, L. W. Science, 1968, 160:537.
1122.   Herriman, I. D., and Hunter, G. D. J. Neurochem., 1965, 12:937-947.
1123.   Herschkowitz, N., Shooter, E. M., and McKahn, G. M. In: First International Meeting of the International Society for Neurochemistry, Strasbourg, 1967, p. 94.
1124.   Hess, A. Nature, 1955, 175:387-388.

1125. Himwich, H. E. Brain Metabolism and Cerebral Disorders. Williams and Wilkins Company, Baltimore, 1951.

1126. Himwich, H. E., and Himwich, W. A. In: Biochemistry of the Developing Nervous System, Academic Press, New York, 1955, pp. 202-207.

1127. Himwich, H. E., and Himwich, W. A. J. Chronic Diseases, 1956, 3:486-498.

1128. Himwich, W. A., and Himwich, H. E. Geriatrics, 1957, 12:19-27.

1129. Hirsch, H. E., and Robins, E. J. Neurochem., 1962, 9:63-70.

1130. Hoch-Ligeti, C., Stutzman, E., and Gruenwald, C. J. Neurochem., 1968, 15:417-425.

1131. Hofmann, G., and Schinko, H. Klin. Wochenschr., 1956, 34:86-90.

1132. Hogness, D. S., Cohn, M., and Monod, J. Biochim. Biophys. Acta, 1955, 16:99-116.

1133. Hokfelt, T., Jonson, G., and Ljungdahl, A. Life Sci., Part 1, 1970, 9:203-212.

1134. Hokin, L. E., and Hokin, M. R. Ann. Rev. Biochem., 1963, 32:553-578.

1135. Hokin, L. E., and Hokin, M. R. Federat. Proc., 1963, Part I, 22:8-18.

1136. Honegger, C. G. Ann. New York Acad. Sci., 1965, 122:199-208.

1137. Hooper, K. C. Biochem. J., 1964, 90:584-587.

1138. Hopkins, J. Proc. Nat. Acad. Sci. (Washington), 1959, 45:1461-1470.

1139. Horrocks, L. A. J. Lipid Res., 1967, 8:569-576.

1140. Horrocks, L. A. J. Neurochem., 1968, 15:483-488.

1141. Hsu, L., and Tappel, A. L. Nature, 1965, 207:1200.

1142. Hulcher, F. H. Arch. Biochem. Biophys., 1963, 100:237-244.

1143. Von Hungen, K., Mahler, H., and Moore, W. J. In: First International Meeting of the International Society for Neurochemistry, Strasbourg, 1967, p. 102.

1144. Von Hungen, K., Mahler, H. R., and Moore, W. J. J. Biol. Chem., 1968, 243:1415-1423.

1145. Hunter, G. D., and Millson, G. C. J. Neurochem., 1966, 13:375-383.

1146. Hydén, H. In: Proc. IV Internat. Congr. Biochem., 1959, 3:64-89.

1147. Hydén, H. Nature, 1959, 184:433-435.

1148. Hydén, H. In: Structure and Function of the Cerebral Cortex, Elsevier, Amsterdam, 1960, pp. 348-356.

1149. Hydén, H., and Egyhazi, E. Proc. Nat. Acad. Sci. (Washington), 1962, 48:1366-1373.

1150. Hydén, H., and Egyhazi, E. Proc. Nat. Acad. Sci. (Washington), 1963, 49:618-624.

1151. Hydén, H., and Lange, P. W. Science, 1968, 159:1370-1373.

1152. Hydén, H., and Lange, P. W. Proc. Nat. Acad. Sci. (Washington), 1970, 65:898-904.

1153. Hydén, H., and McEwen, B. Proc. Nat. Acad. Sci. (Washington), 1966, 55:354-358.

1154. Hydén, H., and Pigon, A. J. Neurochem., 1960, 6:57-72.

1155. Imanishi, A., Momotani, Y., and Isemura, T. J. Biochem., 1965, 57:417-429.

1156. Inesi, G., and Dessi, P. Boll. Soc. Ital. Biol. Sperim., 1957, 33:7-9.

1157. Irwin, L. N. Brain Res., 1969, 15:518-521.

1158. Iton, T., and Quastel, J. H. Biochem. J., 1970, 116:641-655.

1159. Iversen, L. L., and Snyder, S. H. Nature, 1968, 220:796-798.

1160. Jacob, M., et al. J. Neurochem., 1967, 14:169-178.

1161.   Jakoubek, B., et al. Physiol. Bohemosl., 1963, 12:553-561.
1162.   Jakoubek, B., et al., J. Neurochem., 1968, 15:633-641.
1163.   Janković, B. D., et al., Nature, 1968, 218:270-271.
1164.   Jeffay, H. J. Biol. Chem., 1960, 235-2352-2356.
1165.   Jenner, R., and Szafarz, D. Arch. Biochem. Biophys., 1950, 26:54-67.
1166.   Jibril, A. O., and McCay, P. B. Nature, 1965, 205:1214-1215.
1167.   Jöbsis, F. F. Biochim. Biophys. Acta, 1963, 74:60-68.
1168.   Johnson, T. C. J. Neurochem., 1967, 14:1075-1081.
1169.   Johnson, T. C. J. Neurochem., 1968, 15:1189-1194.
1170.   Johnson, T. C. J. Neurochem., 1969, 16:1125-1131.
1171.   Johnson, T. C., and Luttges, M. W. J. Neurochem., 1966, 13:545-552.
1172.   Johnston, G. A. R., de Groat, W. C., and Curtis, D. R. J. Neurochem., 1969, 16:797-800.
1173.   Johnston, R. B., Mycek, M. I., and Fruton, J. S. J. Biol. Chem., 1950, 185:629-641.
1174.   Johnstone, R. M., and Scholefield, P. G. J. Biol. Chem., 1961, 236:1419-1424.
1175.   Jones, M. E., et al. J. Biol. Chem., 1952, 195:645-656.
1176.   Jordan, W. K., and March, R. J. Histochem. Cytochem., 1956, 4:301-311.
1177.   Kadenbach, B. Biochim. Biophys. Acta, 1967, 134:430-442.
1178.   Kadenbach, B. Biochim. Biophys. Acta, 1967, 138:651-654.
1179.   Kalf, G. F. Biochemistry, 1964, 3:1702-1706.
1180.   Kamin, H., and Handler, P. J. Biol. Chem., 1951, 188:193-205.
1181.   Kamrin, R. P., and Kamrin, A. A. J. Neurochem., 1961, 6:219-225.
1182.   Kaps, G. Arch. Psychiatr. Z. Neurol., 1954, 192, 115-129.
1183.   Karcher, D., van Sande, M., and Lowenthal, I. J. Neurochem., 1959, 4:135-140.
1184.   Katz, R. I., Chase, T. N., and Kopin, I. J. J. Neurochem., 1969, 16:961-967.
1185.   Kelley, B. Amer. J. Physiol., 1956, 185:299-301.
1186.   Van Kempen, G. M. J., et al. J. Neurochem., 1965, 12:581-588.
1187.   Kessler, D., Levine, L., and Fasman, G. Biochemistry, 1968, 7:758-764.
1188.   Kety, S. S. In: Metabolism of the Nervous System, Pergamon Press, London, 1957, pp. 221-262.
1189.   Kety, S. S. Science, 1959, 129:1528-1532; 1590-1596.
1190.   Keup, W. Confinia Neurologica, 1955, 15:73-83.
1191.   Kibler, R. F., et al. Science, 1969, 164:577-580.
1192.   Kibler, R. F., Fox, R. H., and Shapira, R. Nature, 1964, 204:1273-1275.
1193.   Kibler, R. F., and Shapira, R. J. Biol. Chem., 1968, 243:281-286.
1194.   Kies, M. W. In: Protein Metabolism of the Nervous System, Plenum Press, New York, 1970:659-666.
1195.   Kies, M. W., et al. Science, 1966, 151:821-822.
1196.   Kies, M. W., Alvord, E. C., and Roboz, E. J. Neurochem., 1958, 2:261-264.
1197.   Kies, M. W., and Schwimmer, S. J. Biol. Chem., 1942, 145:685-691.
1198.   Kiyota, K. J. Neurochem., 1957, 1:301-302.
1199.   Kiyota, K. J. Neurochem., 1959, 4:202-208.
1200.   Klee, C. B., and Sokoloff, L. J. Neurochem., 1964, 11:709-716.
1201.   Klee, C. B., and Sokoloff, L. Proc. Nat. Acad. Sci. (Washington), 1965, 53:1014-1020.

1202.   Kleine, R., and Hanson, H. Acta Biol. Med. Germ., 1962, 9:606-622.
1203.   Klug, H. Z. Mikrosk.-Anat. Forsch., 1966, 75:109-122.
1204.   Knauff, H. G., and Böck, F. J. Neurochem., 1961, 6:171-182.
1205.   Knauff, H. G., Gottstein, U., and Miller, B. Klin. Wochenschr., 1964, 42:27-39.
1206.   Koch, W. J., and Koch, M. L. J. Biol. Chem., 1913, 15:423-448.
1207.   Koch, W. J. Amer. J. Physiol., 1904, 11:303.
1208.   Koenig, E. J. Neurochem., 1965, 12:343-355; 357-361.
1209.   Koenig, E. In: First International Meeting of the International Society for Neurochemistry, Strasbourg, 1967, p. 118.
1210.   Koenig, E., and Koelle, G. B. Science, 1960, 132:1249-1250.
1211.   Koenig, H. J. Histochem. Cytochem., 1963, 11:556-557.
1212.   Koenig, H. J. Histochem. Cytochem., 1965, 13:411-413.
1213.   Koenig, H., et al. J. Neurochem., 1964, 11:729-743.
1214.   Kom, E. D. Science, 1966, 153:1491-1498.
1215.   Komer, A., and Tarver, H. J. Gen. Physiol., 1957, 41:219-231.
1216.   Komguth, S. E., and Thompson, H. G. Arch. Biochem. Biophys., 1964, 105:308-314.
1217.   Kosinski, E., and Grabar, P. J. Neurochem., 1967, 14:273-281.
1218.   Koszalka, T. R., and Miller, L. L. Federat. Proc., 1958, 17:257.
1219.   Koszalka, T. R., and Miller, L. L. J. Biol. Chem., 1960, 235:665-668.
1220.   Koszalka, T. R., and Miller, L. L. J. Biol. Chem., 1960, 235:669-672.
1221.   Krawczynski, J. J. Neurochem., 1961, 7:1-4.
1222.   Krawczynski, J. J. Neurochem., 1961, 8:50-54.
1223.   Krawczynski, J., et al. Acta Biochim. Polon., 1958, 5:139-154.
1224.   Krawczynski, J., Wiszniowska, S., and Drewnowska, J. J. Neurochem., 1960, 5:109-113.
1225.   Krebs, H. A. Biochem. J., 1935, 29:1951-1969.
1226.   Krebs, H. A., Eggleston, L. V., and Hems, R. Biochem. J., 1949, 44:159-163.
1227.   Krimsky, I., and Racker, E. J. Biol. Chem., 1949, 179:903-914.
1228.   Křivánek, J. J. Neurochem., 1970, 17:531-538.
1229.   Kroon, A. M. Biochim. Biophys. Acta, 1963, 72:391-402.
1230.   Kroon, A. M. Biochim. Biophys. Acta, 1964, 91:145-154.
1231.   Kroon, A. M. Biochim. Biophys. Acta, 1965, 108:275-284.
1232.   Kržalić, Lj., Cupić, D., and Mihailović, Lj. T. Arch. Internat. Physiol. Biochim., 1965, 73:817-825.
1233.   Kržalić, Lj., Mandić, V., and Mihailović, Lj. T. Experientia, 1962, 18:368-369.
1234.   Kubler, H., and Frieden, E. Biochim. Biophys. Acta, 1964, 93:635-643.
1235.   Kuff, E. L., and Schneider, W. C. J. Biol. Chem., 1954, 206:677-675.
1236.   Kuhne, W., and Chittenden, R. H. Z. Biol., 1890, 26:291-323.
1237.   Kukral, J. G., et al. Amer. J. Physiol., 1963, 204:262-264.
1238.   Kun, E., and Abood, L. G. Science, 1949, 109:144-146.
1239.   Kurokawa, M., Sakamoto, T., and Katom, M. Biochim. Biophys. Acta, 1965, 94:307-309.
1240.   Kuttner, R., Sims, J. A., and Gordon, M. W. J. Neurochem., 1961, 6:311-317.
1241.   Laatsch, R. H., et al. J. Exptl. Med., 1962, 115:777-788.

1242. La Bella, F. S., and Brown, I. H. U. J. Biophys. Biochem. Cytol., 1959, 5:17-23.

1243. La Belle, F., Vivian, S., and Queen, G. Biochim. Biophys. Acta, 1968, 158:286-288.

1244. Lahiri, S., and Lajtha, A. J. Neurochem., 1964, 11:77-86.

1245. Lajtha, A. J. Neurochem., 1957, 1:216-227.

1246. Lajtha, A. J. Neurochem., 1957, 2:209-215.

1247. Lajtha, A. J. Neurochem., 1959, 3:358-365.

1248. Lajtha, A. In: Regional Neurochemistry, Proceedings of the Fourth International Neurochemical Symposium, Pergamon Press, New York, 1961, pp. 19-24.

1249. Lajtha, A. Ibid. pp. 25-36.

1250. Lajtha, A. Intern. Rev. Neurobiol., 1964, 6:1-98.

1251. Lajtha, A. Intern. Rev. Neurobiol., 1964, 7:1-40.

1252. Lajtha, A. In: Problems in Biochemistry of the Brain [in Russian], Vol. 3, Izd. Akad. Nauk Arm. SSR, Erevan, 1967, pp. 31-44.

1253. Lajtha, A., et al. J. Neurochem., 1957, 1:289-300.

1254. Lajtha, A., Berl, S., and Waelsch, H. J. Neurochem., 1959, 3:322-332.

1255. Lajtha, A., Furst, S., and Waelsch, H. Experientia, 1957, 13:168-172.

1256. Lajtha, A., Lahiri, S., and Toth, J. J. Neurochem., 1963, 10:765-773.

1257. Lajtha, A., and Marks, N. Diseases Nerv. Syst., 1969, 30:36-43.

1258. Lajtha, A., and Mella, P. J. Neurochem., 1961, 7:210-217.

1259. Lajtha, A., and Toth, J. J. Neurochem., 1961, 8:216-225.

1260. Lajtha, A., and Toth, J. J. Neurochem., 1962, 9:199-212.

1261. Lajtha, A., and Toth, J. J. Neurochem., 1963, 10:909-920.

1262. Lajtha, A., and Toth, J. Biochem. Pharmacol.., 1965, 14:729-738.

1263. Lajtha, A., and Toth, J. Biochem. Biophys. Res. Commun., 1966, 23:294-298.

1264. Laki, K., and Gladner, J. A. Physiol. Revs., 1964, 44:127-160.

1265. Landolt, R., and Hess, H. H. J. Neurochem., 1966, 13:1453-1459.

1266. Lapetina, E. G., Soto, E. F., and de Robertis, E. In: First International Meeting of the International Society for Neurochemistry, Strasbourg, 1967, p. 128.

1267. Lapetina, E. G., Soto, E. F., and de Robertis, E. J. Neurochem., 1968, 15:437-445.

1268. Lapetina, E. G., et al. J. Neurochem., 1969, 16:101-106.

1269. Lapham, L. W. Science, 1968, 159:310-312.

1270. Lapresle, C., and Webb, T. Biochem. J., 1962, 84:455-462.

1271. Lazarow, A., and Cooperstein, S. J. Exptl. Cell. Res., 1953, 5:56-61.

1272. Leake, Ch. D., et al., Science, 1958, 127:162-163.

1273. Le Baron, F. N. Ann. Rev. Biochem., 1959, 28:579-604.

1274. Le Baron, F. N., and Folch, J. J. Neurochem., 1956, 1:101-108.

1275. Le Baron, F. N., and Folch, J. J. Neurochem., 1959, 4:1-8.

1276. Leblond, C. P., and Walker, B. E. Physiol. Revs., 1956, 36:255-276.

1277. Ledig, M., Palat, L., and Mandel, P. Compt. Rend. Soc. Biol., 1966 (1967), 160:2455-2458.

1278. Lees, M. B., Amaducci, L., and Waksman, B. N. J. Neurochem., 1961, 8:285-298.

1279. Lehninger, A. L. Proc. Nat. Acad. Sci. (Washington), 1968, 60:1069-1080.
1280. Lehr, P., and Gayet, J. J. Neurochem., 1966, 13:805-810.
1281. Lehr, P., and Gayet, J. J. Neurochem., 1967, 14:927-936.
1282. Levene, P. Arch. Neurol. Psychopath., 1899, 2:1.
1283. Levi, G., and Amaducci, L. J. Neurochem., 1968, 15:459-469.
1284. Levi, G., Blasberg, R., and Lajtha, A. Arch. Biochem. Biophys., 1966, 114:339-351.
1285. Levi, G., Cherayil, A., and Lajtha, A. J. Neurochem., 1965, 12:757-770.
1286. Levi, G., Kandera, J., and Lajtha, A. Arch. Biochem. Biophys., 1967, 119:303-311.
1287. Levi, G., and Lajtha, A. J. Neurochem., 1965, 12:639-648.
1288. Levin, E., Nogueira, G. J., and Garcia Argiz, C. A. J. Neurochem., 1966, 13:761-767.
1289. Levine, L., and Moore, B. W. Neurosciences Res. Prog. Bull., 1965, 3:18. Cited by (1400).
1290. Lewin, E., and Hess, H. H. J. Neurochem., 1967, 14:71-80.
1291. Li Pen Chao, and Einstein, E. R. J. Biol. Chem., 1968, 243:6050-6055.
1292. Li Tsai-ping, and Sheng Pei-ken, Acta Physiol. Sinica, 1957, 21:292-302.
1293. Liakopoulou, A., and MacPherson, C. F. C. J. Immunol., 1970, 105:512-520.
1294. Libenson, L., and Jena, M. Arch. Biochem. Biophys., 1964, 104:292-296.
1295. Lim, L., and Adams, D. H. Biochem. J., 1967, 104:229-238.
1296. Lim, R., and Agranoff, B. W. J. Neurochem., 1969, 16:431-445.
1297. Lim, R., and Tadayyon, E. Analyt. Biochem., 1970, 34:9-15.
1298. Lipmann, F. In: Metabolism of the Nervous System, Pergamon Press, New York, 1957, pp. 329-339.
1299. Lisowski, J. Arch. Immunol. Therap. Exptl., 1964, 12:645-656.
1300. Livett, B. G., Geffen, L. B., and Austin, L. J. Neurochem., 1968, 15:931-939.
1301. Lodin, Z., and Kolousek, J. Physiol. Bohemosl., 1956, Suppl. 5:43-46.
1302. Loftfield, R. B. In: Amino Acid Pools, Elsevier, Amsterdam, 1962, pp. 732-737.
1303. Loftfield, R. B., Grover, J. W., and Stephenson, M. L. Nature, 1953, 171:1024-1025.
1304. Lombardo, G., and Tamburino, G. Z. Naturforsch., 1963, 18b:776-777.
1305. Lombardo, G., and Tamburino, G. Z. Naturforsch., 1964, 19b:267-268.
1306. Løvtrup, S. in: Progress in Brain Research, 4. Growth and Maturation of the Brain, Elsevier, Amsterdam, 1964, pp. 237-253.
1307. Lowden, J. A., Moscarello, M. I., and Morecki, R. Canad. J. Biochem., 1966, 44:567-577.
1308. Lowden, J. A., and Wolfe, L. S. Canad. J. Biochem., 1964, 12:1587-1594.
1309. Lowry, O. H., et al. J. Biol. Chem., 1951, 193:265-275.
1310. Lubinska, L., et al. J. Neurochem., 1963, 10:25-41.
1311. Lumsden, C. E., Robertson, D. M., and Blight, R. J. Neurochem., 1966, 13:127-162.
1312. Luscombe, M. Nature, 1963, 197:1010.
1313. Luse, S. A. J. Histochem. Cytochem., 1960, 8:398-411.
1314. Luttges, M., et al. Science, 1966, 151:834-837.
1315. Luxoro, M. Nature, 1960, 188:1119-1120.

1316.   McCaman, R. E., and Robins, E. J. Neurochem., 1959, 5:32-42.
1317.   McEwen, B. S. and Hydén, H. J. Neurochem., 1966, 13:823-833.
1318.   McFarlane, A. S. In: Mammalian Protein Metabolism, Academic Press, New York, 1964, 297-341.
1319.   McGregor, H. H. J. Biol. Chem., 1917, 28:403-427.
1320.   McIlwain, H. Biochem. J., 1952, 52:289-295.
1321.   McIlwain, H. Biochem. J., 1953, 55:618-624.
1322.   McIlwain, H. Biochem. J., 1961, 78:24-32.
1323.   McIlwain, H. Brit. Med. Bull., 1968, 24:174-178.
1324.   MacInnes, J. W., McConkey, E. H., and Schlessinger, K. J. Neurochem., 1970, 17:457-460.
1325.   MacPherson, C. F. C., and Liakopoulou, A. Federat. Proc., 1965, 24:176.
1326.   MacPherson, C., and Liakopoulou, A. J. Immunol., 1966, 97:450-457.
1327.   Mammalian Protein Metabolism, Parts I-II. Academic Press, New York, 1964.
1328.   Mandel, P., et al. J. Neurochem., 1966, 13:533-536.
1329.   Mandel, P., Edel, S., and Poirel, G. J. Neurochem., 1966, 13:885-886.
1330.   Mandel, P., and Edel-Harth, S. J. Neurochem., 1966, 13:591-595.
1331.   Mandel, P., and Ledig, M. Biochem. Biophys. Res. Commun., 1966, 24:275-279.
1332.   Mandel, P., and Mark, J. J. Neurochem., 1965, 12:987-992.
1333.   Mandel, P., and Jacob, M. In: Protein Metabolism of the Nervous System, Plenum Press, New York, 1970, pp. 129-148.
1334.   Mangan, J. L., and Whittaker, V. P. Biochem. J., 1966, 98:128-137.
1335.   Mans, R. J., and Novelli, G. D. Arch. Biochem. Biophys., 1961, 94:48-52.
1336.   Manzoli, F. A., and Wegelin, I. J. Neurochem., 1969, 16:829-831.
1337.   Maraini, G., et al. Exptl. Eye Res., 1967, 6:299-302.
1338.   Marchbanks, R. M. J. Neurochem., 1966, 13:1481-1493.
1339.   Marcucci, F., and Airoldi, L. J. Neurochem., 1969, 16:673-674.
1340.   Marcucci, F., Airoldi, L., and Mussini, E. J. Neurochem., 1969, 16:272-273.
1341.   Margolis, F. L. J. Neurochem., 1969, 16:447-456.
1342.   Margolis, R. K., and Lajtha, A. Biochim. Biophys. Acta, 1968, 163:374-385.
1343.   Margolis, R. U., Barkulis, S. S., and Geiger, A. J. Neurochem., 1960, 5:379-382.
1344.   Mark, J., and Mandel, P. Compt. Rend. Soc. Biol., 1964 (1965), 158:2478-2481.
1345.   Marks, N. Internat. Rev. Neurobiol., 1968, 11:57-97.
1346.   Marks, N., et al. Brain Res., 1970, 18:309-324.
1347.   Marks, N., Datta, R. K., and Lajtha, A. In: Macromolecules and the Function of the Neuron. Excerpta Medica Foundation, Amsterdam, 1968, pp. 220-230.
1348.   Marks, N., Datta, R. K., and Lajtha, A. J. Neurochem., 1970, 17:53-63.
1349.   Marks, N., and Lajtha, A. Biochem. J., 1963, 89:438-447.
1350.   Marks, N., and Lajtha, A. Biochem. J., 1965, 97:74-83.
1351.   Marks, N., and Lajtha, A. in: Protein Metabolism of the Nervous System, Plenum Press, New York, 1970, 39-73.
1352.   Marks, N., and McIlwain, H. Biochem. J., 1959, 73:401-410.

1353. Martenson, R. E., Deibler, G. E., and Kies, M. W. J. Biol. Chem., 1969, 244:4261-4267.
1354. Martenson, R. E., Deibler, G. E., and Kies, M. W. J. Biol. Chem., 1969, 244:4268-4272.
1355. Martenson, R. E., Deibler, G. E., and Kies, M. W. Biochim. Biophys. Acta, 1970, 200:353-362.
1356. Martenson, R. E., Deibler, G., and Kies, M. W. J. Neurochem., 1970, 17:1329-1330.
1357. Martenson, R. E., and Gaitonde, M. K. J. Neurochem., 1969, 16:333-347; 889-898.
1358. Martenson, R. E., Gaitonde, M. K., and Richter, D. In: First International Meeting of the International Society for Neurochemistry. Strasbourg, 1967, p. 151.
1359. Martenson, R. E., and Le Baron, F. N. Federat. Proc., 1965, 24:360.
1360. Martenson, R. E., and Le Baron, F. N. J. Neurochem., 1966, 13:1469-1479.
1361. Martin, C. J. Federat. Proc., 1954, 13:260.
1362. Martin, C. J. J. Biol. Chem., 1961, 236:2673-2676.
1363. Martin, C. J., and Axelrod, A. E. Biochim. Biophys. Acta, 1957, 26:490-501.
1364. Martin, C. J., and Axelrod, A. E. J. Biol. Chem., 1957, 224:309-321.
1365. Martin, C. J., and Axelrod, A. E. Biochim. Biophys. Acta, 1958, 27:52-62; 532-538.
1366. Martini, E. Experientia, 1959, 15:182-183.
1367. Mase, K., Takahashi, Y., and Ogata, K. J. Neurochem., 1962, 9:281-288.
1368. Massey, V. Biochem. J., 1953, 53:67-71.
1369. Massieu, G. H., et al. J. Neurochem., 1962, 9:143-151.
1370. Matheson, D. F. J. Neurochem., 1969, 16:215-223.
1371. Matheson, D. F., and Gavanach, J. B. Nature, 1967, 214:721-722.
1372. May, L., and Grenell, R. G. Proc. Soc. Exptl. Biol. (New York), 1959, 102:235-239.
1373. Mehl, E. In: First International Meeting of the International Society for Neurochemistry. Strasbourg, 1967, p. 154.
1374. Mehl, E., and Wolfgram, F. J. Neurochem., 1969, 16:1091-1097.
1375. Merei, F. T., and Gallyas, F. J. Neurochem., 1964, 11:251-256; 257-264.
1376. Methods of Separation of the Subcellular Structural Components, Univ. Press, Cambridge, 1963.
1377. Metrione, R. M., Neves, A. G., and Fruton, J. S. Biochemistry, 1966, 5:1597-1604.
1378. Metzger, H. P., Guenod, M., Grynbaum, A., and Waelsch, H. J. Neurochem., 1967, 14:183-187.
1379. Miani, N., Rizzoli, A., and Bucciante, G. J. Neurochem., 1961, 7:161-173.
1380. Michels, R., Cason, J., and Sokoloff, L. Science, 1963, 140:1417-1418.
1381. Mihailović, Lj. T., et al. Experientia, 1965, 21:100-101.
1382. Mihailović, Lj. T., et al. Exptl. Neurol., 1969, 24:325.
1383. Mihailović, Lj. T., and Janković, B. D. Nature, 1961, 192:665-666.
1384. Mihailović, Lj. T., and Kržalić, Lj. Experientia, 1964, 20:262-263.
1385. Mihailović, Lj. T., Kržalić, Lj., and Cupic, D. Experientia, 1965, 21:709-710.

1386.   Miller, L. L., et al. J. Exptl. Med., 1949, 90:297-313.

1387.   Millson, G. C., and Hunter, G. D. J. Neurochem., 1968, 15:447-453.

1388.   Minard, F. N., and Richter, D. J. Neurochem., 1968, 15:1463-1468.

1389.   Mirsky, A., Perisutti, G., and Dixon, F. J. J. Biol. Chem., 1955, 214:397-408.

1390.   Mirsky, A. E., and Pollister, A. W. J. Gen. Physiol., 1946, 30:117-147.

1391.   Missere, G., Tonini, G., and De Risio, C. Boll. Soc. Ital. Biol. Sperim.,
         1957, 33:491-493.

1392.   Moissker, W. W. J. Psychiatr. Res., 1966, 4:235-304.

1393.   Mokrasch, L. C. J. Neurochem., 1966, 13:49-58.

1394.   Mokrasch, L. C., and Manner, Ph. J. Neurochem., 1963, 10:541-547.

1395.   Moldave, K. J. Biol. Chem., 1956, 221:543-553.

1396.   Moore, B. W. Biochem. Biophys. Res. Commun., 1965, 19:739-744.

1397.   Moore, B. W., and McGregor, D. J. Biol. Chem., 1965, 240:1647-1653.

1398.   Moore, B. W., Peña-Ramos, A., and Perez, V. J. In: First International
         Meeting of the International Society for Neurochemistry. Strasbourg, 1967,
         p. 158.

1399.   Moore, B. W., and Perez, V. J. In: Physiological and Biochemical Aspects
         of Nervous Integration, Prentice-Hall, New Jersey, 1968, pp. 343-360.

1400.   Moore, B. W., Perez, V. J., and Gehring, M. J. Neurochem., 1968, 15:265-272.

1401.   Moore, S., Spackman, D. H., and Stein, W. H. Analyt. Chem., 1958, 30:1185-
         1189.

1402.   Moore, S., and Stein, W. H. J. Biol. Chem., 1948, 176:367-388.

1403.   Morgan, I. G., and Austin, L. J. Neurochem., 1968, 15:41-51.

1404.   Morgan, I. G., and Austin, L. Life Sci., 1969, Part II, 8:79-84.

1405.   Morrison, W. L., and Neurath, H. J. Biol. Chem., 1953, 200:39-51.

1406.   Mourek, J., et al. Physiol. Bohemosl., 1968, 17:104-112.

1407.   Muir, H. M., Neuberger, A., and Perrone, J. C. Biochem. J., 1951, 49:IV.

1408.   Mullins, L. J. Amer. J. Physiol., 1953, 175:358-362.

1409.   Murthy, M. R. V. Biochim. Biophys. Acta, 1966, 119:586-598; 599-613.

1410.   Murthy, M. R. V. Canad. J. Biochem., 1969, 47:75-78.

1411.   Murthy, M. R. V. In: Protein Metabolism of the Nervous System, Plenum
         Press, New York, 1970, pp. 109-125.

1412.   Murthy, M. R. V., and Rappoport, D. A. Biochim. Biophys. Acta, 1965,
         95:121-131.

1413.   Murthy, M. R. V., and Rappoport, D. A. Biochim. Biophys. Acta, 1965,
         95:132-145.

1414.   Mussini, E., and Marcucci, F. In: Amino Acid Pools. Elsevier, Amsterdam,
         1962, pp. 486-492.

1415.   Myrbäck, K. Hoppe-Seyler's Z. Physiol. Chem., 1926, 159:1-84.

1416.   Nachlas, M. M., Goldstein, Th. P., and Seligman, A. M. Arch. Biochem.
         Biophys., 1962, 97:223-231.

1417.   Nagel, W., and Willig, F. Nature, 1964, 201:617-618.

1418.   Nagel, W., and Willig, F. Naturwissenschaften, 1964, 51:115.

1419.   Nakamura, R., and Nagayama, M. J. Neurochem., 1966, 13:305-313.

1420.   Nakamura, S., Hayashi, Y., and Tanaka, S. J. Biochem., 1954, 41:13-21.

1421.   Nakao, A., Davis, W. J., and Einstein, E. R. Biochim. Biophys. Acta, 1966,
         130:163-170.

1422. Nakao, A., and Roboz-Einstein, E. Federat. Proc., 1965, 24:242.
1423. Navon, S. and Lajtha, A. Biochim. Biophys. Acta, 1969, 173:518-531.
1424. Neame, K. D. Nature, 1961, 192:173-174.
1425. Neame, K. D. J. Neurochem., 1961, 6:358-366.
1426. Neame, K. D. J. Neurochem., 1962, 9:321-324.
1427. Neame, K. D. J. Physiol., 1962, 162:1-12.
1428. Neame, K. D. J. Neurochem., 1964, 11:67-76.
1429. Neame, K. D. J. Physiol., 1965, 181:114-123.
1430. Neame, K. D., and Smith, S. E. J. Neurochem., 1965, 12:87-91.
1431. Neidle, A., van den Berg, C. J., and Grynbaum, A. J. Neurochem., 1969,
       16:225-234.
1432. Neuberger, A., and Richards, F. F. In: Mammalian Protein Metabolism, 1.
       Academic Press, New York, 1964, 243-296.
1433. Neubert, D., Bass, R., and Helge, H. Naturwissenschaften, 1966, 53:23-24.
1434. Neubert, D., and Helge, H. Biochem. Biophys. Res. Commun., 1965, 18:600-
       605.
1435. Neumann, H. et al. Biochem. J., 1959, 73:33-41.
1436. Neumann, H., and Sharon, N. Biochim. Biophys. Acta, 1960, 41:370-371.
1437. Neurath, H. Advances Protein Chem., 1957, 12:319-386.
1438. Neurath, H., and Dixon, G. H. Federat. Proc., 1957, 16:791-801.
1439. Neville, D. M. Biochim. Biophys. Acta, 1967, 133:168-170.
1440. Nievel, J. G., and Cumings, J. N. Nature, 1967, 214:1123-1124.
1441. Nievel, J. G., Robinson, N., and Eayrs, J. T. Experientia, 1968, 24:677-678.
1442. Nilsson, K. K., and Fruton, J. S. Biochemistry, 1964, 3:1220-1224.
1443. Noak, R., et al. Biochem. J., 1966, 100:775-778.
1444. Noale, M. W., et al. Science, 1957, 126:1002-1005.
1445. Van den Noort, S., and Uzman, L. L. Proc. Soc. Exptl. Biol. Med., 1961,
       108:32-34.
1446. Nordmann, Jo., Nordman, R., and Gauchery, M. O. Bull. Soc. Chim. Biol.,
       1951, 33:1826-1836.
1447. Norton, W. T., and Autilio, L. A. J. Neurochem., 1966, 13:213-222.
1448. Nukada, T. Canad. J. Biochem., 1965, 43:1119-1127.
1449. Ochs, S. In: Protein Metabolism of the Nervous System, Plenum Press,
       New York, 1970, pp. 291-302.
1450. Ochs, S., Johnson, J., and Ng, M.-H. J. Neurochem., 1967, 14:317-331.
1451. Oja, S. S. Ann. Acad. Scient. Fenn., 1967, Ser. A., 131:1-81.
1452. Oja, S. S. In: First International Meeting of the International Society for
       Neurochemistry, Strasbourg, 1967, p. 163.
1453. Oja, S. S., and Oja, H. J. Neurochem., 1970, 17:901-912.
1454. Ojemann, R. G. NRP Bull., 1966, 4:71.
1455. Okumura, N., Otsuki, S., and Aoyama, T. J. Biochem., 1959, 46:207-212.
1456. Okumura, N., Otsuki, S., and Kameyama, A. J. Biochem., 1960, 47:315-320.
1457. Okumura, N., Otsuki, S., and Nasu, H. J. Biochem., 1959, 46:247-252.
1458. Olivo, F. Boll. Soc. Ital. Biol. Sperim., 1959, 35:371-373.
1459. O'Neal, R. M., and Koeppe, R. E. J. Neurochem., 1966, 13:835-847.
1460. O'Neal, R. M., Koeppe, R. E., and Williams, E. I. Biochem. J., 1966, 101:591-
       597.

1461.   O'Neill, J. J., and Duffy, T. E. Life Sci., 1966, 5:1849-1857.
1462.   Orechowitsch, W. N. (Orekhovich, V. N.). Biochem. Z., 1936, 286:91-92.
1463.   Orechowitsch, W. N. (Orekhovich, V. N.), Bromley, N. W., and Kuzmina,
         N. A. Biochem. Z., 1935, 277:186-190.
1464.   Orrego, F., and Lipmann, F. J. Biol. Chem., 1967, 242:665-671.
1465.   Owman, C., and Rosengren, E. J. Neurochem., 1967, 14:547-550.
1466.   Palladin, A. V. In: Biochemistry of the Developing Nervous System, Aca-
         demic Press, New York, 1955, pp. 177-183.
1467.   Palladin, A. V. In: Metabolism of the Nervous System, Pergamon Press,
         London, 1957, pp. 456-458.
1468.   Palladin, A. V. In: Radioisotopes in Scientific Research, Pergamon Press,
         London, 1958, 403-415.
1469.   Palladin, A. V. In: Regional Neurochemistry, Pergamon Press, New York,
         1961, pp. 8-18.
1470.   Palladin, A. V. In: Comparative Neurochemistry, Pergamon Press, Oxford,
         1964, 131-137.
1471.   Palladin, A. V. In: Second International Meeting of the International Society
         for Neurochemistry, Milan, 1969, 312-313.
1472.   Palladin, A. V., et al. In: Problems of the Biochemistry of the Nervous
         System, Pergamon Press, New York, 1964, pp. 3-17.
1473.   Palladin, A. V., and Belik, Ya. V. In: Protein Metabolism of the Nervous
         System. Plenum Press, New York, 1970, pp. 77-91.
1474.   Palladin, A. V., Belik, Ya. V., and Kraciko, L. S. (Krachko, L. S.). Probl.
         Biochim. Seria Med., 1958, 6:3-17.
1475.   Palladin, A. V., and Polyakova, N. M. In: Proceedings of the 4th Interna-
         tional Congress of Biochemistry, 3, Symp. III. Vienna, 1958, pp. 185-189.
1476.   Palladin, A. V., and Polyakova, N. M. In: Biochemistry of the Central Ner-
         vous System, Pergamon Press, 1959, pp. 185-189.
1477.   Palladin, A. V., and Polyakova, N. M. In: Handbook of Neurochemistry, 5.
         Plenum Press, New York, 1971.
1478.   Palladin, A. V., Polyakova, N. M., and Lischko, V. K. (Lijhko, V. K.).
         J. Neurochem., 1963, 10:187-194.
1479.   Patel, A. J., and Balázs, R. J. Neurochem., 1970, 17:955-971.
1480.   Paterson, P. Y. Ann. New York Sci., 1958, 73:811-818.
1481.   Patterson, J. D. E., and Finean, J. B. J. Neurochem., 1961, 7:251-258.
1482.   Penn, N. W. Biochim. Biophys. Acta, 1960, 37:55-63.
1483.   Penn, N. W. Biochim. Biophys. Acta, 1961, 53:490-494.
1484.   Perez, V. J., et al. J. Neurochem., 1970, 17:511-519.
1485.   Perez, V. J., and Moore, B. W. J. Neurochem., 1968, 15:971-977.
1486.   Peterson, J. A., Bray, J. J., and Austin, L. J. Neurochem., 1968, 15:741-745.
1487.   Petrowsky, D. Pflug. Arch., 1873, 7:367-370.
1488.   Phillips, D. M. P., and Johns, E. W. Biochem. J., 1959, 72:538-544.
1489.   Piha, R. S., Cuénod, M., and Waelsch, H. J. Biol. Chem., 1966, 241:2397- 2404.
1490.   Piha, R. S., and Lahdesmaki, P. In: Second International Meeting of the
         International Society for Neurochemistry. Milan, 1969, 231-232.

1491. Pisano, J. J., Abraham, D., and Udenfriend, S. Arch. Biochem. Biophys., 1963, 100:323-329.
1492. Pope, A. J. Neurophysiol., 1952, 15:115-130.
1493. Pope, A. In: Biochemistry of the Developing Nervous System, Academic Press, New York, 1955, pp. 341-399.
1494. Pope, A. J. Neurochem., 1959, 4:31-41.
1495. Pope, A., and Anfinsen, C. B. J. Biol. Chem., 1948, 173:305-311.
1496. Porcellati, G., and Curti, B. J. Neurochem., 1960, 5:277-282.
1497. Porcellati, G., Millo, A., and Monocchio, I. J. Neurochem., 1961, 7:317-320.
1498. Porter, H., and Folch, J. J. Neurochem., 1957, 1:260-271.
1499. Porter, H., and Folch-Pi, J. In: Progress in Neurobiology, 1. Neurochemistry. Hoeber-Harper Books, New York, 1956, pp. 40-50.
1500. Portugalov, V. V., Dovedova, E. L., and Skrebitsky, V. G. J. Histochem. Cytochem., 1962, 10:213-221.
1501. Poulik, M. D. Nature, 1957, 180:1477-1479.
1502. Prensky, A. L., and Moser, H. W. J. Neurochem., 1966, 13:863-874.
1503. Prensky, A. L., and Moser, H. W. J. Neurochem., 1967, 14:117-121.
1504. Price, S. A. P., and West, G. B. Nature, 1960, 185:80-81.
1505. Putnam, F. W. Advances Protein Chem., 1948, 4:80-118.
1506. Quastel, J. H. In: Structure and Function of the Cerebral Cortex, Elsevier, Amsterdam, 1960, pp. 374-384.
1507. Quastel, J. H. Proc. Roy. Soc., 1965, B, 163:169-196.
1508. Rafelson, M. E., Winzler, R. J., and Pearson, H. E. J. Biol. Chem., 1951, 193: 205-217.
1509. Reed, D. J., and Woodbury, D. M. J. Physiol., 1963, 169:816-850.
1510. Reichel, W., et al. J. Gerontol., 1968, 23:71-78.
1511. Reichelt, K. L., and Kvamme, E. FEBS Abstracts. Prague, 1968, p. 203.
1512. Reid, B. R., and Cole, R. D. Proc. Nat. Acad. Sci. (Washington), 1964, 51:1044-1050.
1513. Reinis, S. Nature, 1968, 220:177-178.
1514. Reiss, J. M., Reiss, M., and Wyatt, A. Proc. Soc. Exptl. Biol. (New York), 1956, 93:19-22.
1515. Reiss, M., et al. J. Neurochem., 1963, 10:851-857.
1516. Richard, J., and Neimarevic, D. Acta Neurol. Psychiatr. Belg., 1962, 62:1079-1086.
1517. Richardson, S. H., Hultin, H. O., and Green, D. E. Proc. Nat. Acad. Sci. (Washington), 1963, 50:821-827.
1518. Richlik, I., Dancheva, K. I., and Cerhova, M. Collect. Czechosl. Chem. Commun., 1965, 30:138-145.
1519. Richter, D. In: Biochemistry of the Developing Nervous System, Academic Press, New York, 1955, pp. 225-247.
1520. Richter, D. In: Proceedings of the 4th International Congress of Biochemistry, Symp. III. Vienna, 1958, pp. 173-184.
1521. Richter, D. Brit. Med. J., 1959, 5132:1255-1259.
1522. Richter, D. Brit. Med. Bull., 1965, 21:76-80.

1523.   Richter, D. In: Protein Metabolism of the Nervous System, Plenum Press, New York, 1970, pp. 241-254.

1524.   Richter, D., Gaitonde, M. K., and Cohn, P. In: Structure and Function of the Cerebral Cortex. Elsevier, Amsterdam, 1960, pp. 340-347.

1525.   Riekkinen, P. J., et al., Enzymologia, 1967, 32:97-109.

1526.   Riekkinen, P. J., Clausen, J., and Arstila, A. U. Brain Res., 1970, 19:213-227.

1527.   Riekkinen, P. J., Ekfors, T. O., and Hopsu-Havu, V. K. Enzymologia, 1967, 32:110-127.

1528.   De Risio, C., Tonini, G., and Missers, G. Boll. Soc. Ital. Biol., Sperim., 1957, 33:204.

1529.   Rittenberg, D., Sproul, E. E., and Shemin, D. Federat. Proc., 1948, 7:180.

1530.   De Robertis, E. Science, 1967, 156:907-914.

1531.   De Robertis, E., et al. J. Biophys. Biochem. Cytol., 1961, 9:229-237.

1532.   De Robertis, E., et al. J. Neurochem., 1962, 9:23-35.

1533.   De Robertis, E., et al. Nature, 1962, 194:794-795.

1534.   De Robertis, E., et al. J. Neurochem., 1963, 10:225-235.

1535.   Roberts, E., Frankel, S., and Harman, P. J. Proc. Soc. Exptl. Biol. (New York), 1950, 74:383-387.

1536.   Roberts, E., Harman, P. J., and Frankel, S. Proc. Soc. Exptl. Biol. (New York), 1951, 78:799-803.

1537.   Roberts, E., et al. J. Exptl. Zool., 1958, 138:313-328.

1538.   Roberts, E., Rethstein, M., and Baxter, C. F. Proc. Soc. Exptl. Biol. (New York), 1958, 97:796-802.

1539.   Roberts, E., and Simonsen, D. G. In: Amino Acid Pools, Elsevier, Amsterdam, 1962, pp. 284-349.

1540.   Roberts, E. D., and Bennett, H. S. Exper. Cell. Res., 1954, 6:543-545. Cited by (310).

1541.   Roberts, R. B., Flexner, J. B., and Flexner, L. B. J. Neurochem., 1959, 4:78-90.

1542.   Roberts, S. J. Neurochem., 1963, 10:931-940.

1543.   Roberts, S., and Morelos, B. S. J. Neurochem., 1965, 12:373-387.

1544.   Roberts, S., Zomzely, C. E., and Bondy, S. C. In: Protein Metabolism of the Nervous System, Plenum Press, New York, 1970, pp. 3-35.

1545.   Robertson, D. In: Progress in Neurobiology, II. Ultrastructure and Cellular Chemistry of Neutral Tissue. Hoeber-Harper Books, 1957, pp. 1-22.

1546.   Robertson, D. M. J. Neurochem., 1957, 1:358-363.

1547.   Robertson, D. M. J. Neurochem., 1960, 5:145-149.

1548.   Robins, E., et al. J. Neurochem., 1956, 1:68-76.

1549.   Robins, E., Smith, D. E., and Eydt, K. M. J. Neurochem., 1956, 1:54-67.

1550.   Roboz, E., Henderson, N., and Kies, M. W. J. Neurochem., 1958, 2:254-260.

1551.   Roboz-Einstein, E., et al. J. Neurochem., 1962, 9:353-361.

1552.   Roboz-Einstein, E., Robertson, D., and DeCaprio, J. Federat. Proc, 1960, 19:331.

1553.   Robuschi, L., and Benassi, G. Giorn. Psichiatr. Neuropatol., 1957, 85:183-191.

1554.   Rojas, E., and Atwater, I. Nature, 1967, 215:850-852.

1555.   Roodyn, D. B. Biochem. J., 1962, 85:177-189.

1556.   Roodyn, D. B. Biochem. J., 1965, 97:782-793.

1557.	Roodyn, D. B., Reis, P. J., and Work, T. S. In: Protein Biosynthesis, Academic Press, London, 1961, pp. 37-46.
1558.	Roodyn, D. B., Reis, P. J., and Work, T. S. Biochem. J., 1961, 80:9-21.
1559.	De Ropp, R. S., and Snedeker, E. H. J. Neurochem., 1961, 7:128-134.
1560.	De Ropp, R. S., and Snedeker, E. H. Proc. Soc. Exptl. Biol. (New York), 1961, 106:696-700.
1561.	Rose, S. P. R. FEBS Letters, 1969, 5:305-312.
1562.	Rose, S. P. R., and Sinha, A. K. Life Sci., Part II, 1970, 9:907-915.
1563.	Rothfield, L., and Finkelstein, A. Ann. Rev. Biochem., 1968, 37:463-496.
1564.	Rotshild, H., and Junqueira, L. C. M. Arch. Biochem. Biophys., 1951, 34:453-456.
1565.	Rous, P. J. Exptl. Med., 1925, 16:399-411.
1566.	Rubin, A. L., and Stenzel, K. H. Proc. Nat. Acad. Sci. (Washington), 1965, 53:963-968.
1567.	Rudnick, D., and Waelsch, H. J. J. Exptl. Zool., 1955, 129:309-326.
1568.	Rusca, G., and Calissano, P. Biochim. Biophys. Acta, 1970, 221:74-86.
1569.	Ruščak, M. J. Neurochem., 1961, 7:305-307.
1570.	Ruščak, M. Biologia (Bratislava), 1962, 17:81-88.
1571.	Ruščak, M. Physiol. Bohemosl., 1962, 11:192-198.
1572.	Ruščak, M. Physiol. Bohemosl., 1962, 11:199-205.
1573.	Salganicoff, L., and de Robertis, E. J. Neurochem., 1965, 12:287-309.
1574.	Salzmann, R., and Weber, R. Experientia, 1963, 19:352-354.
1575.	Samec, J., Mandel, P., and Jacob, M. J. Neurochem., 1967, 14:887-892.
1576.	Sarcar, N. K., and Dounce, A. L. Arch. Biochem. Biophys., 1961, 92:321-328.
1577.	Satake, M., et al. J. Biochem., 1964, 56:504-511.
1578.	Satake, M., et al. J. Biochem., 1965, 57:184-191.
1579.	Satake, M., and Abe, S. J. Biochem., 1966, 59:72-75.
1580.	Satake, M., et al. Biochim. Biophys. Acta, 1960, 41:366-367.
1581.	Sawant, P. L., et al. Biochim. Biophys. Acta, 1964, 85:82-92.
1582.	Sawant, P. L., Desai, D., and Tappel, A. L. Biochim. Biophys. Acta, 1964, 85:93-102.
1583.	Schade, J. P., van Backer, H., and Colon, E. In: Progress in Brain Research. 4. Growth and Maturation of the Brain, Elsevier, Amsterdam, 1964, pp. 150-175.
1584.	Schain, R. J., Copenhaver, J. H., and Carver, M. J. J. Neurochem., 1967, 14:195-201.
1585.	Schanberg, S., and Giarman, N. J. Biochim. Biophys. Acta, 1960, 41:556-558.
1586.	Schapira, G., Dreyfus, J. C., and Kruh, J. Biochem. J., 1962, 82:290-297.
1587.	Scharrer, E., and Scharrer, B. Physiol. Revs., 1945, 25:171-181.
1588.	Schepartz, B. J. Neurochem., 1963, 10:825-829.
1589.	Schilling, E. D., Burchill, P. I., and Clayton, R. A. Analyt. Biochem., 1963, 5:1-6.
1590.	Schmidt, G., and Thannhauser, S. J. J. Biol. Chem., 1945, 161:83-89.
1591.	Schmitt, F, O. In: Metabolism of the Nervous System, Pergamon Press, London, 1957, 35:47.
1592.	Schneider, D., and Roberts, S. J. Neurochem., 1968, 15:1469-1474.

1593.   Schoenheimer, R. The Dynamic State of Body Constituents, Harvard University Press, 1946.

1594.   Schoenheimer, R., Ratner, S., and Rittenberg, D. J. Biol. Chem., 1942, 144:541-544.

1595.   Schoenheimer, R., and Rittenberg, D. Physiol. Revs., 1940, 20:218-248.

1596.   Schreier, K. Monatschr. Kinderheilkunde, 1962, 110:290-296.

1597.   Schurr, P. E., et al. J. Biol. Chem., 1950, 182:39-45.

1598.   Schwerin, P., Bessman, S. P., and Waelsch, H. J. Biol. Chem., 1950, 184:37-44.

1599.   Sellinger, O. Z., and Ohlsson, W. G. Life Sci., 1969, Part II, 8:1083-1088.

1600.   Sellinger, O. Z., and Rucker, G. Life Sci., 1964, Part II, 3:1097-1102.

1601.   Sellinger, O. Z., and Rucker, G. D. Life Sci., 1966, Part II, 5:163-167.

1602.   Sellinger, O. Z., Rucker, D. L., and Vebster, F. B. J. Neurochem., 1964, 11:271-280.

1603.   Sellinger, O. Z., and Vebster, F. J. Biol. Chem., 1962, 237:2836-2844.

1604.   Seminario, L. M., Hren, N., and Gomez, C. J. J. Neurochem., 1964, 11:197-207.

1605.   Shapot, V. S. In: Metabolism of the Nervous System, Pergamon Press, London, 1957, 257-262.

1606.   Shaw, R. K., and Heine, J. D. J. Neurochem., 1965, 12:151-155.

1607.   Sheng Pei-ken, Li Tsai-ping, and Tsao Tien-chin, Scientia Sinica, 1957, 6:309-316.

1608.   Siebert, G. Experientia, 1958, 14:65-66.

1609.   Siebert, G., et al., Naturwissenschaften, 1955, 42:156-159.

1610.   Siebert, G., Schmidtt, A., and Malortie, R. Hoppe-Seyler's Z. Physiol. Chem., 1965, 342:20-39.

1611.   Sikdar, K., and Ghosh, J. J. J. Neurochem., 1964, 11:545-549.

1612.   Simon, G., Drori, J. B., and Cohen, M. M. Biochem. J., 1967, 102:153-162.

1613.   Simpson, M. V. J. Biol. Chem., 1953, 201:143-154.

1614.   Simpson, M. V., and Velick, S. F. J. Biol. Chem., 1954, 208:61-71.

1615.   Singh, S. I., and Malhotra, C. L. J. Neurochem., 1962, 9:37-42; 585-588.

1616.   Singh, S. I., and Malhotra, C. L. J. Neurochem., 1964, 11:865-872.

1617.   Singh, S. I., and Malhotra, C. L. J. Neurochem., 1967, 14:135-140.

1618.   Singh, U. B., and Talwar, G. P. J. Neurochem., 1967, 14:675-680.

1619.   Singh, U. B., and Talwar, G. P. J. Neurochem., 1969, 16:951-959.

1620.   Sky-Peck, H. H., Pearson, H. E., and Visser, D. W. J. Biol. Chem., 1956, 223:1033-1041.

1621.   Sky-Peck, H. H., Rosenbloom, C., and Winzler, R. J. J. Neurochem., 1966, 13:223-228.

1622.   Slonimski, P. Biochem. J., 1967, 105:38p.

1623.   Smith, E. L. J. Biol. Chem., 1948, 173:553-569.

1624.   Smith, J. C., et al. Biochim. Biophys. Acta, 1966, 115:81-87.

1625.   Smith, M. E. J. Neurochem., 1969, 16:83-92.

1626.   Smith, L. C., and Nelson, S. R. Proc. Soc. Exptl. Biol. (New York), 1957, 94:644-646.

1627.   Snoke, J. E., and Neurath, H. J. Biol. Chem., 1950, 187:127-135.

1628.   Sokoloff, L. Neurology, 1961, 11:34-40.

1629.   Sokoloff, L. In: Protein Metabolism of the Nervous System, Plenum Press, New York, 1970, 367-382.

1630.   Sokoloff, L., et al. J. Biol. Chem., 1963, 238:1432-1437.

1631.   Sokoloff, L., and Kaufman, S. Science, 1959, 129:569-570.

1632.   Sokoloff, L., and Kaufman, S. J. Biol. Chem., 1961, 236:795-803.

1633.   Somogyi, J., Fonyo, A., and Vincze, J. Acta Physiol. Acad. Sci., 1962, 21:295-300.

1634.   Soto, E. F., Seminario de Bohner, and del Carmen Calvino, M. J. Neurochem., 1966, 13:989-998.

1635.   Steinberg, D., and Vaughan, M. Arch. Biochem. Biophys., 1956, 65:93-105.

1636.   Steinberg, D., Vaughan, M., and Anfinsen, Ch. B. Science, 1956, 124:389-395.

1637.   Steinzel, K. H., Aronson, R. F., and Rubin, A. L. Biochemistry, 1966, 5:930-936.

1638.   Stern, J. R., et al. Biochem. J., 1949, 44:410-418.

1639.   Strecker, H. J. In: Metabolism of the Nervous System, Pergamon Press, New York, 1957, pp. 459-473.

1640.   Sung Shan-Ching. Canad. J. Biochem., 1969, 47:47-50.

1641.   Suttie, J. W. Biochem. J., 1962, 84:382-386.

1642.   Suzuki, K., Korey, S. R., and Terry, R. D. J. Neurochem., 1964, 11:403-412.

1643.   Svet-Moldavsky, G. J., et al. Bull. World Health Org., 1965, 32:47-58.

1644.   Swaiman, K. F., and Nelson, C. E. J. Neurochem., 1967, 14:905-910.

1645.   Swaiman, K. F., and Wolfe, R. N. In: First International Meeting of the International Society for Neurochemistry, Strasbourg, 1967, p. 194.

1646.   Swaiman, K. F., and Wolfe, R. N. Proc. Soc. Exptl. Biol. (New York), 1968, 127:411-414.

1647.   Swanborg, R. H., and Shulman, S. Immunology, 1970, 19:31-40.

1648.   Syrový, J. Ceskosl. Fysiol., 1966, 15:217-224.

1649.   Syrový, J., Hajek, I., and Gutmann, E. Physiol. Bohemosl., 1965, 14:12-16.

1650.   Syrový, J., Hajek, I., and Gutmann, E. Physiol. Bohemosl., 1966, 15:7-13.

1651.   Szafranski, P., and Bagdasarian, M. Post. Biochem., 1961, 7:49-62

1652.   Szafranski, P., and Bagdasarian, M. Nature, 1961, 190:719-720.

1653.   Takahashi, Y., and Abe, S. Experientia, 1963, 19:186-187.

1654.   Takahashi, Y., and Akabane, Y. Canad. J. Biochem. Physiol., 1960, 38:1149-1157.

1655.   Takahasi, Y., and Akabane, Y. J. Neurochem., 1961, 7:89-96.

1656.   Takahashi, Y., Mase, K., and Abe, S. J. Biochem., 1966, 60:363-371.

1657.   Takahashi, Y., Mase, K., and Sugano, H. Biochim. Biophys. Acta, 1966, 119:627-629.

1658.   Takahashi, Y., Nomura, M., and Furusawa, S. J. Neurochem., 1961, 7:97-102.

1659.   Tallan, H. H. In: Amino Acid Pools, Elsevier, Amsterdam, 1962, pp. 471-485.

1660.   Tallan, H. H., Jones, M. E., and Fruton, J. S. J. Biol. Chem., 1952, 194:793-805.

1661.   Tallan, H. H., Moore, S., and Stein, W. H. J. Biol. Chem., 1954, 211-927-939.

1662.   Tallan, H. H., Moore, S., and Stein, W. H. J. Biol. Chem., 1956, 219-257-264.

1663.   Tamburino, G., et al. Radiobiol. Latina, 1962, 5:423-426.

1664.   Tanaka, R., and Abood, L. G. J. Neurochem., 1963, 10:571-576.

1665.   Tanaka, R., and Abood, L. G. Arch. Biochem. Biophys., 1964, 105:554-562.

1666.   Tappel, A. L., et al. Arch. Biochem. Biophys., 1962, 96:340-346.

1667.   Tardy, J., et al. J. Physiol. (Paris), 1967, 59-510.

1668.   Tardy, J., et al. C. R. Acad. Sci. (Paris), 1968, Ser. D., 267:669-672.

1669.   Tarver, H., and Morse, L. M. J. Biol. Chem., 1948, 173:53-61.

1670.   Tarver, H., and Schmidt, L. A. J. Biol. Chem., 1942, 146:69-84.

1671.   Tata, J. R., et al. Biochem. J., 1963, 86:408-428.

1672.   Tata, J. R., and Windnell, C. C. Biochem. J., 1964, 92:26p.

1673.   Terlizzi, L., and Mitolo, M. Boll. Soc. Ital. Biol. Sperim., 1955, 31:1487-1489.

1674.   Tewari, S., and Baxter, C. F. J. Neurochem., 1969, 16:171-180.

1675.   Tews, J. K., et al. J. Neurochem., 1963, 10:641-653.

1676.   Tews, J. K., and Lowell, R. A. J. Neurochem., 1967, 14:1-7.

1677.   Tews, J. K., and Stone, W. E. Biochem. Pharmacol., 1964, 13:543-545.

1678.   Thompson, R. C., and Ballou, J. E. J. Biol. Chem., 1956, 223:795-809.

1679.   Thudichum, J. L. W. Die chemische Konstitution des Gehirns des Menschen und der Tiere, Tubingen, 1901.

1680.   Tiplady, B., and Rose, S. P. R. Biochem. J., 1970, 117:65p.

1681.   Todd, P. E. E., and Trikojus, V. M. Biochim. Biophys. Acta, 1960, 45:234-242.

1682.   Tolani, A. J., and Mokrasch, L. C. Life Sci., 1967, 6:1771-1774.

1683.   Tolani, A. J., and Talwar, G. P. Biochem. J., 1963, 88:357-362.

1684.   Tomasi, L. G., and Kornguth, S. E. J. Biol. Chem., 1967, 242:4933-4938.

1685.   Tonini, G., and Missere, G. Boll. Soc. Ital. Biol. Sperim., 1957, 33:1169-1171.

1686.   Tower, D. B. In: Proceedings of the 4th International Congress of Biochemistry, 3. Symp. III. Vienna, 1958, pp. 213-250.

1687.   Tower, D. B. Amer. J. Clin. Nutr., 1963, 12:308-320.

1688.   Truman, D. E. S., and Korner, A. Biochem. J., 1962, 85:154-158.

1689.   Tsukada, Y., et al. J. Neurochem., 1963, 10:241-256.

1690.   Tsukada, Y., Nagata, Y., and Hirano, S. Nature, 1960, 186:474-475.

1691.   Tsukada, Y., Uyemura, K., and Noduchi, T. In: First International Meeting of the International Society for Neurochemistry, Strasbourg, 1967, p. 205.

1692.   Tugan, N. A., and Adams, C. W. M. J. Neurochem., 1961, 6:327-333.

1693.   Turk, V., et al. Z. Naturforsch., 1967, 226:561.

1694.   Tyce, G. M., Flock, E. V., and Owen, C. A. In: Progress in Brain Research, 9. The Developing Brain, Elsevier, Amsterdam, 1964, pp. 198-203.

1695.   Udenfriend, S. Amer. J. Clin. Nutr., 1963, 12:287-290.

1696.   Umana, R. Arch. Biochem. Biophys., 1967, 119:526-535.

1697.   Ungar, G. Lancet, 1952, 11:742-746.

1698.   Ungar, G. Nature, 1965, 207:419-420.

1699.   Ungar, G., et al. J. Gen. Physiol., 1957, 40:635-652.

1700.   Ungar, G., and Damgaard, E. Proc. Soc. Exptl. Biol. (New York), 1954, 87:378-383.

1662.   Tallan, H. H., Moore, S., and Stein, W. H. J. Biol. Chem., 1956, 219-257-264.

1663.   Tamburino, G., et al. Radiobiol. Latina, 1962, 5:423-426.

1664.   Tanaka, R., and Abood, L. G. J. Neurochem., 1963, 10:571-576.

1665.   Tanaka, R., and Abood, L. G. Arch. Biochem. Biophys., 1964, 105:554-562.

1666.   Tappel, A. L., et al. Arch. Biochem. Biophys., 1962, 96:340-346.

1667.   Tardy, J., et al. J. Physiol. (Paris), 1967, 59-510.

1668.   Tardy, J., et al. C. R. Acad. Sci. (Paris), 1968, Ser. D., 267:669-672.

1669.   Tarver, H., and Morse, L. M. J. Biol. Chem., 1948, 173:53-61.

1670.   Tarver, H., and Schmidt, L. A. J. Biol. Chem., 1942, 146:69-84.

1671.   Tata, J. R., et al. Biochem. J., 1963, 86:408-428.

1672.   Tata, J. R., and Windnell, C. C. Biochem. J., 1964, 92:26p.

1673.   Terlizzi, L., and Mitolo, M. Boll. Soc. Ital. Biol. Sperim., 1955, 31:1487-1489.

1674.   Tewari, S., and Baxter, C. F. J. Neurochem., 1969, 16:171-180.

1675.   Tews, J. K., et al. J. Neurochem., 1963, 10:641-653.

1676.   Tews, J. K., and Lowell, R. A. J. Neurochem., 1967, 14:1-7.

1677.   Tews, J. K., and Stone, W. E. Biochem. Pharmacol., 1964, 13:543-545.

1678.   Thompson, R. C., and Ballou, J. E. J. Biol. Chem., 1956, 223:795-809.

1679.   Thudichum, J. L. W. Die chemische Konstitution des Gehirns des Menschen und der Tiere, Tubingen, 1901.

1680.   Tiplady, B., and Rose, S. P. R. Biochem. J., 1970, 117:65p.

1681.   Todd, P. E. E., and Trikojus, V. M. Biochim. Biophys. Acta, 1960, 45:234-242.

1682.   Tolani, A. J., and Mokrasch, L. C. Life Sci., 1967, 6:1771-1774.

1683.   Tolani, A. J., and Talwar, G. P. Biochem. J., 1963, 88:357-362.

1684.   Tomasi, L. G., and Kornguth, S. E. J. Biol. Chem., 1967, 242:4933-4938.

1685.   Tonini, G., and Missere, G. Boll. Soc. Ital. Biol. Sperim., 1957, 33:1169-1171.

1686.   Tower, D. B. In: Proceedings of the 4th International Congress of Biochemistry, 3. Symp. III. Vienna, 1958, pp. 213-250.

1687.   Tower, D. B. Amer. J. Clin. Nutr., 1963, 12:308-320.

1688.   Truman, D. E. S., and Korner, A. Biochem. J., 1962, 85:154-158.

1689.   Tsukada, Y., et al. J. Neurochem., 1963, 10:241-256.

1690.   Tsukada, Y., Nagata, Y., and Hirano, S. Nature, 1960, 186:474-475.

1691.   Tsukada, Y., Uyemura, K., and Noduchi, T. In: First International Meeting of the International Society for Neurochemistry, Strasbourg, 1967, p. 205.

1692.   Tugan, N. A., and Adams, C. W. M. J. Neurochem., 1961, 6:327-333.

1693.   Turk, V., et al. Z. Naturforsch., 1967, 226:561.

1694.   Tyce, G. M., Flock, E. V., and Owen, C. A. In: Progress in Brain Research, 9. The Developing Brain, Elsevier, Amsterdam, 1964, pp. 198-203.

1695.   Udenfriend, S. Amer. J. Clin. Nutr., 1963, 12:287-290.

1696.   Umana, R. Arch. Biochem. Biophys., 1967, 119:526-535.

1697.   Ungar, G. Lancet, 1952, 11:742-746.

1698.   Ungar, G. Nature, 1965, 207:419-420.

1699.   Ungar, G., et al. J. Gen. Physiol., 1957, 40:635-652.

1700.   Ungar, G., and Damgaard, E. Proc. Soc. Exptl. Biol. (New York), 1954, 87:378-383.

1736.   Waelsch, H. Schweiz. med. Wochenschr., 1963, 93:1289-1293.
1737.   Waelsch, H., and Lajtha, A. In: The Neurochemistry of Nucleotides and Amino Acids, New York, 1960, pp. 205-214.
1738.   Waelsch, H., and Lajtha, A. Physiol. Revs., 1961, 41:709-736.
1739.   Waksman, B. H., et al. J. Exptl. Med., 1954, 100:451-457.
1740.   Waley, S. G., and van Heyningen, R. Biochem. J., 1962, 83:274-283.
1741.   Waley, S. G., and Watson, J. Biochem. J., 1954, 57:529-538.
1742.   Wallach, D. F. H., and Gordon, A. Federat. Proc., 1968, 27:1263-1268.
1743.   Walsh, K. A., et al. Proc. Nat. Acad. Sci. (Washington), 1964, 51:301-308.
1744.   Walter, H. Nature, 1960, 188:643-645.
1745.   Wang, T.-Y. Biochim. Biophys. Acta, 1963, 68:52-61.
1746.   Wang, T.-Y. Biochim. Biophys. Acta, 1966, 114:620-627.
1747.   Warecka, K. Life Sci., 1967, 6:1999-2002.
1748.   Warecka, K. In: Second International Meeting of International Society for Neurochemistry, Milan, 1969, p. 413.
1749.   Warecka, K. J. Neurochem., 1970, 17:829-830.
1750.   Warecka, K., and Bauer, H. In: First International Meeting of the International Society for Neurochemistry, Strasbourg, 1967, p. 215.
1751.   Warecka, K., and Bauer, H. J. Neurochem., 1967, 14:783-787.
1752.   Warecka, K., and Bauer, H. J. Dtsch. Z. Nervenheilkunde, 1968, 194:66-75.
1753.   Wattiaux-de Coninck, S., Rutgeerts, M.-J., and Wattiaux, R. Biochim. Biophys. Acta, 1965, 105:446-459.
1754.   Webster, G. C. Biochem. Biophys. Res. Commun., 1960, 2:56-58.
1755.   Wegelin, I., and Manzolli, F. A. J. Neurochem., 1967, 14:1161-1165.
1756.   Weil-Malherbe, H., and Green, R. H. Biochem. J., 1955, 61-210-218.
1757.   Weinbach, E. C., and Garbus, J. J. Biol. Chem., 1959, 234:412-417.
1758.   Weinstock, I. M., Epstein, S., and Milhorat, A. T. Proc. Soc. Exptl. Biol. (New York), 1958, 99:272-276.
1759.   Weiss, P., and Hiscoe, H. B. J. Exptl. Zool., 1948, 107:315-395.
1760.   Weiss, P., Taylor, A., and Pillai, P. Science, 1962, 136:330-331.
1761.   Weissman, G., and Thomas, L. J. Exptl. Med., 1962, 116:433-450.
1762.   Weissmann, G., and Thomas, L. J. Clin. Investig., 1963, 42:661-669.
1763.   Wells, M. A., and Dittmer, J. C. Biochemistry, 1967, 6:3169-3175.
1764.   Wender, M., and Hierowski, M. In: Progress in Brain Research. 4. Growth and Maturation of the Brain, Elsevier, Amsterdam, 1964, pp. 273-280.
1765.   Wender, M., and Waligora, Z. J. Neurochem., 1961, 7:259-263.
1766.   Wender, M., and Waligora, Z. J. Neurochem., 1962, 9:115-118.
1767.   Wender, M., and Waligora, Z. J. Neurochem., 1964, 11:243-246; 583-588.
1768.   Whittaker, V. P. Biochem. Pharmacol., 1959, 1:351-352.
1769.   Whittaker, V. P. Biochem. J., 1959, 72:694-706.
1770.   Whittaker, V. P. In: Methods of Separation of Subcellular Structural Components, University Press, Cambridge, 1963, pp. 109-126.
1771.   Whittaker, V. P., and Dowe, G. H. C. Biochem. Pharmacol., 1965, 14:194-196.
1772.   Wiechert, P. Acta Biol. Med. German, 1963, 10:305-310.
1773.   Wiechert, P. Acta Biol. Med. German, 1963, 11:68-76.

1774. Wiechert, P., and Schröter, P. Acta Biol. Med. German, 1964, 12:475-480.
1775. Wilbrandt, W. Experientia, 1969, 25:673-677.
1776. Williams, J. N. J. Nutr., 1961, 73:199-209.
1777. Wills, E. D., and Wilkinson, A. E. Biochem. J., 1966, 99:657-666.
1778. Wilson, C. W. M., and Brodie, B. B. J. Pharmacol. Exptl. Therap., 1961, 133:332-334.
1779. Windmueller, H. G., and Kaplan, N. O. Biochim. Biophys. Acta, 1962, 56:388-391.
1780. Winnick, T., Friedberg, F., and Greenberg, D. M. J. Biol. Chem., 1948, 173:189-197.
1781. Winzler, R. J., et al. J. Biol. Chem., 1952, 199:485-492.
1782. Withrow, C. D., and Woodbury, D. M. In: Progress in Brain Research, 9. The Developing Brain, Elsevier, Amsterdam, 1964, pp. 204-206.
1783. Woessner, J. F. Biochem. J., 1962, 83:304-314.
1784. Woessner, J. F., and Brewer, T. H. Federat. Proc., 1960, 19:335.
1785. Wolfgram, F. J. Neurochem., 1966, 13:461-470.
1786. Wolfgram, F., and Kotorii, K. J. Neurochem., 1968, 15:1281-1290; 1291-1295.
1787. Wolfgram, F., and Rose, A. S. J. Neurochem., 1961, 8:161-168.
1788. Wolfgram, F., and Rose, A. S. J. Neurochem., 1962, 9:623-627.
1789. Woodman, R. J., and McIlwain, H. Biochem. J., 1961, 81:83-93.
1790. Work, T. S. Biochem. J., 1967, 105:38p-39p.
1791. Wunderly, Ch., and Bustamante, V. Klin. Wochenschr., 1957, 35:578-760.
1792. Yamagami, S., Fritz, R. R., and Rappoport, D. A. Biochim. Biophys. Acta, 1966, 129:532-547.
1793. Yamagami, S., Masui, M., and Kawakita, Y. J. Neurochem., 1963, 10:849-850.
1794. Yamagami, S., and Mori, K. J. Neurochem., 1970, 17:721-731.
1795. Yordanov, B. Compt. Rend. Acad. Bulg. Sci., 1966, 19:867-870.
1796. Yoshino, Y., and Elliott, K. A. C. Canad. J. Biochem., 1970, 48:228-235.
1797. Yoshino, Y., and Elliott, K. A. C. Canad. J. Biochem., 1970, 48:236-243.
1798. Yoshino, Y., Mozai, T., and Nakao, K. J. Neurochem., 1966, 13:1223-1230.
1799. Yphantis, D. A. Biochemistry, 1964, 3:297-317.
1800. Zachar, F., Skupenová, A., and Turský, T. Biologiya (Bratislava), Ser. C., 1969, 24:229-235.
1801. Zalkin, H., et al. Federat. Proc., 1961, 20:303.
1802. Zamenhof, S. van Martens, E., and Margolis, F. L. Science, 1968, 160:322-323.
1803. Zeller, E. A., and Shock, D. Federat. Proc., 1959, 18:358.
1804. Zomzely, C. E., et al. J. Molec. Biol., 1966, 19:455-468.
1805. Zomzely, C. E., Roberts, S., and Rapaport, D. J. Neurochem., 1964, 11:567-582.
1806. Zuckerman, J. E., Hershman, H. R., and Levine, L. J. Neurochem., 1970, 17:247-251.